Universitext

Universitext is a series of textbooks that presents material from a wide variety of mathematical disciplines at master's level and beyond. The books, often well class-tested by their author, may have an informal, personal, or even experimental approach to their subject matter. Some of the most successful and established books in the series have evolved through several editions, always following the evolution of teaching curricula, into very polished texts.

Thus as research topics trickle down into graduate-level teaching, first textbooks written for new, cutting-edge courses may find their way into *Universitext*.

Paolo Baldi

Probability

An Introduction Through Theory and Exercises

 Springer

Paolo Baldi
Dipartimento di Matematica
Università di Roma Tor Vergata
Roma, Italy

ISSN 0172-5939   ISSN 2191-6675   (electronic)
Universitext
ISBN 978-3-031-38491-2   ISBN 978-3-031-38492-9   (eBook)
https://doi.org/10.1007/978-3-031-38492-9

Mathematics Subject Classification: 60-XX

This Springer imprint is published by the registered company Springer Nature Switzerland AG
The registered company address is: Gewerbestrasse 11, 6330 Cham, Switzerland

Paper in this product is recyclable

Preface

This book is based on a one-semester basic course on probability with measure theory for students in mathematics at the University of Roma "Tor Vergata".

The main objective is to provide the necessary notions required for more advanced courses in this area (stochastic processes and statistics, mainly) that students might attend later.

This explains some choices:

- Random elements in spaces more general than the finite dimensional Euclidean spaces are considered: in the future the student might be led to consider r.v.'s with values in Banach spaces, the sphere, a group of rotations...
- Some classical finer topics (e.g. around the Law of Large numbers and the Central Limit Theorem) are omitted. This has made it possible to devote some time to other topics more essential to the objective indicated above (e.g. martingales).

  It is assumed that students

- Are already familiar with elementary notions of probability and in particular know the classical distributions and their use
- Are acquainted with the manipulations of basic calculus and linear algebra and the main definitions of topology
- Already know measure theory or are following simultaneously a course on measure theory

The book consists of six chapters and an additional chapter of solutions to the exercises.

The first is a recollection of the main topics of measure theory. Here "recollection" means that only the more significant proofs are given, skipping the more technical points, the important thing being to become comfortable with the tools and the typical ways of reasoning of this theory.

The second chapter develops the main core of probability theory: independence, laws and the computations thereof, characteristic functions and the complex Laplace transform, multidimensional Gaussian distributions.

The third chapter concerns convergence, the fourth is about conditional expectations and distributions and the fifth is about martingales.

Chapters 1 to 5 can be covered in a 64-hour course with some time included for exercises.

The sixth chapter develops two subjects that regretfully did not fit into the time schedule above: simulation and tightness (the last one without proofs).

Most of the material is, of course, classical and appears in many of the very good textbooks already available. However, the present book also includes some topics that, in my experience, are important in view of future study and which are seldom developed elsewhere: the behavior of Gaussian laws and r.v.'s concerning convergence (Sect. 3.7) and conditioning (Sect. 4.4), quadratic functionals of Gaussian r.v.'s (Sects. 2.9 and 3.9) and the complex Laplace transform (Sect. 2.7), which is of constant use in stochastic calculus and the gateway to changes of probability.

Particular attention is devoted to the exercises: detailed solutions are provided for all of them in the final chapter, possibly making these notes useful for self study.

In the preparation of this book, I am indebted to B. Pacchiarotti and L. Caramellino, of my University, whose lists of exercises have been an important source, and P. Priouret, who helped clarify a few notions that were a bit misty in my head.

Roma, Italy   Paolo Baldi
April 2023

Contents

Notation

Real, Complex Numbers, $\mathbb{R}^m$

$x \vee y$	$= \max(x, y)$ the largest of the real numbers x and y
$x \wedge y$	$= \min(x, y)$ the smallest of the real numbers x and y
$\langle x, y \rangle$	the scalar product of $x, y \in \mathbb{R}^m$ or $x, y \in \mathbb{C}^m$
x^+, x^-	the positive and negative parts of $x \in \mathbb{R}$: $x^+ = \max(x, 0)$, $x^- = \max(-x, 0)$
$\|x\|$	according to the context, the absolute value of the real number x, the modulus of the complex number x or the norm of the vector x
$\Re z, \Im z$	the real and imaginary parts of $z \in \mathbb{C}$
$B_R(x)$	$= \{y \in \mathbb{R}^m, |y - x| < R\}$, the open ball centered at x with radius R
$A^*, \operatorname{tr} A, \det A$	the transpose, trace, determinant of matrix A

Functional Spaces

$M_b(E)$	real bounded measurable functions on the topological space E
$\|f\|_\infty$	the sup norm$= \sup_{x \in E} |f(x)|$ if $f \in M_b(E)$
$C_b(E)$	the Banach space of real bounded continuous functions on the topological space E endowed with the norm $\| \ \|_\infty$
$C_0(E)$	the subspace of $C_b(E)$ of the functions f vanishing at infinity, i.e. such that for every $\varepsilon > 0$ there exists a compact set K_ε such that $|f| \leq \varepsilon$ outside K_ε
$C_K(E)$	the subspace of $C_b(E)$ of the continuous functions with compact support. It is dense in $C_0(E)$

To be Precise

Throughout this book, "positive" means ≥ 0, "strictly positive" means > 0. Similarly "increasing" means $\geq$, "strictly increasing" $>$.

Chapter 1
Elements of Measure Theory

The building block of probability is the triple $(\Omega, \mathcal{F}, P)$, where $\mathcal{F}$ is a σ-algebra of subsets of a set Ω and P a probability.

This is the typical setting of measure theory. In this first chapter we shall peruse the main points of this theory. We shall skip the more technical proofs and focus instead on the results, their use and the typical ways of reasoning.

In the next chapters we shall see how measure theory allows us to deal with many, often difficult, problems in probability. For more information concerning measure theory in view of probability and of further study see in the references the books [3], [5], [11], [12], [17], [19], [24], [20].

1.1 Measurable Spaces, Measurable Functions

Let E be a set and $\mathcal{E}$ a family of subsets of E.

$\mathcal{E}$ is a σ-*algebra* (resp. an *algebra*) if

- $E \in \mathcal{E}$,
- $\mathcal{E}$ is stable with respect to set complementation;
- $\mathcal{E}$ is stable with respect to countable (resp. finite) unions.

This means that if $A \in \mathcal{E}$ then also $A^c \in \mathcal{E}$ and that if $(A_n)_n \subset \mathcal{E}$ then also $\bigcup_n A_n \in \mathcal{E}$. Of course $\emptyset = E^c \in \mathcal{E}$.

Actually a σ-algebra is also stable with respect to countable intersections: if $(A_n)_n \subset \mathcal{E}$ then we can write

$$\bigcap_{n=1}^{\infty} A_n = \Big(\bigcup_{n=1}^{\infty} A_n^c \Big)^c ,$$

© The Author(s), under exclusive license to Springer Nature Switzerland AG 2023
P. Baldi, *Probability*, Universitext, https://doi.org/10.1007/978-3-031-38492-9_1

so that also $\bigcap_{n=1}^{\infty} A_n \in \mathscr{E}$.

A pair $(E, \mathscr{E})$, where $\mathscr{E}$ is a σ-algebra on E, is a *measurable space*.

Of course the family $\mathscr{P}(E)$ of all subsets of E is a σ-algebra and it is immediate that the intersection of *any* family of σ-algebras is a σ-algebra. Hence, given a class of sets $\mathscr{C} \subset \mathscr{P}(E)$, we can consider the smallest σ-algebra containing $\mathscr{C}$: it is the intersection of all σ-algebras containing $\mathscr{C}$ (such a family is non-empty as certainly $\mathscr{P}(E)$ belongs to it). It is the σ-*algebra generated* by $\mathscr{C}$, denoted $\sigma(\mathscr{C})$.

Definition 1.1 A *monotone class* is a family $\mathscr{M}$ of subsets of E such that

- $E \in \mathscr{M}$,
- $\mathscr{M}$ is stable with respect to relative complementation, i.e. if $A, B \in \mathscr{M}$ and $A \subset B$, then $B \setminus A \in \mathscr{M}$.
- $\mathscr{M}$ is stable with respect to increasing limits: if $(A_n)_n \subset \mathscr{M}$ is an increasing sequence of sets, then $A = \bigcup_n A_n \in \mathscr{M}$.

Note that a σ-algebra is a monotone class. Actually, if $\mathscr{E}$ is a σ-algebra, then

- if $A, B \in \mathscr{E}$ and $A \subset B$, then $A^c \in \mathscr{E}$ hence also $B \setminus A = B \cap A^c \in \mathscr{E}$;
- if $(A_n)_n \subset \mathscr{E}$ then $A = \bigcup_n A_n \in \mathscr{E}$, whether the sequence is increasing or not.

On the other hand, to prove that a family of sets is a monotone class may turn out to be easier than to prove that it is a σ-algebra. For this reason the next result will be useful in the sequel (for a proof, see e.g. [24], p. 39).

Theorem 1.2 (The Monotone Class Theorem) *Let $\mathscr{C} \subset \mathscr{P}(E)$ be a family of sets that is stable with respect to finite intersections and let $\mathscr{M}$ be a monotone class containing $\mathscr{C}$. Then $\mathscr{M}$ also contains $\sigma(\mathscr{C})$.*

Note that in the literature the definition of "monotone class" may be different and the statement of Theorem 1.2 modified accordingly (see e.g. [2], p. 43).

The next definition introduces an important class of σ-algebras.

Definition 1.3 Let E be a topological space and $\mathscr{O}$ the class of all open sets of E. The σ-algebra $\sigma(\mathscr{O})$ (i.e. the smallest one containing all open sets) is the *Borel σ-algebra* of E, denoted $\mathscr{B}(E)$.

Of course $\mathscr{B}(E)$ is also the smallest σ-algebra containing all closed sets. Actually the latter also contains all open sets, which are the complements of closed sets, hence also contains $\mathcal{B}(E)$, that is the smallest σ-algebra containing the open sets. By the same argument (closed sets are the complements of open sets) the σ-algebra generated by all closed sets is contained in $\mathcal{B}(E)$ hence the two σ-algebras coincide.

If E is a separable metric space, then $\mathcal{B}(E)$ is also generated by smaller families of sets.

Example 1.4 Assume that E is a separable metric space and let $D \subset E$ be a dense subset. Then the Borel σ-algebra $\mathcal{B}(E)$ is also generated by the family $\mathcal{D}$ of the balls centered at D with rational radius. Actually every open set is a countable union of these balls. Hence $\mathcal{B}(E) \subset \sigma(\mathcal{D})$ and, as the opposite inclusion is obvious, $\mathcal{B}(E) = \sigma(\mathcal{D})$.

Let $(E, \mathscr{E})$ and $(G, \mathscr{G})$ be measurable spaces. A map $f : E \to G$ is said to be *measurable* if, for every $A \in \mathscr{G}$, $f^{-1}(A) \in \mathscr{E}$.

It is immediate that if g is measurable from $(E, \mathscr{E})$ to $(G, \mathscr{G})$ and h is measurable from $(G, \mathscr{G})$ to $(H, \mathscr{H})$ then $h \circ g$ is measurable from $(E, \mathscr{E})$ to $(H, \mathscr{H})$.

Remark 1.5 (A very useful criterion) In order for f to be measurable it suffices to have $f^{-1}(A) \in \mathscr{E}$ for every $A \in \mathscr{C}$, where $\mathscr{C} \subset \mathscr{G}$ is such that $\sigma(\mathscr{C}) = \mathscr{G}$.

Indeed the class $\widetilde{\mathscr{G}}$ of the sets $A \subset G$ such that $f^{-1}(A) \in \mathscr{E}$ is a σ-algebra, thanks to the easy relations

$$\bigcup_{n=1}^{\infty} f^{-1}(A_n) = f^{-1}\left(\bigcup_{n=1}^{\infty} A_n\right), \qquad (1.1)$$
$$f^{-1}(A)^c = f^{-1}(A^c).$$

As $\widetilde{\mathscr{G}}$ contains the class $\mathscr{C}$, it also contains the whole σ-algebra $\mathscr{G}$ that is generated by $\mathscr{C}$. Therefore $f^{-1}(A) \in \mathscr{E}$ also for every $A \in \mathscr{G}$.

The criterion of Remark 1.5 is very useful as often one knows explicitly the sets of a class $\mathscr{C}$ generating $\mathscr{G}$, but not those of $\mathscr{G}$.

For instance, if $\mathscr{G}$ is the Borel σ-algebra of a topological space G, in order to establish the measurability of f it is sufficient, for instance, to check that $f^{-1}(A) \in \mathscr{E}$ for every open set A.

In particular, if E, G are topological spaces, a continuous map $f : E \to G$ is measurable with respect to the respective Borel σ-algebras.

1.2 Real Measurable Functions

If the target space is $\mathbb{R}, \overline{\mathbb{R}}, \overline{\mathbb{R}}^+, \mathbb{R}^d, \mathbb{C}$, we shall always understand that it is endowed with the respective Borel σ-algebra. Here $\overline{\mathbb{R}} = \mathbb{R} \cup \{+\infty, -\infty\}$ and $\overline{\mathbb{R}}^+ = \mathbb{R}^+ \cup \{+\infty\}$.

Let $(E, \mathscr{E})$ be a measurable space. In order for a numerical map (i.e. $\overline{\mathbb{R}}$-valued) to be measurable it is sufficient to have, for every $a \in \mathbb{R}$, $\{f > a\} = \{x, \ f(x) > a\} = f^{-1}(]a, +\infty]) \in \mathscr{E}$, as the sets of the form $]a, +\infty]$ generate the Borel σ-algebra (Exercise 1.2) and we can apply the criterion of Remark 1.5. Generating families of sets are also those of the form $\{f < a\}, \{f \le a\}, \{f \ge a\}$ (see Exercise 1.2).

Many natural operations are possible on numerical measurable functions. Are linear combinations, products, limits ...of measurable functions still measurable?

These properties are easily proved: for instance if $(f_n)_n$ is a sequence of measurable numerical functions and $h = \sup_n f_n$, then, for every $a \in \mathbb{R}$, the sets $\{f_n \le a\} = f_n^{-1}([-\infty, a])$ are measurable and

$$\{h \le a\} = \bigcap_{n=1}^{\infty} \{f_n \le a\},$$

hence $\{h \le a\}$ is measurable, being the countable intersection of measurable sets.

Similarly, if $g = \inf_n f_n$, then

$$\{g \ge a\} = \bigcap_{n=1}^{\infty} \{f_n \ge a\},$$

hence $\{g \ge a\}$ is also measurable.

Recall that

$$\overline{\lim_{n\to\infty}} f_n(x) = \lim_{n\to\infty} \downarrow \sup_{k\ge n} f_k(x), \quad \underline{\lim_{n\to\infty}} f_n(x) = \lim_{n\to\infty} \uparrow \inf_{k\ge n} f_k(x), \tag{1.2}$$

where these quantities are $\overline{\mathbb{R}}$-valued. If the f_n are measurable, then also $\overline{\lim}_{n\to\infty} f_n$, $\underline{\lim}_{n\to\infty} f_n$, $\lim_{n\to\infty} f_n$ (if it exists) are measurable: actually, for the $\overline{\lim}$ for instance, the functions $g_n = \sup_{k\ge n} f_k$ are measurable, being the supremum of measurable functions, and then also $\overline{\lim}_{n\to\infty} f_n$, being the infimum of the g_n.

As a consequence, if $(f_n)_n$ is a sequence of measurable real functions and $f_n \to_{n \to \infty} f$ pointwise, then f is measurable. This is true also for sequences of measurable functions with values in a separable metric space, see Exercise 1.6.

The same argument gives that if $f, g : E \to \overline{\mathbb{R}}$ are measurable then also $f \vee g$ and $f \wedge g$ are measurable. In particular

$$f^+ = f \vee 0 \quad \text{and} \quad f^- = -f \vee 0$$

are measurable functions. f^+ and f^- are the *positive and negative parts* of f and we have

$$f = f^+ - f^- ,$$
$$|f| = f^+ + f^- .$$

Note that *both* f^+ and f^- are positive functions.

Let f_1, f_2 be real measurable maps defined on the measurable space $(E, \mathscr{E})$. Then the map $f = (f_1, f_2)$ is measurable with values in $(\mathbb{R}^2, \mathscr{B}(\mathbb{R}^2))$. Indeed, if $A_1, A_2 \in \mathscr{B}(\mathbb{R})$, then $f^{-1}(A_1 \times A_2) = f_1^{-1}(A_1) \cap f_2^{-1}(A_2) \in \mathscr{E}$. Moreover, it is easy to prove, with the argument of Example 1.4, that every open set of $\mathbb{R}^2$ is a countable union of rectangles $A_1 \times A_2$, so that they generate $\mathscr{B}(\mathbb{R}^2)$ and we can apply the criterion of Remark 1.5.

As $(x, y) \mapsto x + y$ is a continuous map $\mathbb{R}^2 \to \mathbb{R}$, it is also measurable. It follows that $f_1 + f_2$ is also measurable, being the composition of the measurable maps $f = (f_1, f_2)$ and $(x, y) \mapsto x + y$. In the same way one can prove the measurability of the maps $f_1 f_2$ and $\frac{f_1}{f_2}$ (if defined). Similar results hold for numerical functions f_1 and f_2, provided that we ensure that indeterminate forms such as $+\infty - \infty, 0/0, \infty/\infty \dots$ are not possible.

> As these examples suggest, in order to prove the measurability of a real function f, one will seldom try to use the definition, but rather apply the criterion of Remark 1.5, investigating $f^{-1}(A)$ for A in a class of sets generating the σ-algebra of the target space, or, for real or numerical functions, writing f as the sum, product, limit, ... of measurable functions.

If $A \subset E$, the *indicator function of A*, denoted 1_A, is the function that takes the value 1 on A and 0 on A^c. We have the obvious relations

$$1_{A^c} = 1 - 1_A, \quad 1_{\cap A_n} = \prod_n 1_{A_n} = \inf_n 1_{A_n}, \quad 1_{\cup A_n} = \sup_n 1_{A_n}.$$

It is immediate that, if $A \in \mathscr{E}$, then 1_A is measurable. A function $f : (E, \mathscr{E}) \to \mathbb{R}$ is *elementary* if it is of the form $f = \sum_{k=1}^n a_k 1_{A_k}$ with $A_k \in \mathscr{E}$ and $a_k \in \mathbb{R}$.

The following result is fundamental, as it allows us to approximate every positive measurable function with elementary functions. It will be of constant use.

Proposition 1.6 *Every positive measurable function f is the limit of an increasing sequence of elementary positive functions.*

Proof Just consider

$$f_n(x) = \sum_{k=0}^{n2^n-1} \frac{k}{2^n} 1_{\{x; \frac{k}{2^n} \le f(x) < \frac{k+1}{2^n}\}}(x) + n 1_{\{f(x) \ge n\}},$$ (1.3)

i.e.

$$f_n(x) = \begin{cases} \dfrac{k}{2^n} & \text{if } f(x) < n \text{ and } \dfrac{k}{2^n} \le f(x) < \dfrac{k+1}{2^n} \\ n & \text{if } f(x) \ge n . \end{cases}$$

Clearly the sequence $(f_n)_n$ is increasing. Moreover, as $f(x) - \frac{1}{2^n} \le f_n(x) \le f(x)$ if $f(x) < n$, $f_n(x) \to_{n \to \infty} f(x)$. ∎

Let f be a map from E into another measurable space $(G, \mathcal{G})$. We denote by $\sigma(f)$ the σ-algebra generated by f, i.e. the smallest σ-algebra on E such that $f : (E, \sigma(f)) \to (G, \mathcal{G})$ is measurable.

It is easy to check that the family $\mathcal{E}_f = \{f^{-1}(A), \ A \in \mathcal{G}\}$ is a σ-algebra of subsets of E (use the relations (1.1)). Hence it coincides with $\sigma(f)$.

Actually $\sigma(f)$ must contain every set of the form $f^{-1}(A)$, so that $\sigma(f) \supset \mathcal{E}_f$. Conversely, f is obviously measurable with respect to $\mathcal{E}_f$, so that $\mathcal{E}_f$ must contain $\sigma(f)$, which, by definition, is the smallest σ-algebra enjoying this property.

More generally, if $(f_i, \ i \in I)$ is a family of maps on E with values respectively in the measurable spaces $(G_i, \mathcal{G}_i)$, we denote by $\sigma(f_i, \ i \in I)$ the smallest σ-algebra on E with respect to which all the f_i are measurable. We shall call $\sigma(f_i, \ i \in I)$ the σ-algebra generated by the f_i.

Proposition 1.7 (Doob's Criterion) *Let f be a map from E to some measurable space $(G, \mathcal{G})$ and let $h : E \to \overline{\mathbb{R}}$. Then h is $\sigma(f)$-measurable if and only if there exists a $\mathcal{G}$-measurable function $g : G \to \overline{\mathbb{R}}$ such that $h = g \circ f$ (see Fig. 1.1).*

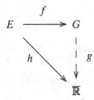

Fig. 1.1 Proposition 1.7 states the existence of a g such that $h = g \circ f$

Proof Of course if $h = g \circ f$ with g measurable, then h is $\sigma(f)$-measurable, being the composition of measurable maps.

Conversely, let us assume first that h is $\sigma(f)$-measurable, positive and elementary. Then h is of the form $h = \sum_{k=1}^{n} a_k 1_{B_k}$ with $B_k \in \sigma(f)$ and therefore $B_k = f^{-1}(A_k)$ for some $A_k \in \mathcal{G}$. As $1_{B_k} = 1_{f^{-1}(A_k)} = 1_{A_k} \circ f$, we can write $h = g \circ f$ with $g = \sum_{k=1}^{n} a_k 1_{A_k}$.

Let us drop the assumption that h is elementary and assume h positive and $\sigma(f)$-measurable. Then $h = \lim_{n \to \infty} \uparrow h_n$ for an increasing sequence $(h_n)_n$ of elementary positive functions (Proposition 1.6). Thanks to the first part of the proof, h_n is of the form $h_n = g_n \circ f$ with g_n positive and $\mathcal{G}$-measurable. We deduce that $h = g \circ f$ with $g = \overline{\lim}_{n \to \infty} g_n$, which is a positive $\mathcal{G}$-measurable function.

Let now h be $\sigma(f)$-measurable (not necessarily positive). It can be decomposed into the difference of its positive and negative parts, $h = h^+ - h^-$, and we know that we can write $h^+ = g^+ \circ f$, $h^- = g^- \circ f$ for some positive $\mathcal{G}$-measurable functions g^+, g^-. The function h being $\sigma(f)$-measurable, we have $\{h \geq 0\} = f^{-1}(A_1)$ and $\{h < 0\} = f^{-1}(A_2)$ for some $A_1, A_2 \in \mathcal{G}$. Therefore $h = g \circ f$ with $g = g^+ 1_{A_1} - g^- 1_{A_2}$. There is no danger of encountering a form $+\infty - \infty$ as the sets A_1, A_2 are disjoint. ∎

1.3 Measures

Let $(E, \mathcal{E})$ be a measurable space.

> **Definition 1.8** A measure on $(E, \mathcal{E})$ is a map $\mu : \mathcal{E} \to \overline{\mathbb{R}}^+$ (it can also take the value $+\infty$) such that
>
> (a) $\mu(\emptyset) = 0$,
> (b) for every sequence $(A_n)_n \subset \mathcal{E}$ of *pairwise disjoint* sets
>
> $$\mu\left(\bigcup_{n \geq 1} A_n\right) = \sum_{n=1}^{\infty} \mu(A_n).$$
>
> The triple $(E, \mathcal{E}, \mu)$ is a *measure space*.

Some terminology.

- If $E = \bigcup_n E_n$ for $E_n \in \mathscr{E}$ with $\mu(E_n) < +\infty$, μ is said to be σ-*finite*.
- If $\mu(E) < +\infty$, μ is said to be *finite*.
- If $\mu(E) = 1$, μ is a *probability*, or also a *probability measure*.

As we shall see, the assumption that μ is σ-finite will be necessary in most statements.

Remark 1.9 Property (b) of Definition 1.8 is called σ-*additivity*. If in Definition 1.8 we assume that $\mathscr{E}$ is only an algebra and we add to (b) the condition that $\bigcup_n A_n \in \mathscr{E}$, then we have the notion of a *measure on an algebra*.

Remark 1.10 (A Few Properties of a Measure as a Consequence of the Definition) (a) If $A, B \in \mathscr{E}$ and $A \subset B$, then $\mu(A) \leq \mu(B)$. Indeed A and $B \cap A^c$ are disjoint measurable sets and their union is equal to B. Therefore $\mu(B) = \mu(A) + \mu(B \cap A^c) \geq \mu(A)$.

(b) If $(A_n)_n \subset \mathscr{E}$ is a sequence of measurable sets *increasing* to A, i.e. such that $A_n \subset A_{n+1}$ and $A = \bigcup_n A_n$, then $\mu(A_n) \uparrow \mu(A)$ as $n \to \infty$.

Indeed let $B_1 = A_1$ and recursively define $B_n = A_n \setminus A_{n-1}$. The B_n are pairwise disjoint ($B_{n-1} \subset A_{n-1}$ whereas $B_n \subset A_{n-1}^c$) and, clearly, $B_1 \cup \cdots \cup B_n = A_n$. Hence

$$A = \bigcup_{n=1}^{\infty} A_n = \bigcup_{n=1}^{\infty} B_n$$

and, as the B_n are pairwise disjoint,

$$\mu(A) = \mu\Big(\bigcup_{k=1}^{\infty} B_k \Big) = \sum_{k=1}^{\infty} \mu(B_k) = \lim_{n \to \infty} \sum_{k=1}^{n} \mu(B_k) = \lim_{n \to \infty} \mu(A_n).$$

(c) If $(A_n)_n \subset \mathscr{E}$ is a sequence of measurable sets decreasing to A (i.e. such that $A_{n+1} \subset A_n$ and $\bigcap_{n=1}^{\infty} A_n = A$) and if, for some n_0, $\mu(A_{n_0}) < +\infty$, then $\mu(A_n) \downarrow \mu(A)$ as $n \to \infty$.

Indeed we have $A_{n_0} \setminus A_n \uparrow A_{n_0} \setminus A$ as $n \to \infty$. Hence, using the result of (b) on the increasing sequence $(A_{n_0} \setminus A_n)_n$,

$$\mu(A) \stackrel{\downarrow}{=} \mu(A_{n_0}) - \mu(A_{n_0} \setminus A) = \mu(A_{n_0}) - \lim_{n \to \infty} \mu(A_{n_0} \setminus A_n)$$

$$= \lim_{n \to \infty} \left(\mu(A_{n_0}) - \mu(A_{n_0} \setminus A_n) \right) = \lim_{n \to \infty} \mu(A_n)$$

($\downarrow$ denotes the equality where the assumption $\mu(A_{n_0}) < +\infty$ is necessary). In general, a measure does not necessarily pass to the limit along decreasing sequences of events (we shall see examples). Note, however, that the condition $\mu(A_{n_0}) < +\infty$ for some n_0 is always satisfied if μ is finite.

The next, very important, statement says that if two measures coincide on a class of sets that is large enough, then they coincide on the whole generated σ-algebra.

Proposition 1.11 (Carathéodory's Criterion) *Let μ, ν be measures on the measurable space $(E, \mathscr{E})$ and let $\mathscr{C} \subset \mathscr{E}$ be a class of sets which is stable with respect to finite intersections and such that $\sigma(\mathscr{C}) = \mathscr{E}$. Assume that*

- *for every $A \in \mathscr{C}$, $\mu(A) = \nu(A)$;*
- *there exists an increasing sequence of sets $(E_n)_n \subset \mathscr{C}$ such that $E = \bigcup_n E_n$ and $\mu(E_n) < +\infty$ (hence also $\nu(E_n) < +\infty$) for every n.*

Then $\mu(A) = \nu(A)$ for every $A \in \mathscr{E}$.

Proof Let us assume first that μ and ν are finite. Let $\mathscr{M} = \{A \in \mathscr{E}, \ \mu(A) = \nu(A)\}$ (the family of sets of $\mathscr{E}$ on which the two measures coincide) and let us check that $\mathscr{M}$ is a monotone class. We have

- $\mu(E) = \lim_{n \to \infty} \mu(E_n) = \lim_{n \to \infty} \nu(E_n) = \nu(E)$, so that $E \in \mathscr{M}$.
- If $A, B \in \mathscr{M}$ and $A \subset B$ then, as A and $B \setminus A$ are disjoint sets and their union is equal to B,

$$\mu(B \setminus A) \stackrel{\downarrow}{=} \mu(B) - \mu(A) = \nu(B) - \nu(A) = \nu(B \setminus A)$$

and therefore $\mathscr{M}$ is stable with respect to relative complementation ($\downarrow$: here we use the assumption that μ and ν are finite).
- If $(A_n)_n \subset \mathscr{M}$ is an increasing sequence of sets and $A = \bigcup_n A_n$, then (Remark 1.10 (b))

$$\mu(A) = \lim_{n \to \infty} \mu(A_n) = \lim_{n \to \infty} \nu(A_n) = \nu(A) \, ,$$

so that also $A \in \mathcal{M}$. By Theorem 1.2, the Monotone Class Theorem, $\mathcal{E} = \sigma(\mathcal{E}) \subset \mathcal{M}$, hence μ and ν coincide on $\mathcal{E}$.

In order to deal with the general case (i.e. μ and ν not necessarily finite), let, for $A \in \mathcal{E}$,

$$\mu_n(A) = \mu(A \cap E_n) , \qquad \nu_n(A) = \nu(A \cap E_n) .$$

It is easy to check that μ_n, ν_n are measures on $\mathcal{E}$ and as $\mu_n(E) = \mu(E_n) < +\infty$ and $\nu_n(E) = \nu(E_n) < +\infty$ they are finite. They obviously coincide on $\mathcal{E}$ (which is stable with respect to finite intersections) and, thanks to the first part of the proof, also on $\mathcal{E}$. Now, if $A \in \mathcal{E}$, as $A \cap E_n \uparrow A$, we have

$$\mu(A) = \lim_{n \to \infty} \mu(A \cap E_n) = \lim_{n \to \infty} \nu(A \cap E_n) = \nu(A) .$$

■

Remark 1.12 If μ and ν are finite measures, the statement of Proposition 1.11 can be simplified: if μ and ν coincide on a class $\mathcal{E}$ which is stable with respect to finite intersections, containing E and generating $\mathcal{E}$, then they coincide on $\mathcal{E}$.

An interesting, and natural, problem is the construction of measures satisfying particular properties. For instance, such that they take given values on some classes of sets. The key tool in this direction is the following theorem. We shall skip its proof.

Theorem 1.13 (Carathéodory's Extension Theorem) *Let μ be a measure on an algebra $\mathcal{A}$ (see Remark 1.9). Then μ can be extended to a measure on $\sigma(\mathcal{A})$. Moreover, if μ is σ-finite this extension is unique.*

Let us now introduce a particular class of measures.

A *Borel measure* on a topological space E is a measure on $(E, \mathcal{B}(E))$ such that $\mu(K) < +\infty$ for every compact set $K \subset E$.

Let us have a closer look at the Borel measures on $\mathbb{R}$. Note first that the class $\mathcal{E} = \{]a, b], -\infty < a < b < +\infty\}$ (the half-open intervals) is stable with respect to finite intersections and that $\sigma(\mathcal{E}) = \mathcal{B}(\mathbb{R})$ (Exercise 1.2). Thanks to Proposition 1.11 (Carathéodory's criterion), a Borel measure μ on $\mathscr{B}(\mathbb{R})$ is determined by the values $\mu(]a, b])$, $a, b \in \mathbb{R}$, $a < b$, which are finite, as μ is finite on compact sets. Given such a measure let us define a function F by setting $F(0) = 0$ and

$$F(x) = \begin{cases} \mu(]0, x]) & \text{if } x > 0 \\ -\mu(]x, 0]) & \text{if } x < 0 . \end{cases} \tag{1.4}$$

Then F is right-continuous, as a consequence of Remark 1.10 (c): if $x > 0$ and $x_n \downarrow x$, then $]0, x] = \bigcap_n]0, x_n]$ and, as the sequence $(]0, x_n])_n$ is decreasing and $(\mu(]0, x_n]))_n$ is bounded by $\mu(]0, x_1])$, we have $F(x_n) = \mu(]0, x_n]) \downarrow \mu(]0, x]) = F(x)$. If $x < 0$ or $x = 0$ the argument is the same. F is obviously increasing and we have

$$\mu(]a, b]) = F(b) - F(a) . \tag{1.5}$$

A right-continuous increasing function F satisfying (1.5) is a *distribution function* (d.f.) of μ. Of course the d.f. of a Borel measure on $\mathbb{R}$ is not unique: $F + c$ is again a d.f. for every $c \in \mathbb{R}$.

Conversely, let $F : \mathbb{R} \to \mathbb{R}$ be an increasing right-continuous function, does a measure μ on $\mathscr{B}(\mathbb{R})$ exist such that $\mu(]a, b]) = F(b) - F(a)$? Such a μ would be a Borel measure, of course.

Let us try to apply Theorem 1.13, Carathéodory's existence theorem. Let $\mathscr{C}$ be the family of sets formed by the half-open intervals $]a, b]$. It is immediate that the algebra $\mathscr{A}$ generated by $\mathscr{C}$ is the family of finite disjoint unions of these intervals, i.e.

$$\mathscr{A} = \left\{ A = \bigcup_{k=1}^{n}]a_k, b_k], -\infty \le a_1 < b_1 < a_2 < \cdots < b_{n-1} < a_n < b_n \le +\infty \right\}$$

with the understanding $]a_n, b_n] =]a_n, +\infty[$ if $b_n = +\infty$.

Let us define μ on $\mathscr{A}$ by setting $\mu(A) = \sum_{k=1}^{n}(F(b_k) - F(a_k))$, with $F(+\infty) = \lim_{x \to +\infty} F(x)$, $F(-\infty) = \lim_{x \to -\infty} F(x)$. It is easy to prove that μ is additive on $\mathscr{A}$; a bit more delicate is to prove that μ is σ-additive on $\mathscr{A}$, and we shall skip the proof of this fact. As $\sigma(\mathscr{A}) = \mathscr{B}(\mathbb{R})$, we have therefore, thanks to Theorem 1.13, the following result that characterizes the Borel measures on $\mathbb{R}$.

Theorem 1.14 *Let $F : \mathbb{R} \to \mathbb{R}$ be a right-continuous increasing function. Then there exists a unique Borel measure μ on $\mathscr{B}(\mathbb{R})$ such that, for every $a < b$, $\mu(]a, b]) = F(b) - F(a)$.*

Uniqueness, of course, is a consequence of Proposition 1.11, Carathéodory's criterion, as the class $\mathscr{C}$ of the half-open intervals is stable with respect to finite intersections and generates $\mathscr{B}(\mathbb{R})$ (Exercise 1.2).

Borel measures on $\mathbb{R}$ are, of course, σ-finite: the sets $]-n, n]$ have finite measure equal to $F(n) - F(-n)$, and their union is equal to $\mathbb{R}$. The property of σ-finiteness of Borel measures holds in more general topological spaces: actually it is sufficient for the space to be σ-compact (i.e. a countable union of compact sets), which is the case, for instance, if it is locally compact and separable (see Lemma 1.26 below).

If we choose $F(x) = x$, we obtain existence and uniqueness of a measure λ on $\mathscr{B}(\mathbb{R})$ such that $\lambda(I) = |I| = b - a$ for every interval $I =]a, b]$. This is the *Lebesgue measure* of $\mathbb{R}$.

Let $(E, \mathscr{E}, \mu)$ be a measure space. A subset $A \in \mathscr{E}$ is said to be *negligible* if $\mu(A) = 0$. We say that a property is true *almost everywhere* (a.e.) if it is true outside a negligible set. For instance, $f = g$ a.e. means that the set $\{x \in E, f(x) \neq g(x)\}$ is negligible. If μ is a probability, we say *almost surely* (a.s.) instead of a.e.

Beware that in the literature sometimes a slightly different definition of negligible set can be found.

Note that if $(A_n)_n$ is a sequence of negligible sets, then their union is also negligible (Exercise 1.7).

Remark 1.15 If $(f_n)_n$ is a sequence of real measurable functions such that $f_n \to_{n\to\infty} f$, then we know that f is also measurable. But what if the convergence only takes place a.e.?

Let N be the negligible set outside which the convergence takes place and let $\widetilde{f_n} = f_n 1_N$. Then the $\widetilde{f_n}$ are also measurable and converge, everywhere, to $\widetilde{f} := f 1_N$, which is therefore measurable.

In conclusion, we can state that there exists at least one measurable function which is the a.e. limit of $(f_n)_n$.

Using Exercise 1.6, this remark also holds for sequences $(f_n)_n$ of functions with values in a separable metric space.

1.4 Integration

Let $(E, \mathscr{E}, \mu)$ be a measure space. In this section we define the integral with respect to μ. As above we shall be more interested in ideas and tools and shall skip the more technical proofs.

Let us first define the integral with respect to μ of a measurable function f : $E \to \overline{\mathbb{R}}^+$. If f is positive elementary then $f = \sum_{k=1}^n a_k 1_{A_k}$, with $A_k \in \mathscr{E}$, and $a_k \geq 0$ and we can define

$$\int_E f \, d\mu := \sum_{k=1}^n a_k \mu(A_k) \, .$$

Some simple remarks show that this number (which can turn out to be $= +\infty$) does not depend on the representation of f (different numbers a_k and sets A_k can define the same function). If f, g are positive and elementary, we have easily

(a) if $a, b > 0$ then $\int (af + bg) \, d\mu = a \int f \, d\mu + b \int g \, d\mu$,
(b) if $f \leq g$, then $\int f \, d\mu \leq \int g \, d\mu$.

The following technical result is the key to the construction.

Lemma 1.16 *If $(f_n)_n$, $(g_n)_n$ are increasing sequences of positive elementary functions such that $\lim_{n \to \infty} \uparrow f_n = \lim_{n \to \infty} \uparrow g_n$, then also*

$$\lim_{n \to \infty} \uparrow \int_E f_n \, d\mu = \lim_{n \to \infty} \uparrow \int_E g_n \, d\mu \, .$$

Let now $f : E \to \overline{\mathbb{R}}^+$ be a positive $\mathscr{E}$-measurable function. Thanks to Proposition 1.6 there exists a sequence $(f_n)_n$ of elementary positive functions such that $f_n \uparrow f$ as $n \to \infty$; then the sequence $(\int f_n \, d\mu)_n$ of their integrals is increasing thanks to (b) above; let us define

$$\int_E f \, d\mu := \lim_{n \to \infty} \uparrow \int_E f_n \, d\mu \, . \tag{1.6}$$

By Lemma 1.16, this limit does not depend on the particular approximating sequence $(f_n)_n$, hence (1.6) is a good definition. Taking the limit, we obtain immediately that, if f, g are positive measurable, then

- for every $a, b > 0$, $\int (af + bg) \, d\mu = a \int f \, d\mu + b \int g \, d\mu$;
- if $f \leq g$, $\int f \, d\mu \leq \int g \, d\mu$.

In order to define the integral of a numerical $\mathscr{E}$-measurable function, let us write the decomposition $f = f^+ - f^-$ of f into positive and negative parts. The simple idea is to define

$$\int_E f \, d\mu := \int_E f^+ \, d\mu - \int_E f^- \, d\mu$$

. provided that at least one of the quantities $\int f^+ d\mu$ and $\int f^- d\mu$ is finite.

- f is said to be *lower semi-integrable* (l.s.i.) if $\int f^- d\mu < +\infty$. In this case the integral of f is well defined (but can take the value $+\infty$).
- f is said to be *upper semi-integrable* (u.s.i.) if $\int f^+ d\mu < +\infty$. In this case the integral of f is well defined (but can take the value $-\infty$).
- f is said to be *integrable* if both f^+ and f^- have finite integral.

Clearly a function is l.s.i. if and only if it is bounded below by an integrable function. A positive function is always l.s.i. and a negative one is always u.s.i. Moreover, as $|f| = f^+ + f^-$, f is integrable if and only if $\int |f| d\mu < +\infty$.

If f is semi-integrable (upper or lower) we have the inequality

$$\left| \int_E f \, d\mu \right| = \left| \int_E f^+ \, d\mu - \int_E f^- \, d\mu \right|$$

$$\leq \int_E f^+ \, d\mu + \int_E f^- \, d\mu = \int_E |f| \, d\mu .$$

(1.7)

Note the difference of the integral just defined (the Lebesgue integral) with respect to the Riemann integral: in both of them the integral is first defined for a class of elementary functions. But for the Riemann integral the elementary functions are piecewise constant and defined by splitting the *domain* of the function. Here the elementary functions (have a look at the proof of Proposition 1.6) are obtained by splitting its *co-domain*.

The integral is easily defined also for complex-valued functions. If $f : E \to \mathbb{C}$, and $f = f_1 + if_2$, then it is immediate that if $\int |f| d\mu < +\infty$ (here $| \ |$ denotes the complex modulus), then both f_1 and f_2 are integrable, as both $|f_1|$ and $|f_2|$ are majorized by $|f|$. Thus we can define

$$\int_E f \, d\mu = \int_E f_1 \, d\mu + i \int_E f_2 \, d\mu .$$

Also (a bit less obvious) (1.7) still holds, with $| \ |$ meaning the complex modulus.

It is easy to deduce from the properties of the integral of positive functions that

- (linearity) if $a, b \in \mathbb{C}$ and f and g are both integrable, then also $af + bg$ is integrable and $\int (af + bg)\, d\mu = a \int f\, d\mu + b \int g\, d\mu$;
- (monotonicity) if f and g are real and semi-integrable and $f \leq g$, then $\int f\, d\mu \leq \int g\, d\mu$.

The following properties are often very useful (see Exercise 1.9).

(a) If f is positive measurable and if $\int f\, d\mu < +\infty$, then $f < +\infty$ a.e. (recall that we consider numerical functions that can take the value $+\infty$)
(b) If f is positive measurable and $\int f\, d\mu = 0$ then $f = 0$ a.e.

The reader is encouraged to write down the proofs: it is important to become acquainted with the simple arguments they use.

If f is positive measurable (resp. integrable) and $A \in \mathscr{B}$, then $f 1_A$ is itself positive measurable (resp. integrable). We define then

$$\int_A f\, d\mu := \int_E f 1_A\, d\mu \,.$$

The following are the three classical convergence results.

Theorem 1.17 (Beppo Levi's Theorem or the Monotone Convergence Theorem) *Let $(f_n)_n$ be an increasing sequence of measurable functions bounded from below by an integrable function and $f = \lim_{n \to \infty} \uparrow f_n$. Then*

$$\lim_{n \to \infty} \uparrow \int_E f_n\, d\mu = \int_E f\, d\mu \,.$$

We already know (Remark 1.10 (b)) that if $f_n = 1_{A_n}$ where $(A_n)_n$ is an increasing sequence of measurable sets, then $f_n \uparrow f = 1_A$ where $A = \bigcup_n A_n$ and

$$\int_E f_n\, d\mu = \mu(A_n) \uparrow \mu(A) = \int_E f\, d\mu \,.$$

Hence Beppo Levi's Theorem is an extension of the property of passing to the limit of a measure on increasing sequences of sets.

Proposition 1.18 (Fatou's Lemma) *Let $(f_n)_n$ be a sequence of measurable functions bounded from below (resp. from above) by an integrable function, then*

$$\lim_{n\to\infty} \int_E f_n \, d\mu \geq \int_E \lim_{n\to\infty} f_n \, d\mu$$

$$\left(resp. \ \overline{\lim_{n\to\infty}} \int_E f_n \, d\mu \leq \int_E \overline{\lim_{n\to\infty}} f_n \, d\mu \right).$$

Fatou's Lemma and Beppo Levi's Theorem are most frequently applied to sequences of positive functions.

Fatou's Lemma implies

Theorem 1.19 (Lebesgue's Theorem) *Let $(f_n)_n$ be a sequence of integrable functions such that $f_n \to_{n\to\infty} f$ a.e. and such that, for every n, $|f_n| \leq g$ for some integrable function g. Then*

$$\lim_{n\to\infty} \int_E f_n \, d\mu = \int_E f \, d\mu.$$

Lebesgue's Theorem has a useful "continuous" version.

Corollary 1.20 *Let $(f_t)_{t\in U}$ be a family of integrable functions, where $U \subset \mathbb{R}^d$ is an open set. Assume that $\lim_{t\to t_0} f_t = f$ a.e. and that, for every $t \in U$, $|f_t| \leq g$ for some integrable function g. Then $\lim_{t\to t_0} \int f_t \, d\mu = \int f \, d\mu$.*

Proof Just note that $\lim_{t\to t_0} \int f_t \, d\mu = \int f \, d\mu$ if and only if, for every sequence $(t_n)_n \subset U$ converging to t_0, $\lim_{n\to\infty} \int f_{t_n} \, d\mu = \int f \, d\mu$, which holds thanks to Theorem 1.19. ∎

This corollary has an important application.

Proposition 1.21 (Derivation Under the Integral Sign) *Let* $(E, \mathscr{E}, \mu)$ *be a measure space,* $I \subset \mathbb{R}$ *an open interval and* $(f(t, x), t \in I)$ *a family of integrable functions* $E \to \mathbb{C}$. *Let, for every* $t \in I$,

$$\phi(t) = \int_E f(t, x) \, d\mu(x) .$$

Let us assume that there exists a negligible set $N \in \mathscr{E}$ *such that*

- *for every* $x \in N^c$, $t \mapsto f(t, x)$ *is differentiable on* I;
- *there exists an integrable function* g *such that, for every* $t \in I$, $x \in N^c$, $|\frac{\partial f}{\partial t}(t, x)| \leq g(x)$.

Then ϕ *is differentiable in the interior of* I *and*

$$\phi'(t) = \int_E \frac{\partial f}{\partial t}(t, x) \, d\mu(x) . \tag{1.8}$$

Proof Let $t \in I$. The idea is to write, for $h > 0$,

$$\frac{1}{h}\big(\phi(t + h) - \phi(t)\big) = \int \frac{1}{h}\big(f(t + h, x) - f(t, x)\big) \, d\mu(x) \tag{1.9}$$

and then to take the limit as $h \to 0$. We have for every $x \in N^c$

$$\frac{1}{h}\big(f(t + h, x) - f(t, x)\big) \underset{h \to 0}{\to} \frac{\partial f}{\partial t}(t, x)$$

and by the mean value theorem, for $x \in N^c$,

$$\left|\frac{1}{h}\big(f(t + h, x) - f(t, x)\big)\right| = \left|\frac{\partial f}{\partial t}(\tau, x)\right| \leq g(x)$$

for some τ, $t \leq \tau \leq t + h$ (τ possibly depending on x). Hence by Lebesgue's Theorem in the version of Corollary 1.20

$$\int_E \frac{1}{h}\big(f(t + h, x) - f(t, x)\big) \, d\mu(x) \underset{h \to 0}{\to} \int_E \frac{\partial f}{\partial t}(t, x) \, d\mu(x) .$$

Going back to (1.9), this proves that ϕ is differentiable and that (1.8) holds. ∎

Another useful consequence of the "three convergence theorems" is the following result of integration by series.

Corollary 1.22 *Let $(E, \mathscr{E}, \mu)$ be a measure space.*

(a) *Let $(f_n)_n$ be a sequence of positive measurable functions. Then*

$$\sum_{k=1}^{\infty} \int_E f_k \, d\mu = \int_E \sum_{k=1}^{\infty} f_k \, d\mu \,. \tag{1.10}$$

(b) *Let $(f_n)_n$ be a sequence of real measurable functions such that*

$$\sum_{k=1}^{\infty} \int_E |f_k| \, d\mu < +\infty \,. \tag{1.11}$$

Then (1.10) holds.

Proof

(a) As the partial sums increase to the sum of the series, (1.10) follows as

$$\sum_{k=1}^{\infty} \int_E f_k \, d\mu = \lim_{n\to\infty} \sum_{k=1}^{n} \int_E f_k \, d\mu = \lim_{n\to\infty} \int_E \sum_{k=1}^{n} f_k \, d\mu \overset{\downarrow}{=} \int_E \sum_{k=1}^{\infty} f_k \, d\mu \,,$$

where the equality indicated by $\downarrow$ is justified by Beppo Levi's Theorem.

(b) Thanks to (a) we have

$$\sum_{k=1}^{\infty} \int_E |f_k| \, d\mu = \int_E \sum_{k=1}^{\infty} |f_k| \, d\mu$$

so that by (1.11) the sum $\sum_{k=1}^{\infty} |f_k|$ is integrable. Then, as above,

$$\sum_{k=1}^{\infty} \int_E f_k \, d\mu = \lim_{n\to\infty} \sum_{k=1}^{n} \int_E f_k \, d\mu = \lim_{n\to\infty} \int_E \sum_{k=1}^{n} f_k \, d\mu \overset{\downarrow}{=} \int_E \sum_{k=1}^{\infty} f_k \, d\mu \,,$$

where now $\downarrow$ follows by Lebesgue's Theorem, as

$$\left| \sum_{k=1}^{n} f_k \right| \leq \sum_{k=1}^{\infty} |f_k| \qquad \text{for every } n \,,$$

so that the partial sums are bounded in modulus by an integrable function. ∎

Example 1.23 Let us compute

$$\int_{-\infty}^{+\infty} \frac{x}{\sinh x} \, dx \, .$$

Recall the power series expansion $\frac{1}{1-x} = \sum_{k=0}^{\infty} x^k$ (for $|x| < 1$), so that, for $x > 0$,

$$\frac{1}{1 - e^{-2x}} = \sum_{k=0}^{\infty} e^{-2kx} \, .$$

As $x \mapsto \frac{x}{\sinh x}$ is an even function we have

$$\int_{-\infty}^{+\infty} \frac{x}{\sinh x} \, dx = 4 \int_{0}^{+\infty} \frac{x}{e^x - e^{-x}} \, dx = 4 \int_{0}^{+\infty} \frac{xe^{-x}}{1 - e^{-2x}} \, dx$$

$$= 4 \int_{0}^{+\infty} \sum_{k=0}^{\infty} xe^{-(2k+1)x} \, dx = 4 \sum_{k=0}^{\infty} \int_{0}^{+\infty} xe^{-(2k+1)x} \, dx$$

$$= 4 \sum_{k=0}^{\infty} \frac{1}{(2k+1)^2} = \frac{\pi^2}{2} \, .$$

Integration by series is authorized here, everything being positive.

Let us denote by $\mathscr{E}^+$ the family of positive measurable functions. If we write, for simplicity, $I(f) = \int_E f \, d\mu$, the integral is a functional $I : \mathscr{E}^+ \to \overline{\mathbb{R}}^+$. We know that I enjoys the properties:

(a) (positive linearity) if $f, g \in \mathscr{E}^+$ and $a, b \in \overline{\mathbb{R}}^+$, $I(af + bg) = aI(f) + bI(g)$ (with the understanding $0 \cdot +\infty = 0$);
(b) if $(f_n)_n \subset \mathscr{E}^+$ and $f_n \uparrow f$, then $I(f_n) \uparrow I(f)$ (this is Beppo Levi's Theorem).

We have seen how, given a measure, we can define the integral I with respect to it and that I is a functional $\mathscr{E}^+ \to \overline{\mathbb{R}}^+$ satisfying (a) and (b) above. Let us see now how it is possible to reverse the argument, i.e. how, starting from a given functional $I : \mathscr{E}^+ \to \overline{\mathbb{R}}^+$ enjoying the properties (a) and (b) above, a measure μ on $(E, \mathscr{E})$ can be defined such that I is the integral with respect to μ.

Proposition 1.24 *Let $(E, \mathscr{E})$ be a measurable space and $I : \mathscr{E}^+ \to \overline{\mathbb{R}}^+$ a functional enjoying the properties (a) and (b) above. Then $\mu(A) := I(1_A)$, $A \in \mathscr{E}$, defines a measure on $\mathscr{E}$ and, for every $f \in \mathscr{E}^+$, $I(f) = \int f \, d\mu$.*

Proof Let us prove that μ is a measure. Let $f_0 \equiv 0$, then $\mu(\emptyset) = I(f_0) = I(0 \cdot f_0) = 0 \cdot I(f_0) = 0$.

As for σ-additivity: let $(A_n)_n \subset \mathscr{E}$ be a sequence of pairwise disjoint sets whose union is equal to A; then $1_A = \sum_{k=1}^{\infty} 1_{A_k} = \lim_{n \to \infty} \uparrow \sum_{k=1}^{n} 1_{A_k}$ and, thanks to the properties (a) and (b) above,

$$\mu(A) = I(1_A) = I\left(\lim_{n \to \infty} \uparrow \sum_{k=1}^{n} 1_{A_k} \right) = \lim_{n \to \infty} \uparrow \sum_{k=1}^{n} I(1_{A_k})$$

$$= \lim_{n \to \infty} \uparrow \sum_{k=1}^{n} \mu(A_k) = \sum_{k=1}^{\infty} \mu(A_k) \, .$$

Hence μ is a measure on $(E, \mathscr{E})$. Moreover, by (a) above, for every positive elementary function $f = \sum_{k=1}^{m} a_k 1_{A_k}$,

$$\int_E f \, d\mu = \sum_{k=1}^{m} a_k \mu(A_k) = I(f) \, ,$$

hence the integral with respect to μ coincides with the functional I on positive elementary functions. Proposition 1.6 and (b) above give that $\int f \, d\mu = I(f)$ for every $f \in \mathscr{E}^+$. ∎

Thanks to Proposition 1.24, measure theory and integration can be approached in two different ways.

- The first approach is to investigate and construct measures, i.e. set functions μ satisfying Definition 1.8, and then construct the integral of measurable functions with respect to measures (thus obtaining functionals on positive functions).
- The second approach is to directly construct functionals on $\mathscr{E}^+$ satisfying properties (a) and (b) above and then obtain measures by applying these functionals to functions that are indicators of sets, as in Proposition 1.24.

  These two points of view are equivalent but, according to the situation, one of them may turn out to be significantly simpler. So far we have followed the first one but we shall see situations where the second one turns out to be much easier.

A question that we shall often encounter in the sequel is the following: assume that we know that the integrals with respect to two measures μ and ν coincide for every function in some class $\mathcal{D}$, for example continuous functions in a topological space setting. Can we deduce that $\mu = \nu$?

With this goal in mind it is useful to have results concerning the approximation of indicator functions by means of "regular" functions. If E is a metric space and $G \subset E$ is an open set, let us consider the sequence of continuous functions $(f_n)_n$ defined as

$$f_n(x) = n\,d(x, G^c) \wedge 1\,. \tag{1.12}$$

A quick look shows immediately that f_n vanishes on G^c whereas $f_n(x)$ increases to 1 if $x \in G$. Therefore $f_n \uparrow 1_G$ as $n \to \infty$.

> **Proposition 1.25** *Let (E, d) be a metric space.*
>
> *(a) Let μ, ν be finite measures on $\mathcal{B}(E)$ such that*
>
> $$\int_E f\,d\mu = \int_E f\,d\nu \tag{1.13}$$
>
> *for every bounded continuous function $f : E \to \mathbb{R}$. Then $\mu = \nu$.*
> *(b) Assume that E is also separable and locally compact and that μ and ν are Borel measures on E (not necessarily finite). Then if (1.13) holds for every function f which is continuous and compactly supported we have $\mu = \nu$.*

Proof (a) Let $G \subset E$ be an open set and let f_n be as in (1.12). As $f_n \uparrow 1_G$ as $n \to \infty$, by Beppo Levi's Theorem

$$\mu(G) = \lim_{n \to \infty} \int_E f_n\,d\mu = \lim_{n \to \infty} \int_E f_n\,d\nu = \nu(G)\,, \tag{1.14}$$

hence μ and ν coincide on open sets. Taking $f \equiv 1$ we have also $\mu(E) = \nu(E)$: just take $f \equiv 1$ in (1.13). As the class of open sets is stable with respect to finite intersections the result follows thanks to Carathéodory's criterion, Proposition 1.11.

(b) Let $G \subset E$ be a relatively compact open set and f_n as in (1.12). f_n is continuous and compactly supported (its support is contained in $\overline{G}$) and, by (1.14), $\mu(G) = \nu(G)$.

Hence μ and ν coincide on the class $\mathscr{C}$ of relatively compact open subsets of E. Thanks to Lemma 1.26 below, there exists a sequence $(W_n)_n$ of relatively compact open sets increasing to E. Hence we can apply Carathéodory's criterion, Proposition 1.11, and μ and ν also coincide on $\sigma(\mathscr{C})$. Moreover, every open set G belongs to $\sigma(\mathscr{C})$ as

$$G = \bigcup_{n=1}^{\infty} G \cap W_n$$

and $G \cap W_n$ is a relatively compact open set. Hence $\sigma(\mathscr{C})$ contains all open sets and also the Borel σ-algebra $\mathscr{B}(E)$, completing the proof. ∎

Lemma 1.26 *Let E be a locally compact separable metric space.*

(a) *E is the countable union of an increasing sequence of relatively compact open sets. In particular, E is σ-compact, i.e. the union of a countable family of compact sets.*

(b) *There exists an increasing sequence of compactly supported continuous functions $(h_n)_n$ such that $h_n \uparrow 1$ as $n \to \infty$.*

Proof (a) Let $\mathscr{D}$ be the family of open balls with rational radius centered at the points of a given countable dense subset D. $\mathscr{D}$ is countable and every open set of E is the union (countable of course) of elements of $\mathscr{D}$.

Every $x \in E$ has a relatively compact neighborhood U_x, E being assumed to be locally compact. Then $V \subset U_x$ for some $V \in \mathscr{D}$. Such balls V are relatively compact, as $V \subset U_x$. The balls V that are contained in some of the U_x's as above are countably many as $\mathscr{D}$ is itself countable, and form a countable covering of E that is comprised of relatively compact open sets. If we denote them by $(\widetilde{V}_n)_n$ then the sets

$$W_n = \bigcup_{k=1}^{n} \widetilde{V}_k \tag{1.15}$$

form an increasing sequence of relatively compact open sets such that $W_n \uparrow E$ as $n \to \infty$.

(b) Let

$$h_n(x) := n \, d(x, W_n^c) \wedge 1$$

with W_n as in (1.15). The sequence $(h_n)_n$ is obviously increasing and, as the support of h_n is contained in W_n, each h_n is also compactly supported. As $W_n \uparrow E$, for every $x \in E$ we have $h_n(x) = 1$ for n large enough. ∎

Note that if E is not locally compact the relation $\int f\,d\mu = \int f\,d\nu$ for every compactly supported continuous function f does not necessarily imply that $\mu = \nu$ on $\mathscr{B}(E)$. This should be kept in mind, as it can occur when considering measures on, e.g., infinite-dimensional Banach spaces, which are not locally compact.

In some sense, if the space is not locally compact, the class of compactly supported continuous functions is not "large enough".

1.5   Important Examples

Let us present some examples of measures and some ways to construct new measures starting from given ones.

• (Dirac masses) If $x \in E$ let us consider the measure on $\mathscr{P}(E)$ (all subsets of E) that is defined as

$$\mu(A) = 1_A(x) . \tag{1.16}$$

This is the measure that gives to a set A the value 0 or 1 according as $x \in A$ or not. It is immediate that this is a measure; it is denoted δ_x and is called the *Dirac mass* at x. We have the formula

$$\int_E f\,d\delta_x = f(x) ,$$

which can be easily proved by the same argument as in the forthcoming Propositions 1.27 or 1.28.

• (Countable sets) If E is a countable set, a measure on $(E, \mathscr{P}(E))$ can be constructed in a simple (and natural) way: let us associate to every $x \in E$ a number $p_x \in \mathbb{R}^+$ and let, for $A \subset E$, $\mu(A) = \sum_{x \in A} p_x$. The summability properties of positive series imply that μ is a measure: actually, if $A_1, A_2, \ldots$ are pairwise disjoint subsets of E, and $A = \bigcup_n A_n$, then the σ-additivity relationship

$$\mu(A) = \sum_{n=1}^{\infty} \mu(A_n)$$

is equivalent to

$$\sum_{n=1}^{\infty} \left(\sum_{x \in A_n} p_x \right) = \sum_{x \in A} p_x ,$$

which holds because the sum of a series whose terms are positive does not depend on the order of summation.

A natural example is the choice $p_x = 1$ for every x. In this case the measure of a set A coincides with its cardinality. This is the *counting measure* of E.

• (Image measures) Let $(E, \mathscr{E})$ and $(G, \mathscr{G})$ be measurable spaces, $\Phi : E \to G$ a measurable map and μ a measure on $(E, \mathscr{E})$; we can define a measure ν on $(G, \mathscr{G})$ via

$$\nu(A) := \mu\big(\Phi^{-1}(A)\big) \qquad A \in \mathscr{G} . \tag{1.17}$$

Also here it is immediate to check that ν is a measure (thanks to the relations (1.1)). ν is the *image measure* of μ under Φ and is denoted $\Phi(\mu)$ or $\mu \circ \Phi^{-1}$.

Proposition 1.27 (Integration with Respect to an Image Measure) *Let* $g : G \to \overline{\mathbb{R}}^+$ *be a positive measurable function. Then*

$$\int_G g \, d\nu = \int_E g \circ \Phi \, d\mu . \tag{1.18}$$

A measurable function $g : G \to \overline{\mathbb{R}}$ *is integrable with respect to* ν *if and only if* $g \circ \Phi$ *is integrable with respect to* μ *and also in this case* (1.18) *holds.*

Proof Let, for every positive measurable function $g : G \to \overline{\mathbb{R}}^+$,

$$I(g) = \int_E g \circ \Phi \, d\mu .$$

It is immediate that the functional I satisfies the conditions (a) and (b) of Proposition 1.24. Therefore, thanks to Proposition 1.24,

$$A \mapsto I(1_A) = \int_E 1_A \circ \Phi \, d\mu = \int_E 1_{\Phi^{-1}(A)} \, d\mu = \mu(\Phi^{-1}(A))$$

is a measure on $(G, \mathscr{G})$ and (1.18) holds for every positive function g. The proof is completed taking the decomposition of g into positive and negative parts. ∎

• (Measures defined by a density) Let μ be a σ-*finite* measure on $(E, \mathscr{E})$.

A positive measurable function f is a *density* if there exists a sequence $(A_n)_n \subset \mathscr{E}$ such that $\bigcup_n A_n = E$, $\mu(A_n) < +\infty$ and $f 1_{A_n}$ is integrable for every n.

In particular a positive integrable function is a density.

Theorem 1.28 *Let $(E, \mathscr{E}, \mu)$ be a σ-finite measure space and f a density with respect to μ. Let for $A \in \mathscr{E}$*

$$\nu(A) := \int_E 1_A f \, d\mu = \int_A f \, d\mu . \tag{1.19}$$

Then ν is a σ-finite measure on $(E, \mathscr{E})$ which is called the measure of density f with respect to μ, denoted $d\nu = f \, d\mu$. Moreover, for every positive measurable function $g \colon E \to \mathbb{R}$ we have

$$\int_E g \, d\nu = \int_E g f \, d\mu . \tag{1.20}$$

A measurable function $g \colon E \to \mathbb{R}$ is integrable with respect to ν if and only if gf is integrable with respect to μ and also in this case (1.20) holds.

Proof The functional

$$\mathscr{E}^+ \ni g \mapsto \int_E g f \, d\mu$$

is positively linear and passes to the limit on increasing sequences of positive functions by Beppo Levi's Theorem (recall $\mathscr{E}^+$ = the positive measurable functions). Hence, by Proposition 1.24,

$$\nu(A) := I(1_A) = \int_A f \, d\mu$$

is a measure on $(E, \mathscr{E})$ such that (1.20) holds for every positive function g. ν is σ-finite because if $(A_n)_n$ is a sequence of sets of $\mathscr{E}$ such that $\bigcup_n A_n = E$ and with $f 1_{A_n}$ integrable, then

$$\nu(A_n) = \int_E f 1_{A_n} \, d\mu < +\infty .$$

Finally (1.20) is proved to hold for every ν-integrable function by decomposing g into positive and negative parts. $\blacksquare$

Let μ, ν be σ-finite measures on the measurable space $(E, \mathscr{E})$. We say that ν is *absolutely continuous* with respect to μ, denoted $\nu \ll \mu$, if and only if every μ-negligible set $A \in \mathscr{E}$ (i.e. such that $\mu(A) = 0$) is also ν-negligible.

If ν has density f with respect to μ then clearly $\nu \ll \mu$: if A is μ-negligible then the function $f 1_A$ is $\neq 0$ only on A, hence $\nu(A) = \int_E f 1_A \, d\mu = 0$ (Exercise 1.10). A remarkable and non-obvious result is that the converse is also true.

Theorem 1.29 (Radon-Nikodym) *If μ, ν are σ-finite and $\nu \ll \mu$ then ν has a density with respect to μ.*

A proof of this theorem can be found in almost all the books listed in the references. A proof in the case of probabilities will be given in Example 5.27.

It is often important to establish whether a Borel measure ν on $(\mathbb{R}, \mathcal{B}(\mathbb{R}))$ has a density with respect to the Lebesgue measure λ, i.e. is such that $\nu \ll \lambda$, and to be able to compute it.

First, in order for ν to be absolutely continuous with respect to λ it is necessary that $\nu(\{x\}) = 0$ for every x, as $\lambda(\{x\}) = 0$ and the negligible sets for λ must also be negligible for ν. The distribution function of ν, F, therefore must be continuous, as

$$0 = \nu(\{x\}) = \lim_{n \to \infty} \nu(]x - \tfrac{1}{n}, x]) = F(x) - \lim_{n \to \infty} F(x - \tfrac{1}{n}) \,.$$

Assume, moreover, that F is absolutely continuous, hence a.e. differentiable and such that, if $F'(x) = f(x)$, for every $-\infty < a \leq b < +\infty$,

$$\int_a^b f(x) \, dx = F(b) - F(a) \,. \tag{1.21}$$

In (1.21) the term on the right-hand side is nothing else than $\nu(]a, b])$, whereas the left-hand term is the value on $]a, b]$ of the measure $f \, d\lambda$. The two measures ν and $f \, d\lambda$ therefore coincide on the half-open intervals and by Theorem 1.14 (Carathéodory's criterion) they coincide on the whole σ-algebra $\mathcal{B}(\mathbb{R})$.

Note that, to be precise, it is not correct to speak of "the" density of ν with respect to μ: if f is a density, then so is every function g that is μ-equivalent to f (i.e. such that $f = g$ μ-a.e.).

1.6 L^p Spaces

Let $(E, \mathcal{E}, \mu)$ be a measure space, V a normed vector space and $f : E \to V$ a measurable function. Let, for $1 \leq p < +\infty$,

$$\|f\|_p = \left(\int_E |f|^p \, d\mu \right)^{\frac{1}{p}}$$

and, for $p = +\infty$,

$$\|f\|_\infty = \inf\{M; \mu(|f| > M) = 0\} \,.$$

In particular the set $\{|f| > \|f\|_\infty\}$ is negligible. $\| \ \|_p$ and $\| \ \|_\infty$ can of course be $+\infty$. Let, for $1 \le p \le +\infty$,

$$\mathcal{L}^p = \{f; \|f\|_p < +\infty\} \,.$$

Let us state two fundamental inequalities: if $f, g : E \to V$ are measurable functions then

$$\|f + g\|_p \le \|f\|_p + \|g\|_p \,, \quad 1 \le p \le +\infty \,, \tag{1.22}$$

which is *Minkowski's inequality* and

$$\big\| |f| |g| \big\|_1 \le \|f\|_p \|g\|_q \,, \quad 1 \le p \le +\infty, \ \frac{1}{p} + \frac{1}{q} = 1 \,, \tag{1.23}$$

which is *Hölder's inequality*.

Thanks to Minkowski's inequality, $\mathcal{L}^p$ is a vector space and $\| \ \|_p$ a seminorm. It is not a norm as it is possible for a function $f \ne 0$ to have $\|f\|_p = 0$ (this happens if and only if $f = 0$ a.e.). Let us define an equivalence relation on $\mathcal{L}^p$ by setting $f \sim g$ if $f = g$ a.e. and then let $L^p = \mathcal{L}^p / \sim$, the quotient space with respect to this equivalence. Then L^p is a normed space. Actually, $f = g$ a.e. implies $\int |f|^p \, d\mu = \int |g|^p \, d\mu$, and we can define, for $f \in L^p$, $\|f\|_p$ without ambiguity.

Note however that L^p *is not a space of functions*, but of equivalence classes of functions; this distinction is seldom important and in the sequel we shall often identify a function f and its equivalence class. But sometimes it will be necessary to pay attention.

If the norm of V is associated to a scalar product $\langle \cdot, \cdot \rangle$, then, for $p = q = 2$, Hölder's inequality (1.23) gives the *Cauchy-Schwarz inequality*

$$\left(\int_E \langle f, g \rangle \, d\mu \right)^2 \le \int_E |f|^2 \, d\mu \int_E |g|^2 \, d\mu \,. \tag{1.24}$$

It can be proved that if the target space V is complete, i.e. a Banach space, then the normed space L^p is itself a Banach space and therefore also complete. In this case L^2 is a Hilbert space with respect to the scalar product

$$\langle f, g \rangle_2 = \int_E \langle f, g \rangle \, d\mu \,.$$

Note that, if $V = \mathbb{R}$, then

$$\langle f, g \rangle_2 = \int_E fg \, d\mu$$

and, if $V = \mathbb{C}$,

$$\langle f, g \rangle_2 = \int_E f \overline{g} \, d\mu \,.$$

A sequence of functions $(f_n)_n \subset L^p$ is said to converge to f in L^p if $\| f_n - f \|_p \to 0$ as $n \to \infty$.

Remark 1.30 Let $f, g \in L^p$, $p \geq 1$. Then by Minkowski's inequality we have

$$\| f \|_p \leq \| f - g \|_p + \| g \|_p \,,$$
$$\| g \|_p \leq \| f - g \|_p + \| f \|_p$$

from which we obtain both inequalities

$$\| g \|_p - \| f \|_p \leq \| f - g \|_p \quad \text{and} \quad \| f \|_p - \| g \|_p \leq \| f - g \|_p \,,$$

hence

$$\left| \| f \|_p - \| g \|_p \right| \leq \| f - g \|_p \,,$$

so that $f \to \| f \|_p$ is a continuous map $L^p \to \mathbb{R}^+$ and L^p-convergence implies convergence of the L^p norms.

1.7 Product Spaces, Product Measures

Let $(E_1, \mathscr{E}_1), \ldots, (E_m, \mathscr{E}_m)$ be measurable spaces. On the product set $E := E_1 \times \cdots \times E_m$ let us define the *product σ-algebra* $\mathscr{E}$ by setting

$$\mathscr{E} := \mathscr{E}_1 \otimes \cdots \otimes \mathscr{E}_m := \sigma(A_1 \times \cdots \times A_m; \ A_1 \in \mathscr{E}_1, \ldots, A_m \in \mathscr{E}_m) \,. \tag{1.25}$$

$\mathscr{E}$ is the smallest σ-algebra that contains the "rectangles" $A_1 \times \cdots \times A_m$ with $A_1 \in \mathscr{E}_1, \ldots, A_m \in \mathscr{E}_m$.

Proposition 1.31 *Let* $p_i : E \to E_i$, $i = 1, \ldots, m$, *be the canonical projections*

$$p_i(x_1, \ldots, x_m) = x_i \,.$$

Then p_i *is measurable* $(E, \mathscr{E}) \to (E_i, \mathscr{E}_i)$ *and the product σ-algebra $\mathscr{E}$ is the smallest σ-algebra on the product space E that makes the projections p_i measurable.*

Proof If $A_i \in \mathscr{E}_i$, then, for $1 \le i \le m$,

$$p_i^{-1}(A_i) = E_1 \times \cdots \times E_{i-1} \times A_i \times E_{i+1} \times \cdots \times E_m \,. \qquad (1.26)$$

This set belongs to $\mathscr{E}$ (it is a "rectangle"), hence p_i is measurable $(E, \mathscr{E}) \to (E_i, \mathscr{E}_i)$.

Conversely, let $\widetilde{\mathscr{E}}$ denote a σ-algebra of subsets of $E = E_1 \times \cdots \times E_m$ with respect to which the canonical projections p_i are measurable. $\widetilde{\mathscr{E}}$ must contain the sets $p_i^{-1}(A_i)$, $A_i \in \mathscr{E}_i$, $i = 1, \ldots, m$. Therefore $\widetilde{\mathscr{E}}$ also contains the rectangles, as, recalling (1.26), we can write $A_1 \times \cdots \times A_m = p_1^{-1}(A_1) \cap \cdots \cap p_m^{-1}(A_m)$. Therefore $\widetilde{\mathscr{E}}$ also contains the product σ-algebra $\mathscr{E}$, which is the smallest σ-algebra containing the rectangles. ∎

Let now $(G, \mathscr{G})$ be a measurable space and $f = (f_1, \ldots, f_m)$ a map from $(G, \mathscr{G})$ to the product space $(E, \mathscr{E})$. As an immediate consequence of Proposition 1.31, f is measurable if and only if all its components $f_i = p_i \circ f : (G, \mathscr{G}) \to (E_i, \mathscr{E}_i)$ are measurable. Indeed, if $f : G \to E$ is measurable $(G, \mathscr{G}) \to (E, \mathscr{E})$, then the components $f_i = p_i \circ f$ are measurable, being compositions of measurable maps. Conversely, if the components $f_1, \ldots, f_m$ are measurable, then for every rectangle $A = A_1 \times \cdots \times A_m \in \mathscr{E}$ we have

$$f^{-1}(A) = f_1^{-1}(A_1) \cap \cdots \cap f_m^{-1}(A_m) \in \mathscr{G} \,.$$

Hence the pullback of every rectangle is a measurable set and the claim follows thanks to Remark 1.5, as the rectangles generate the product σ-algebra $\mathscr{E}$.

Given two topological spaces, on their product we can consider

- the product of the respective Borel σ-algebras
- the Borel σ-algebra of the product topology.

Do they coincide?

In general they do not, but the next proposition states that they do coincide under assumptions that are almost always satisfied. Recall that a topological space is said to have a countable basis of open sets if there exists a countable family $(O_n)_n$ of open sets such that every open set is the union of some of the O_n. In particular, every separable metric space has such a basis.

> **Proposition 1.32** *Let $E_1, \ldots, E_m$ be topological spaces. Then*
>
> *(a) $\mathscr{B}(E_1 \times \cdots \times E_m) \supset \mathscr{B}(E_1) \otimes \cdots \otimes \mathscr{B}(E_m)$.*
> *(b) If $E_1, \ldots, E_m$ have a countable basis of open sets, then $\mathscr{B}(E_1 \times \cdots \times E_m) = \mathscr{B}(E_1) \otimes \cdots \otimes \mathscr{B}(E_m)$.*

Proof In order to keep the notation simple, let us assume $m = 2$.

(a) The projections

$$p_1 : E_1 \times E_2 \to E_1, \quad p_2 : E_1 \times E_2 \to E_2$$

are continuous when we consider on $E_1 \times E_2$ the product topology (which, by definition, is the smallest topology on the product space with respect to which the projections are continuous). They are therefore also measurable with respect to $\mathscr{B}(E_1 \times E_2)$. Hence $\mathscr{B}(E_1 \times E_2)$ contains $\mathscr{B}(E_1) \otimes \mathscr{B}(E_2)$, which is the smallest σ-algebra making the projections measurable (Proposition 1.31).

(b) If $(U_{1,n})_n$, $(U_{2,n})_n$ are countable bases of the topologies of E_1 and E_2 respectively, then the sets $V_{n,m} = U_{1,n} \times U_{2,m}$ form a countable basis of the product topology of $E_1 \times E_2$. As $U_{1,n} \in \mathscr{B}(E_1)$ and $U_{2,n} \in \mathscr{B}(E_2)$, we have $V_{n,m} \in \mathscr{B}(E_1) \otimes \mathscr{B}(E_2)$ ($V_{n,m}$ is a rectangle). As all open sets of $E_1 \times E_2$ are countable unions of the open sets $V_{n,m}$, all open sets of the product topology belong to the σ-algebra $\mathscr{B}(E_1) \otimes \mathscr{B}(E_2)$ which therefore contains $\mathscr{B}(E_1 \times E_2)$. ∎

Let μ, ν be finite measures on the product space. Carathéodory's criterion, Proposition 1.11, ensures that if they coincide on rectangles then they are equal. Indeed the class of rectangles $A_1 \times \cdots \times A_m$ is stable with respect to finite intersections.

In order to prove that $\mu = \nu$ it is also sufficient to check that

$$\int_E f_1(x_1) \cdots f_m(x_m) \, d\mu(x) = \int_E f_1(x_1) \cdots f_m(x_m) \, d\nu(x)$$

for every choice of bounded measurable functions $f_i : E_i \to \mathbb{R}$. If the spaces $(E_i, \mathscr{C}_i)$ are metric spaces, a repetition of the arguments of Proposition 1.25 proves the following criterion.

Proposition 1.33 *Assume that* $(E_i, \mathcal{E}_i)$, $i = 1, \ldots, m$, *are metric spaces endowed with their Borel σ-algebras. Let μ, ν be finite measures on the product space.*

(a) Assume that

$$\int_E f_1(x_1) \cdots f_m(x_m) \, d\mu(x) = \int_E f_1(x_1) \cdots f_m(x_m) \, d\nu(x) \qquad (1.27)$$

for every choice of bounded continuous functions $f_i : E_i \to \mathbb{R}$, $i = 1, \ldots, m$. Then $\mu = \nu$.

(b) If, moreover, the spaces E_i, $i = 1, \ldots, m$, are also separable and locally compact and if (1.27) holds for every choice of continuous and compactly supported functions f_i, then $\mu = \nu$.

Let $\mu_1, \ldots, \mu_m$ be σ-finite measures on $(E_1, \mathcal{E}_1), \ldots, (E_m, \mathcal{E}_m)$ respectively. For every rectangle $A = A_1 \times \cdots \times A_m$ let

$$\mu(A) = \mu_1(A_1) \ldots \mu_m(A_m) . \qquad (1.28)$$

Is it possible to extend μ to a measure on the product σ-algebra $\mathcal{E} = \mathcal{E}_1 \otimes \cdots \otimes \mathcal{E}_m$?

In order to prove the existence of this extension it is possible to take advantage of Theorem 1.13, Carathéodory's extension theorem, whose use here however requires some work in order to check that the set function μ defined in (1.28) is σ-additive on the algebra of finite unions of rectangles (recall Remark 1.9).

It is easier to proceed following the idea of Proposition 1.24, i.e. constructing a positively linear functional on the positive functions on $(E, \mathcal{E})$ that passes to the limit on increasing sequences. More precisely the idea is the following. Let us assume for simplicity $m = 2$ and let $f : E_1 \times E_2 \to \overline{\mathbb{R}}^+$ be a positive $\mathcal{E}_1 \otimes \mathcal{E}_2$-measurable function.

(1) First prove that, for every given $x_1 \in E_1$, $x_2 \in E_2$, the functions $f(x_1, \cdot)$ and $f(\cdot, x_2)$ are respectively $\mathcal{E}_2$- and $\mathcal{E}_1$-measurable.

(2) Then prove that, for every $x_1 \in E_1$, $x_2 \in E_2$, the "partially integrated" functions

$$x_1 \mapsto \int_{E_2} f(x_1, x_2) \, d\mu_2(x_2), \qquad x_2 \mapsto \int_{E_1} f(x_1, x_2) \, d\mu_1(x_1)$$

are respectively $\mathcal{E}_1$- and $\mathcal{E}_2$-measurable.

(3) Now let

$$I(f) = \int_{E_2} d\mu_2(x_2) \int_{E_1} f(x_1, x_2) \, d\mu_1(x_1) \qquad (1.29)$$

(i.e. we integrate first with respect to μ_1 the measurable function $x_1 \mapsto f(x_1, x_2)$, the result is a measurable function of x_2 that is then integrated with respect to μ_2). It is immediate that the functional I satisfies assumptions (a) and (b) of Proposition 1.24 (use Beppo Levi's Theorem twice).

It follows (Proposition 1.24) that $\mu(A) := I(1_A)$ defines a measure on $\mathscr{E}_1 \otimes \mathscr{E}_2$. Such a μ satisfies (1.28), as, by (1.29),

$$\mu(A_1 \times A_2) = I(1_{A_1 \times A_2})$$

$$= \int_{E_1} 1_{A_1}(x_1) \, d\mu_1(x_1) \int_{E_2} 1_{A_2}(x_2) \, d\mu_2(x_2)$$

$$= \mu_1(A_1)\mu_2(A_2) \, .$$

This is the extension we were looking for. The measure μ is the *product measure* of μ_1 and μ_2, denoted $\mu = \mu_1 \otimes \mu_2$.

Uniqueness of the product measure follows from Carathéodory's criterion, Proposition 1.11, as two measures satisfying (1.28) coincide on the rectangles having finite measure, which form a class that is stable with respect to finite intersections and, as the measures μ_i are assumed to be σ-finite, generates the product σ-algebra. In order to properly apply Carathéodory's criterion however we also need to prove that there exists a sequence of rectangles of finite measure increasing to the whole product space.

Let, for every $i = 1, \ldots, m$, $C_{i,n} \in \mathscr{E}_i$ be an increasing sequence of sets such that $\mu_i(C_{i,n}) < +\infty$ and $\bigcup_n C_{i,n} = E_i$. Such a sequence exists as the measures $\mu_1, \ldots, \mu_m$ are assumed to be σ-finite. Then the sets $C_n = C_{1,n} \times \cdots \times C_{m,n}$ are increasing, such that $\mu(C_n) < +\infty$ and $\bigcup_n C_n = E$.

The proofs of (1) and (2) above are without surprise: these properties are obvious if f is the indicator function of a rectangle. Let us prove next that they hold if f is the indicator function of a set of $\mathscr{E} = \mathscr{E}_1 \otimes \mathscr{E}_2$: let $\mathscr{M}$ be the class of the sets $A \in \mathscr{E}$ whose indicator functions satisfy 1), i.e. such that $1_A(x_1, \cdot)$ and $1_A(\cdot, x_2)$ are respectively $\mathscr{E}_2$- and $\mathscr{E}_1$-measurable. It is immediate that they form a monotone class. As $\mathscr{M}$ contains the rectangles, a family which is stable with respect to finite intersections, by Theorem 1.2, the Monotone Class Theorem, $\mathscr{M}$ also contains $\mathscr{E}$, which is the σ-algebra generated by the rectangles.

By linearity (1) is also satisfied by the elementary functions on $\mathscr{E}$ and finally by all $\mathscr{E}_1 \otimes \mathscr{E}_2$-positive measurable functions thanks to Proposition 1.6 (approximation with elementary functions).

The argument to prove (2) is similar but requires more care, considering first the case of finite measures and then taking advantage of the assumption of σ-finiteness.

In practice, in order to integrate with respect to a product measure one takes advantage of the following, very important, theorem. We state it with respect to the product of two measures, the statement for the product of m measures being left to the imagination of the reader.

Theorem 1.34 (Fubini-Tonelli) *Let $f : E_1 \times E_2 \to \mathbb{R}$ be an $\mathscr{E}_1 \otimes \mathscr{E}_2$-measurable function and let μ_1, μ_2 be σ-finite measures on $(E_1, \mathscr{E}_1)$ and $(E_2, \mathscr{E}_2)$ respectively. Let $\mu = \mu_1 \otimes \mu_2$ be their product.*

(a) If f is positive, then the functions

$$x_1 \mapsto \int_{E_2} f(x_1, x_2)\, d\mu_2(x_2)$$

$$x_2 \mapsto \int_{E_1} f(x_1, x_2)\, d\mu_1(x_1) \tag{1.30}$$

are respectively $\mathscr{E}_1$- and $\mathscr{E}_2$-measurable. Moreover, we have

$$\int_{E_1 \times E_2} f\, d\mu = \int_{E_2} d\mu_2(x_2) \int_{E_1} f(x_1, x_2)\, d\mu_1(x_1)$$

$$= \int_{E_1} d\mu_1(x_1) \int_{E_2} f(x_1, x_2)\, d\mu_2(x_2)\,. \tag{1.31}$$

(b) If f is real, numerical or complex-valued and integrable with respect to the product measure $\mu_1 \otimes \mu_2$, then the functions in (1.30) are respectively $\mathscr{E}_1$- and $\mathscr{E}_2$-measurable and integrable with respect to μ_1 and μ_2 respectively and (1.31) holds.

For simplicity we shall refer to this theorem as Fubini's Theorem.

The main ideas in the application of Fubini's Theorem for the integration of a function with respect to a product measure are:

- if f is positive everything is allowed (i.e. you can integrate with respect to the variables one after the other in any order) and the result is equal to the integral with respect to the product measure, which can be a real number or possibly $+\infty$ (this is part (a) of Fubini's Theorem);
- if f is real and takes both positive and negative values or is complex-valued, in order for (1.31) to hold f must be integrable with respect to the product measure. In practice one first checks integrability of $|f|$ using part (a) of the theorem and then applies part (b) in order to compute the integral.
- In addition to the two integration results for positive and integrable functions, the measurability and integrability results of the "partially integrated" functions (1.30) is also useful.

Therefore Fubini's Theorem 1.34 contains in fact three different results, all of them very useful.

Remark 1.35 Corollary 1.22 (integration by series) can be seen as a consequence of Fubini's Theorem.

Indeed, let $(E, \mathscr{E}, \mu)$ be a measure space. Given a sequence $(f_n)_n$ of measurable functions $E \to \mathbb{R}$ we can consider the function $\Phi_f : \mathbb{N} \times E \to \mathbb{R}$ defined as $(n, x) \mapsto f_n(x)$. Hence the relation

$$\sum_{n=1}^{\infty} \int_E f_n \, d\mu = \int_E \sum_{n=1}^{\infty} f_n \, d\mu$$

is just Fubini's theorem for the function Φ_f integrated with respect to the product measure $\nu_c \otimes \mu$, ν_c denoting the counting measure of $\mathbb{N}$. Measurability of Φ_f above is immediate.

Let us consider $(\mathbb{R}, \mathscr{B}(\mathbb{R}), \lambda)$ ($\lambda =$ the Lebesgue measure). By Proposition 1.32, $\mathscr{B}(\mathbb{R}) \otimes \ldots \otimes \mathscr{B}(\mathbb{R}) = \mathscr{B}(\mathbb{R}^d)$. Let $\lambda_d = \lambda \otimes \ldots \otimes \lambda$ (d times). We can apply Carathéodory's criterion, Proposition 1.11, to the class of sets

$$\mathscr{C} = \left\{ A; \ A = \prod_{i=1}^{d}]a_i, b_i[, \ -\infty < a_i < b_i < +\infty \right\}$$

and obtain that λ_d is the unique measure on $\mathscr{B}(\mathbb{R}^d)$ such that, for every $-\infty < a_i < b_i < +\infty$,

$$\lambda_d \left(\prod_{i=1}^{d}]a_i, b_i[\right) = \prod_{i=1}^{d} (b_i - a_i) .$$

λ_d is *the Lebesgue's measure* of $\mathbb{R}^d$.

In the sequel we shall also need to consider the product of countably many measure spaces. The theory is very similar to the finite case, at least for probabilities.

Let $(E_i, \mathscr{E}_i, \mu_i)$, $i = 1, 2, \ldots$, be measure spaces. Then the product σ-algebra $\mathscr{E} = \bigotimes_{i=1}^{\infty} \mathscr{E}_i$ is defined as the smallest σ-algebra of subsets of the product $E = \prod_{i=1}^{\infty} E_i$ containing the rectangles $\prod_{i=1}^{\infty} A_i$, $A_i \in \mathscr{E}_i$. The following statement says that on the product space $(E, \mathscr{E})$ there exists a probability that is the product of the μ_i.

Theorem 1.36 *Let $(E_i, \mathscr{E}_i, \mu_i)$, $i = 1, 2, \ldots$, be a countable family of measure spaces such that μ_i is a probability for every i. Then there exists a unique probability μ on $(E, \mathscr{E})$ such that for every rectangle $A = \prod_{i=1}^{\infty} A_i$*

$$\mu(A) = \prod_{i=1}^{\infty} \mu_i(A_i) .$$

For a proof and other details, see Halmos's book [16].

Exercises

1.1 (p. 261) A σ-algebra $\mathscr{F}$ is said to be *countably generated* if there exists a countable subfamily $\mathscr{C} \subset \mathscr{F}$ such that $\sigma(\mathscr{C}) = \mathscr{F}$.

Prove that if E is a separable metric space, then its Borel σ-algebra, $\mathscr{B}(E)$, is countably generated. In particular, so is the Borel σ-algebra of $\mathbb{R}^d$ or, more generally, of any separable Banach space.

1.2 (p. 261) The Borel σ-algebra of $\mathbb{R}$ is generated by each of the following families of sets.

(a) The open intervals $]a, b[$, $a < b$.
(b) The half-open intervals $]a, b]$, $a < b$.
(c) The open half-lines $]a, \infty[$, $a \in \mathbb{R}$.
(d) The closed half-lines $[a, \infty[$, $a \in \mathbb{R}$.

1.3 (p. 261) Let E be a topological space and let us denote by $\mathscr{B}_0(E)$ the smallest σ-algebra of subsets of E with respect to which all real continuous functions are measurable. $\mathscr{B}_0(E)$ is the *Baire σ-algebra*.

(a) Prove that $\mathscr{B}_0(E) \subset \mathscr{B}(E)$.
(b) Prove that if E is metric separable then $\mathscr{B}_0(E)$ and $\mathscr{B}(E)$ coincide.

1.4 (p. 262) Let $(E, \mathscr{E})$ be a measurable space and $S \subset E$ (not necessarily $S \in \mathscr{E}$). Prove that

$$\mathscr{E}_S = \{A \cap S; \, A \in \mathscr{E}\}$$

is a σ-algebra of subsets of S (the *trace σ-algebra* of $\mathscr{E}$ on S).

1.5 (p. 262) Let $(E, \mathscr{E})$ be a measurable space.

(a) Let $(f_n)_n$ be a sequence of real measurable functions. Prove that the set

$$L = \{x; \lim_{n \to \infty} f_n(x) \text{ exists}\}$$

is measurable.

(b) Assume that the f_n take their values in a metric space G. Using unions, intersections, complementation... describe the set of points x such that the Cauchy property for the sequence $(f_n(x))_n$ is satisfied and prove that, if E is complete, L is measurable also in this case.

1.6 (p. 262) Let $(E, \mathcal{E})$ be a measurable space, $(f_n)_n$ a sequence of measurable functions taking values in the metric space (G, d) and assume that $\lim_{n \to \infty} f_n = f$ pointwise. We have seen (p. 4) that if $G = \mathbb{R}$ then f is also measurable. In this exercise we address this question in more generality.

(a) Prove that for every continuous function $\Phi : G \to \mathbb{R}$ the function $\Phi \circ f$ is measurable.
(b) Prove that if the metric space (G, d) is separable, then f is measurable $E \to G$.

Recall that, for $z \in G$, the function $x \mapsto d(x, z)$ is continuous.

1.7 (p. 263) Let $(E, \mathcal{E}, \mu)$ be a measure space.

(a) Prove that if $(A_n)_n \subset \mathcal{E}$ then

$$\mu\left(\bigcup_{n=1}^{\infty} A_n\right) \leq \sum_{n=1}^{\infty} \mu(A_n) . \tag{1.32}$$

(b) Let $(A_n)_n$ be a sequence of negligible events. Prove that $\bigcup_n A_n$ is also negligible.
(c) Let $\mathcal{A} = \{A; A \in \mathcal{E}, \mu(A) = 0 \text{ or } \mu(A^c) = 0\}$. Prove that $\mathcal{A}$ is a σ-algebra.

1.8 (p. 264) (The support of a measure) Let μ be a Borel measure on a separable metric space E. Let us denote by $B_x(r)$ the open ball with radius r centered at x and let

$$F = \{x \in E; \mu(B_x(r)) > 0 \text{ for every } r > 0\}$$

(i.e. F is formed by all $x \in E$ such that all their neighborhoods have strictly positive measure).

(a) Prove that F is a closed set.
(b1) Prove that $\mu(F^c) = 0$.
(b2) Prove that F is the smallest closed subset of E such that $\mu(F^c) = 0$.

- F is the *support* of the measure μ. Note that the support of a measure is always a closed set.

1.9 (p. 264) Let μ be a measure on $(E, \mathscr{E})$ and $f : E \to \overline{\mathbb{R}}$ a measurable function.

(a) Prove that if f is integrable, then $|f| < +\infty$ a.e.

(b) Prove that if f is positive and

$$\int_E f \, d\mu = 0$$

then $f = 0$ μ-a.e.

(c) Prove that if f is semi-integrable and if $\int_A f \, d\mu \geq 0$ for every $A \in \mathscr{E}$, then $f \geq 0$ a.e.

1.10 (p. 265) Let $(E, \mathscr{E}, \mu)$ be a measure space and $f : E \to \mathbb{R}$ a measurable function vanishing outside a negligible set N. Prove that f is integrable and that its integral vanishes.

1.11 (p. 265)

(a) Let $(w_n)_n$ be a bounded sequence of positive numbers and let, for $t > 0$,

$$\phi(t) = \sum_{n=1}^{\infty} w_n \, e^{-tn} . \tag{1.33}$$

Is it true that ϕ is differentiable by series, i.e. that, if $t > 0$, ϕ is differentiable and

$$\phi'(t) = -\sum_{n=1}^{\infty} n w_n \, e^{-tn} \ ? \tag{1.34}$$

(b) Consider the same question where the sequence $(w_n)_n$ is bounded but not necessarily positive.

(c1) And if $w_n = \sqrt{n}$?

(c2) And if $w_n = e^{\sqrt{n}}$?

1.12 (p. 267) (Counterexamples)

(a) Find an example of a measure space $(E, \mathscr{E}, \mu)$ and of a decreasing sequence $(A_n)_n \subset \mathscr{E}$ such that $\mu(A_n)$ does not converge to $\mu(A)$ where $A = \bigcap_n A_n$.

(b) Beppo Levi's Theorem requires the existence of an integrable function f such that $f \leq f_n$ for every n. Give an example where this condition is not satisfied and the statement of Beppo Levi's Theorem is not true.

1.13 (p. 267) Let ν, μ be measures on the measurable space $(E, \mathscr{E})$ such that $\nu \ll \mu$. Let ϕ be a measurable map from E into the measurable space $(G, \mathscr{G})$ and let $\tilde{\nu}, \tilde{\mu}$ be the respective images of ν and μ. Prove that $\tilde{\nu} \ll \tilde{\mu}$.

1.14 (p. 267) Let λ be the Lebesgue measure on $[0, 1]$ and μ the set function on $\mathscr{B}([0, 1])$ defined as

$$\mu(A) = \begin{cases} 0 & \text{if } \lambda(A) = 0 \\ +\infty & \text{if } \lambda(A) > 0 . \end{cases}$$

(a) Prove that μ is a measure on $\mathscr{B}([0, 1])$.

(b) Note that $\lambda \ll \mu$ but the Radon-Nikodym Theorem does not hold here and explain why.

1.15 (p. 267) Let $(E, \mathscr{E}, \mu)$ be a measure space and $(f_n)_n$ a sequence of real functions *bounded* in L^p, $0 < p \le +\infty$, and assume that $f_n \to_{n\to\infty} f$ μ-a.e.

(a1) Prove that $f \in L^p$.

(a2) Does the convergence necessarily also take place in L^p?

(b) Let $g \in L^p$, $0 < p \le +\infty$, and let $g_n = g \wedge n \vee (-n)$. Prove that $g_n \to g$ in L^p as $n \to +\infty$.

1.16 (p. 268) (Do the L^p spaces become larger or smaller as p increases?) Let μ be a *finite* measure on the measurable space $(E, \mathscr{E})$.

(a1) Prove that, if $0 \le p \le q$, then $|x|^p \le 1 + |x|^q$ for every $x \in \mathbb{R}$ and that $L^q \subset L^p$, i.e. the spaces L^p become smaller as p increases (recall that μ is finite).

(a2) Prove that, if $f \in L^q$, then

$$\lim_{p \to q-} \|f\|_p = \|f\|_q . \tag{1.35}$$

(a3) Prove that, if $f \notin L^q$, then

$$\lim_{p \to q-} \int_E |f|^p \, d\mu = +\infty . \tag{1.36}$$

(a4) Prove that

$$\lim_{p \to q+} \|f\|_p \ge \|f\|_q \tag{1.37}$$

but that, if $f \in L^{q_0}$ for some $q_0 > q$, then

$$\lim_{p \to q+} \|f\|_p = \|f\|_q . \tag{1.38}$$

(a5) Give an example of a function that belongs to L^q for a given value of q, but that does not belong to L^p for any $p > q$, so that, in general, $\lim_{p \to q+} \|f\|_p = \|f\|_q$ does not hold.

(b1) Let $f : E \to \mathbb{R}$ be a measurable function. Prove that

$$\varlimsup_{p \to +\infty} \|f\|_p \leq \|f\|_\infty .$$

(b2) Let $M \geq 0$. Prove that, for every $p \geq 0$,

$$\int_E |f|^p \, d\mu \geq M^p \mu(|f| \geq M)$$

and deduce the value of $\lim_{p \to +\infty} \|f\|_p$.

1.17 (p. 269) (Again, do the L^p spaces become larger or smaller as p increases?) Let us consider the set $\mathbb{N}$ endowed with the counting measure: $\mu(\{k\}) = 1$ for every $k \in \mathbb{N}$ (hence not a finite measure). Prove that if $p \leq q$, then $L^p \subset L^q$.

• The L^p spaces with respect to the counting measure of $\mathbb{N}$ are usually denoted ℓ_p.

1.18 (p. 269) The computation of the integral

$$\int_0^{+\infty} \frac{1}{x} e^{-tx} \sin x \, dx \tag{1.39}$$

for $t > 0$ does not look nice. But as

$$\frac{1}{x} \sin x = \int_0^1 \cos(xy) \, dy$$

and Fubinizing... Compute the integral in (1.39) and its limit as $t \to 0+$.

1.19 (p. 270) Let $f, g : \mathbb{R}^d \to \mathbb{R}$ be integrable functions. Prove that

$$x \mapsto \int_{\mathbb{R}^d} f(y)g(x - y) \, dy$$

defines a function in L^1. This is the *convolution* of f and g, denoted $f * g$. Determine a relation between the L^1 norms of f, g and $f * g$.

• Note the following apparently surprising fact: the two functions $y \mapsto f(y)$ and $y \mapsto g(x - y)$ are in L^1 but, in general, the product of functions of L^1 is not integrable.

Chapter 2
Probability

2.1 Random Variables, Laws, Expectation

A probability space is a triple $(\Omega, \mathscr{F}, P)$ where $(\Omega, \mathscr{F})$ is a measurable space and P a probability on $(\Omega, \mathscr{F})$. Other objects of measure theory appear in probability but sometimes they take a new name that takes into account the role they play in relation to random phenomena. For instance the sets of the σ-algebra $\mathscr{F}$ are the *events*.

A *random variable* (r.v.) is a measurable map defined on $(\Omega, \mathscr{F}, P)$ with values in some measurable space $(E, \mathscr{E})$. In most situations $(E, \mathscr{E})$ will be one among $(\mathbb{R}, \mathscr{B}(\mathbb{R}))$, $(\overline{\mathbb{R}}, \mathscr{B}(\overline{\mathbb{R}}))$ (i.e. the values $+\infty$ or $-\infty$ are also possible), $(\mathbb{R}^m, \mathscr{B}(\mathbb{R}^m))$ or $(\mathbb{C}, \mathscr{B}(\mathbb{C}))$ and we shall speak, respectively, of real, numerical, m-dimensional or complex r.v.'s.

It is not unusual, however, to be led to consider more complicated spaces such as, for instance, matrix groups, the sphere $\mathbb{S}_2$, or even function spaces, endowed with their Borel σ-algebras.

R.v.'s are traditionally denoted by capital letters $(X, Y, Z, \dots)$. They of course enjoy all the properties of measurable maps as seen at the beginning of §1.2. In particular, sums, products, limits,... of real r.v.'s are also r.v.'s.

The *law* or *distribution* of the r.v. $X : (\Omega, \mathscr{F}) \to (E, \mathscr{E})$ is the image of P under X, i.e. the probability μ on $(E, \mathscr{E})$ defined as

$$\mu(A) = P(X^{-1}(A)) = P(\{\omega; X(\omega) \in A\}) \qquad A \in \mathscr{E}.$$

We shall write $P(X \in A)$ as a shorthand for $P(\{\omega; X(\omega) \in A\})$ and we shall write $X \sim Y$ or $X \sim \mu$ to indicate that X and Y have the same distribution or that X has law μ respectively.

If X is real, its *distribution function* F is the distribution function of μ (see (1.4)). In this case (i.e. dealing with probabilities) we can take F as the increasing and right continuous function

$$F(x) = \mu(]-\infty, x]) = P(X \leq x).$$

© The Author(s), under exclusive license to Springer Nature Switzerland AG 2023
P. Baldi, *Probability*, Universitext, https://doi.org/10.1007/978-3-031-38492-9_2

If the real or numerical r.v. X is semi-integrable (upper or lower) with respect to P, its *mathematical expectation* (or *mean*), denoted E(X), is the integral $\int X \, dP$. If $X = (X_1, \ldots, X_m)$ is an m-dimensional r.v. we define

$$E(X) := (E[X_1], \ldots, E[X_m]) \, .$$

X is said to be *centered* if E(X) = 0. If X is $(E, \mathcal{E})$-valued, μ its law and $f :$ $E \to \mathbb{R}$ is a measurable function, by Proposition 1.27 (integration with respect to an image measure), $f(X)$ is integrable if and only if

$$\int_E |f(x)| \, d\mu(x) < +\infty$$

and in this case

$$E[f(X)] = \int_E f(x) \, d\mu(x) \, . \tag{2.1}$$

Of course (2.1) holds also if the r.v. $f(X)$ is only semi-integrable (which is always true if f is positive, for instance). In particular, if X is real-valued and semi-integrable we have

$$E(X) = \int_{\mathbb{R}} x \, d\mu(x) \, . \tag{2.2}$$

This is the relation that is used in practice in order to compute the mathematical expectation of an r.v. The equality (2.2) is also important from a theoretical point of view as it shows that the mathematical expectation depends only on the law: different r.v.'s (possibly defined on different probability spaces) which have the same law also have the same mathematical expectation.

Moreover, (2.1) characterizes the law of X: if the probability μ on $(E, \mathcal{E})$ is such that (2.1) holds for every real bounded measurable function f (or for every measurable positive function f), then necessarily μ is the law of X. This is a useful method to determine the law of X, as better explained in §2.3 below.

The following remark provides an elementary formula for the computation of expectations of positive r.v.'s that we shall use very often.

Remark 2.1 (a) Let X be a *positive* r.v. having law μ and $f : \mathbb{R}^+ \to \mathbb{R}$ an absolutely continuous function such that $f(X)$ is integrable. Then

$$E[f(X)] = f(0) + \int_0^{+\infty} f'(y) P(X \geq y) \, dy \, . \tag{2.3}$$

This is a clever application of Fubini's Theorem: actually such an f is a.e. differentiable and

$$f(x) = f(0) + \int_0^x f'(y)\, dy\,,$$

so that

$$
\begin{aligned}
\mathrm{E}[f(X)] &= \int_0^{+\infty} f(x)\, d\mu(x) \\
&= f(0) + \int_0^{+\infty} d\mu(x) \int_0^x f'(y)\, dy \\
&\overset{!}{=} f(0) + \int_0^{+\infty} f'(y)\, dy \int_y^{+\infty} d\mu(x) \\
&= f(0) + \int_0^{+\infty} f'(y)\mathrm{P}(X \geq y)\, dy\,,
\end{aligned}
\tag{2.4}
$$

where ! indicates where we apply Fubini's Theorem, concerning the integral of $(x, y) \mapsto f'(y)$ on the set $\{(x, y); 0 \leq y \leq x\} \subset \mathbb{R}^2$ with respect to the product measure $\mu \otimes \lambda$ (λ =Lebesgue's measure). *Note however* that in order to apply Fubini's theorem the function $(x, y) \mapsto |f'(y)|1_{\{0 \leq y \leq x\}}(x)$ must be integrable with respect to $\mu \otimes \lambda$. For instance (2.4) does not hold for $f(x) = \sin(e^x)$, whose derivative exhibits high frequency oscillations at infinity.

Note also that in (2.3) $\mathrm{P}(X \geq y)$ can be replaced by $\mathrm{P}(X > y)$: the two functions $y \mapsto \mathrm{P}(X \geq y)$ and $y \mapsto \mathrm{P}(X > y)$ are monotone and coincide except at their points of discontinuity, which are countably many at most, hence of Lebesgue measure 0.

Relation (2.4), replacing f with the identity function $x \mapsto x$ and X with $f(X)$ becomes, still for $f \geq 0$,

$$\mathrm{E}[f(X)] = \int_0^{+\infty} \mathrm{P}(f(X) \geq t)\, dt = \int_0^{+\infty} \mu(f \geq t)\, dt\,. \tag{2.5}$$

(b) If X is positive and integer-valued and $f(x) = x$, (2.5) takes an interesting form: as $\mathrm{P}(X \geq t) = \mathrm{P}(X \geq k + 1)$ for $t \in]k, k + 1]$, we have

$$\mathrm{E}(X) = \int_0^{+\infty} \mathrm{P}(X \geq t)\, dt = \sum_{k=0}^{\infty} \int_k^{k+1} \mathrm{P}(X \geq t)\, dt$$

$$= \sum_{k=0}^{\infty} \mathrm{P}(X \geq k + 1) = \sum_{k=1}^{\infty} \mathrm{P}(X \geq k)\,.$$

In the sequel we shall often make a slight abuse: we shall consider some r.v.'s without stating on which probability space they are defined. The justification for this is that, in order to make the computations, often it is only necessary to know the law of the r.v.'s concerned and, anyway, the explicit construction of a probability space on which the r.v.'s are defined is always possible (see Remark 2.13 below).

The model of a random phenomenon will be a probability space $(\Omega, \mathcal{F}, P)$, of an unknown nature, on which some r.v.'s $X_1, \ldots, X_n$ with given laws are defined.

2.2 Independence

In this section $(\Omega, \mathcal{F}, P)$ is a probability space and all the σ-algebras we shall consider are sub-σ-algebras of $\mathcal{F}$.

Definition 2.2 The σ-algebras $\mathcal{B}_i$, $i = 1, \ldots, n$, are said to be *independent* if

$$P\left(\bigcap_{i=1}^{n} A_i\right) = \prod_{i=1}^{n} P(A_i) \tag{2.6}$$

for every choice of $A_i \in \mathcal{B}_i$, $i = 1, \ldots, n$. The σ-algebras of a, possibly infinite, family $(\mathcal{B}_i, i \in I)$ are said to be independent if the σ-algebras of every finite sub-family are independent.

The next remark is obvious but important.

Remark 2.3 If the σ-algebras $(\mathcal{B}_i, i \in I)$ are independent and if, for every $i \in I$, $\mathcal{B}'_i \subset \mathcal{B}_i$ is a sub-σ-algebra, then the σ-algebras $(\mathcal{B}'_i, i \in I)$ are also independent.

The next proposition says that in order to prove the independence of σ-algebras it is sufficient to check (2.6) for smaller classes of events. This is obviously a very useful simplification.

Proposition 2.4 *Let $\mathscr{C}_i \subset \mathscr{B}_i$, $i = 1, \ldots, n$, be families of events that are stable with respect to finite intersections, containing Ω and such that $\mathscr{B}_i = \sigma(\mathscr{C}_i)$.*

Assume that (2.6) holds for every $A_i \in \mathscr{C}_i$, then the σ-algebras $\mathscr{B}_i$, $i = 1, \ldots, n$, are independent.

Proof We must prove that (2.6), which by hypothesis holds for every $A_i \in \mathscr{C}_i$, actually holds for every $A_i \in \mathscr{B}_i$. Let us fix $A_2 \in \mathscr{C}_2, \ldots, A_n \in \mathscr{C}_n$ and on $\mathscr{B}_1$ consider the two finite measures defined as

$$A \mapsto P\left(A \cap \bigcap_{k=2}^{n} A_k\right) \quad \text{and} \quad A \mapsto P(A)P\left(\bigcap_{k=2}^{n} A_k\right).$$

By assumption they coincide on $\mathscr{C}_1$. Thanks to Carathéodory's criterion, Proposition 1.11, they coincide also on $\mathscr{B}_1$. Hence the independence relation (2.6) holds for every $A_1 \in \mathscr{B}_1$ and $A_2 \in \mathscr{C}_2, \ldots, A_n \in \mathscr{C}_n$.

Let us argue by induction: let us consider, for $k = 1, \ldots, n$, the property

$$P\left(\bigcap_{i=1}^{n} A_i\right) = \prod_{i=1}^{n} P(A_i), \text{ for } A_i \in \mathscr{B}_i, i = 1, \ldots, k, \text{ and } A_i \in \mathscr{C}_i, i > k. \quad (2.7)$$

This property is true for $k = 1$; note also that the condition to be proved is simply that this property holds for $k = n$. If (2.7) holds for $k = r - 1$, let $A_i \in \mathscr{B}_i$, $i = 1, \ldots, r - 1$ and $A_i \in \mathscr{C}_i$, $i = r + 1, \ldots, n$ and let us consider on $\mathscr{B}_r$ the two measures

$$\mathscr{B}_r \ni B \mapsto P(A_1 \cap \ldots \cap A_{r-1} \cap B \cap A_{r+1} \cap \ldots \cap A_n)$$

$$\mathscr{B}_r \ni B \mapsto P(A_1) \cdots P(A_{r-1})P(B)P(A_{r+1}) \cdots P(A_n).$$

By the induction assumption they coincide on the events of $\mathscr{C}_r$. Thanks to Proposition 1.11 (Carathéodory's criterion again) they coincide also on $\mathscr{B}_r$, and therefore (2.7) holds for $k = r$. By induction then (2.7) also holds for $k = n$ which completes the proof. ∎

Next let us consider the property of "independence by packets": if $\mathscr{B}_1$, $\mathscr{B}_2$ and $\mathscr{B}_3$ are independent σ-algebras, are the σ-algebras $\sigma(\mathscr{B}_1, \mathscr{B}_2)$ and $\mathscr{B}_3$ also independent? The following proposition gives an answer in a more general setting.

Proposition 2.5 *Let* $(\mathscr{B}_i, i \in \mathscr{I})$ *be independent* σ*-algebras and* $(I_j, j \in J)$ *a partition of* $\mathscr{I}$. *Then the* σ*-algebras* $(\sigma(\mathscr{B}_i, i \in I_j), j \in J)$ *are independent.*

Proof As independence of a family of σ-algebras is by definition the independence of each finite subfamily, it is sufficient to consider the case of a finite J, $J = \{1, \ldots, n\}$ so that the set of indices $\mathscr{I}$ is partitioned into $I_1, \ldots, I_n$. Let, for $j \in J$, $\mathscr{C}_j$ be the family of all finite intersections of events of the σ-algebras $\mathscr{B}_i$ for $i \in I_j$, i.e.

$$\mathscr{C}_j = \{C; \ C = A_{j,i_1} \cap A_{j,i_2} \cap \ldots \cap A_{j,i_\ell}, \ A_{j,i_1} \in \mathscr{B}_{i_1}, \ldots, A_{j,i_k} \in \mathscr{B}_{i_k},$$

$$i_1, \ldots, i_\ell \in I_j, \ell = 1, 2, \ldots\}.$$

The families of events $\mathscr{C}_j$ are stable with respect to finite intersections, generate respectively the σ-algebras $\sigma(\mathscr{B}_i, i \in I_j)$ and contain Ω. As the $\mathscr{B}_i, i \in \mathscr{I}$, are independent, we have, for every choice of $C_j \in \mathscr{C}_j, j \in J$,

$$P\left(\bigcap_{j=1}^n C_j\right) = P\left(\bigcap_{j=1}^n (A_{j,i_1} \cap \ldots \cap A_{j,i_{\ell_j}})\right) = \prod_{j=1}^n \prod_{k=1}^{\ell_j} P(A_{j,i_k})$$

$$= \prod_{j=1}^n P(A_{j,i_1} \cap A_{j,i_2} \cap \ldots \cap A_{j,i_{\ell_j}}) = \prod_{j=1}^n P(C_j),$$

and thanks to Proposition 2.4 the σ-algebras $\sigma(\mathscr{C}_j) = \sigma(\mathscr{B}_i, i \in I_j)$ are independent. ∎

From the definition of independence of σ-algebras we derive the corresponding definitions for r.v.'s and events.

Definition 2.6 The r.v.'s $(X_i)_{i \in \mathscr{I}}$ with values in the measurable spaces $(E_i, \mathscr{E}_i)$ respectively are said to be independent if the generated σ-algebras $(\sigma(X_i))_{i \in \mathscr{I}}$ are independent.

The events $(A_i)_{i \in \mathscr{I}}$ are said to be independent if the σ-algebras $(\sigma(A_i))_{i \in \mathscr{I}}$ are independent.

Besides these formal definitions, let us recall the intuition beyond these notions of independence: independent events should be such that the knowledge that some of them have taken place does not give information about whether the other ones will take place or not.

In a similar way independent σ-algebras are such that the knowledge of whether the events of some of them have occurred or not does not provide useful information concerning whether the events of the others have occurred or not. In this sense a σ-algebra can be seen as a "quantity of information".

This intuition is important when we must construct a model (i.e. a probability space) intended to describe a given phenomenon. A typical situation arises, for instance, when considering events related to subsequent coin or die throws, or to the choice of individuals in a sample.

However let us not forget that when concerned with proofs or mathematical manipulations, only the formal properties introduced by the definitions must be taken into account. Note that independent r.v.'s may take values in different measurable spaces but, of course, must be defined on the same probability space.

Note also that if the events A and B are independent then also A and B^c are independent, as the σ-algebra generated by an event coincides with the one generated by its complement: $\sigma(A) = \{\Omega, A, A^c, \emptyset\} = \sigma(A^c)$. More generally, if $A_1, \ldots, A_n$ are independent events, then also $B_1, \ldots, B_n$ are independent, where $B_i = A_i$ or $B_i = A_i^c$.

This is in agreement with intuition, as A and A^c carry the same information.

Recall (p. 6) that the σ-algebra generated by an r.v. X taking its values in a measurable space $(E, \mathscr{E})$ is formed by the events $X^{-1}(A) = \{X \in A\}$, $A \in \mathscr{E}$. Hence to say that the $(X_i)_{i \in \mathscr{I}}$ are independent means that

$$P(X_{i_1} \in A_{i_1}, \ldots, X_{i_m} \in A_{i_m}) = P(X_{i_1} \in A_{i_1}) \cdots P(X_{i_m} \in A_{i_m}) \qquad (2.8)$$

for every finite subset $\{i_1, \ldots, i_m\} \subset \mathscr{I}$ and for every choice of $A_{i_1} \in \mathscr{E}_{i_1}, \ldots, A_{i_m} \in \mathscr{E}_{i_m}$.

Thanks to Proposition 2.4, in order to prove the independence of $(X_i)_{i \in \mathscr{I}}$, it is sufficient to verify (2.8) for $A_{i_1} \in \mathscr{C}_{i_1}, \ldots, A_{i_n} \in \mathscr{C}_{i_n}$, where, for every i, $\mathscr{C}_i$ is a class of events generating $\mathscr{E}_i$. If these r.v.'s are real-valued, for instance, it is sufficient for (2.8) to hold for every choice of intervals A_{i_k}.

The following statement is immediate.

Lemma 2.7 *If the σ-algebras $(\mathscr{B}_i)_{i \in \mathscr{I}}$ are independent and if, for every $i \in \mathscr{I}$, X_i is $\mathscr{B}_i$-measurable, then the r.v.'s $(X_i)_{i \in \mathscr{I}}$ are independent.*

Actually $\sigma(X_i) \subset \mathcal{B}_i$, hence also the σ-algebras $(\sigma(X_i))_{i \in \mathcal{I}}$ are independent (Remark 2.3).

If the r.v.'s $(X_i)_{i \in \mathcal{I}}$ are independent with values respectively in the measurable spaces $(E_i, \mathcal{E}_i)$ and $f_i : E_i \to G_i$ are measurable functions with values respectively in the measurable spaces $(G_i, \mathcal{G}_i)$, then the r.v.'s $(f_i(X_i))_{i \in \mathcal{I}}$ are also independent as obviously $\sigma(f_i(X_i)) \subset \sigma(X_i)$.

In other words, functions of independent r.v.'s are themselves independent, which agrees with the intuitive meaning described previously: if the knowledge of the values taken by some of the X_i does not give information concerning the values taken by other X_j's, there is no reason why the values taken by some of the $f_i(X_i)$ should give information about the values taken by other $f_j(X_j)$'s.

The next, fundamental, theorem establishes a relation between independence of r.v's. and their joint law.

Theorem 2.8 *Let X_i, $i = 1, \ldots, n$, be r.v.'s with values in the measurable spaces $(E_i, \mathcal{E}_i)$ respectively. Let us denote by μ the law of $(X_1, \ldots, X_n)$, which is an r.v. with values in the product space of the $(E_i, \mathcal{E}_i)$, and by μ_i the law of X_i, $i = 1, \ldots, n$.*

Then $X_1, \ldots, X_n$ are independent if and only if $\mu = \mu_1 \otimes \cdots \otimes \mu_n$.

Proof Let us assume $X_1, \ldots, X_n$ are independent: we have, for every choice of $A_i \in \mathcal{E}_i, i = 1, \ldots, n$,

$$\mu(A_1 \times \cdots \times A_n) = P(X_1 \in A_1, \ldots, X_n \in A_n)$$
$$= P(X_1 \in A_1) \cdots P(X_n \in A_n) = \mu_1(A_1) \cdots \mu_n(A_n) . \tag{2.9}$$

Hence μ coincides with the product measure $\mu_1 \otimes \cdots \otimes \mu_n$ on the rectangles $A_1 \times \cdots \times A_n$. Therefore $\mu = \mu_1 \otimes \cdots \otimes \mu_n$. The converse follows at once by writing (2.9) the other way round: if $\mu = \mu_1 \otimes \cdots \otimes \mu_n$

$$P(X_1 \in A_1, \ldots, X_n \in A_n) = \mu(A_1 \times \cdots \times A_n) = \mu_1(A_1) \cdots \mu_n(A_n) =$$
$$= P(X_1 \in A_1) \cdots P(X_n \in A_n) .$$

so that $X_1, \ldots, X_n$ are independent   ∎

Thanks to Theorem 2.8 the independence of r.v.'s depends only on their joint law: if $X_1, \ldots, X_n$ are independent and $(X_1, \ldots, X_n)$ has the same law as $(Y_1, \ldots, Y_n)$

(possibly defined on a different probability space), then also $Y_1, \ldots, Y_n$ are independent.

The following proposition specializes Theorem 2.8 when the r.v.'s X_i take their values in a metric space.

Proposition 2.9 *Let $X_1, \ldots, X_m$ be r.v.'s taking values in the metric spaces $E_1, \ldots, E_m$. Then $X_1, \ldots, X_m$ are independent if and only if for every choice of bounded continuous functions $f_i : E_i \to \mathbb{R}$, $i = 1, \ldots, m$,*

$$\mathrm{E}[f_1(X_1) \cdots f_m(X_m)] = \mathrm{E}[f_1(X_1)] \cdots \mathrm{E}[f_m(X_m)] . \tag{2.10}$$

If in addition the spaces E_i are also separable and locally compact, then it is sufficient to check (2.10) for compactly supported continuous functions f_i.

Proof In (2.10) we have the integral of $(x_1, \ldots, x_m) \mapsto f_1(x_1) \cdots f_m(x_m)$ with respect to the joint law, μ, of $(X_1, \ldots, X_m)$ on the left-hand side, whereas on the right-hand side appears the integral of the same function with respect to the product of their laws. The statement then follows immediately from Proposition 1.33. ∎

Corollary 2.10 *Let $X_1, \ldots, X_n$ be real integrable independent r.v.'s. Then their product $X_1 \cdots X_n$ is integrable and*

$$\mathrm{E}(X_1 \cdots X_n) = \mathrm{E}(X_1) \cdots \mathrm{E}(X_n) .$$

Proof This result is obviously related to Proposition 2.9, but for the fact that the function $x \mapsto x$ is not bounded. But Fubini's Theorem easily handles this difficulty.

As the joint law of $(X_1, \ldots, X_n)$ is the product $\mu_1 \otimes \cdots \otimes \mu_n$, Fubini's Theorem gives

$$\begin{aligned} \mathrm{E}(|X_1 \cdots X_n|) &= \int |x_1| \, d\mu_1(x_1) \cdots \int |x_n| \, d\mu_n(x_n) \\ &= \mathrm{E}(|X_1|) \cdots \mathrm{E}(|X_n|) < +\infty . \end{aligned} \tag{2.11}$$

Hence the product $X_1 \cdots X_n$ is integrable and, repeating the argument of (2.11) without absolute values, Fubini's Theorem again gives

$$\mathrm{E}(X_1 \cdots X_n) = \int x_1 \, d\mu_1(x_1) \cdots \int x_n \, d\mu_n(x_n) = \mathrm{E}(X_1) \cdots \mathrm{E}(X_n) .$$

∎

Remark 2.11 Let $X_1, \ldots, X_n$ be r.v.'s taking their values in the measurable spaces $E_1, \ldots, E_n$, countable and endowed with the σ-algebra of all subsets respectively. Then they are independent if and only if for every $x_i \in E_i$ we have for every $x_i \in E_i$

$$P(X_1 = x_1, \ldots, X_n = x_n) = P(X_1 = x_1) \cdots P(X_n = x_n) \,. \qquad (2.12)$$

Actually from this relation it is easy to see that the joint law of $(X_1, \ldots, X_n)$ coincides with the product law on the rectangles.

Remark 2.12 Given a family $(X_i)_{i \in \mathcal{I}}$ of r.v.'s, it is possible to have X_i independent of X_j for every $i, j \in \mathcal{I}$, $i \neq j$, without the family being formed of independent r.v.'s, as shown in the following example. In other words, pairwise independence is a (much) weaker property than independence.

Let X and Y be independent r.v.'s such that $P(X = \pm 1) = P(Y = \pm 1) = \frac{1}{2}$ and let $Z = XY$. We have easily that also $P(Z = \pm 1) = \frac{1}{2}$.

X and Z are independent: indeed $P(X = 1, Z = 1) = P(X = 1, Y = 1) = \frac{1}{4} = P(X = 1)P(Z = 1)$ and in the same way we see that $P(X = i, Z = j) = P(X = i)P(Z = j)$ for every $i, j = \pm 1$, so that the criterion of Remark 2.11 is satisfied. By symmetry Y and Z are also independent.

The three r.v.'s X, Y, Z however are not independent: as $X = Z/Y$, X is $\sigma(Y, Z)$-measurable and $\sigma(X) \subset \sigma(Y, Z)$. If they were independent $\sigma(X)$ would be independent of $\sigma(Y, Z)$ and the events of $\sigma(X)$ would be independent of themselves. But if A is independent of itself then $P(A) = P(A \cap A) = P(A)^2$ so that it can only have probability equal to 0 or to 1, whereas here the events $\{X = 1\}$ and $\{X = -1\}$ belong to $\sigma(X)$ and have probability $\frac{1}{2}$.

Note in this example that $\sigma(X)$ is independent of $\sigma(Y)$ and is independent of $\sigma(Z)$, but is not independent of $\sigma(Y, Z)$.

Remark 2.13 (a) Given a probability μ on a measurable space $(E, \mathcal{E})$, it is always possible to construct a probability space $(\Omega, \mathcal{F}, P)$ on which an r.v. X is defined with values in $(E, \mathcal{E})$ and having law μ. It is sufficient, for instance, to set $\Omega = E$, $\mathcal{F} = \mathcal{E}$, $P = \mu$ and $X(x) = x$.

(b) Very often we shall consider sequences $(X_n)_n$ of independent r.v.'s, defined on a probability space $(\Omega, \mathcal{F}, P)$ having given laws, $X_i \sim \mu_i$ say.

Note that such an object always exists. Actually if X_i is $(E_i, \mathscr{C}_i)$-valued and $X_i \sim \mu_i$, let

$\Omega = $ the infinite product set $E_1 \times E_2 \times \cdots$

$\mathscr{F} = $ the product σ-algebra $\mathscr{C}_1 \otimes \mathscr{C}_2 \otimes \cdots$

P $= $ the infinite product probability $\mu_1 \otimes \mu_2 \otimes \cdots$, see Theorem 1.36.

As the elements of the product set Ω are of the form $\omega = (x_1, x_2, \ldots)$ with $x_i \in E_i$, we can define $X_i(\omega) = x_i$. Such a map is measurable $E \to E_i$ (it is a projector, recall Proposition 1.31) and the sequence $(X_n)_n$ defined in this way satisfies the requested conditions. Independence is guaranteed by the fact that their joint law is the product law.

Remark 2.14 Let X_i, $i = 1, \ldots, m$, be real independent r.v.'s with values respectively in the measurable spaces $(E_i, \mathscr{C}_i)$. Let, for every $i = 1, \ldots, m$, ρ_i be a σ-finite measure on $(E_i, \mathscr{C}_i)$ such that the law μ_i of X_i has density f_i with respect to ρ_i. Then the product measure $\mu := \mu_1 \otimes \cdots \otimes \mu_m$, which is the law of $(X_1, \ldots, X_m)$, has density

$$f(x) = f_1(x_1) \cdots f_m(x_m)$$

with respect to the product measure $\rho := \rho_1 \otimes \cdots \otimes \rho_m$.

Actually it is immediate that the two measures $f\, d\rho$ and μ coincide on rectangles.

In this case we shall say that the joint density f is the *tensor* product of the marginal densities f_i.

Theorem 2.15 (Kolmogorov's 0-1 Law) *Let $(X_n)_n$ be a sequence of independent r.v.'s. Let $\mathscr{B}^n = \sigma(X_k, k \geq n)$ and $\mathscr{B}^\infty = \bigcap_n \mathscr{B}^n$ (the tail σ-algebra). Then $\mathscr{B}^\infty$ is P-trivial, i.e. for every $A \in \mathscr{B}^\infty$, we have $\mathrm{P}(A) = 0$ or $\mathrm{P}(A) = 1$. Moreover, if X is an m-dimensional $\mathscr{B}^\infty$-measurable r.v., then X is constant a.s.*

Proof Let $\mathscr{F}_n = \sigma(X_k, \ k \leq n)$, $\mathscr{F}_\infty = \sigma(X_k, \ k \geq 0)$. Thanks to Proposition 2.5 (independence by packets) $\mathscr{F}_n$ is independent of $\mathscr{B}^{n+1}$, which is generated by the X_i with $i > n$. Hence it is also independent of $\mathscr{B}^\infty \subset \mathscr{B}^{n+1}$.

Let us prove that $\mathcal{B}^\infty$ is independent of $\mathcal{F}_\infty$. The family $\mathscr{C} = \bigcup_n \mathcal{F}_n$ is stable with respect to finite intersections and generates $\mathcal{F}_\infty$. If $A \in \bigcup_n \mathcal{F}_n$, then $A \in \mathcal{F}_n$ for some n, hence is independent of $\mathcal{B}^\infty$. Therefore A is independent of $\mathcal{B}^\infty$ and by Proposition 2.4 $\mathcal{B}^\infty$ and $\mathcal{F}_\infty$ are independent.

But $\mathcal{B}^\infty \subset \mathcal{F}_\infty$, so that $\mathcal{B}^\infty$ is independent of itself. If $A \in \mathcal{B}^\infty$, as in Remark 2.12, we have $P(A) = P(A \cap A) = P(A)P(A)$, i.e. $P(A) = 0$ or $P(A) = 1$.

If X is a real $\mathcal{B}^\infty$-measurable r.v., then for every $a \in \mathbb{R}$ the event $\{X \le a\}$ belongs to $\mathcal{B}^\infty$ and its probability can be equal to 0 or to 1 only. Let $c = \sup\{a; P(X \le a) = 0\}$, then necessarily $c < +\infty$, as $1 = P(X < +\infty) = \lim_{a\to\infty} P(X \le a)$, so that $P(X \le a) > 0$ for some k. For every $n > 0$, $P(X \le c + \frac{1}{n}) > 0$, hence $P(X \le c + \frac{1}{n}) = 1$ as $\{X \le c + \frac{1}{n}\} \in \mathcal{B}^\infty$, whereas $P(X \le c - \frac{1}{n}) = 0$. From this we deduce that X takes a.s. only the value c as

$$P(X = c) = P\left(\bigcap_{n=1}^\infty \{c - \tfrac{1}{n} \le X \le c + \tfrac{1}{n}\} \right) = \lim_{n\to\infty} P\left(c - \tfrac{1}{n} \le X \le c + \tfrac{1}{n}\right) = 1 \, .$$

If $X = (X_1, \ldots, X_m)$ is m-dimensional, by the previous argument each of the marginals X_i is a.s. constant and the result follows. ∎

If all events of $\sigma(X)$ have probability 0 or 1 only then X is a.s. constant also if X takes values in a more general space, see Exercise 2.2.

Some consequences of Kolmogorov's 0-1 law are surprising, at least at first sight.

Let $(X_n)_n$ be a sequence of real independent r.v.'s and let $\overline{X}_n = \frac{1}{n}(X_1 + \cdots + X_n)$ (the *empirical means*). Then $\overline{X} = \overline{\lim}_{n\to\infty} \overline{X}_n$ is a tail r.v. Actually we can write, for every integer k,

$$\overline{X}_n = \frac{1}{n}(X_1 + \cdots + X_k) + \frac{1}{n}(X_{k+1} + \cdots + X_n)$$

and as the first term on the right-hand side tends to 0 as $n \to \infty$, $\overline{X}$ does not depend on $X_1, \ldots, X_k$ for every k and is therefore $\mathcal{B}^{k+1}$-measurable. We deduce that $\overline{X}$ is measurable with respect to the tail σ-algebra and is a.s. constant. As the same argument holds for $\underline{\lim}_{n\to\infty} \overline{X}_n$ we also have

$$\{\text{the sequence } (\overline{X}_n)_n \text{ is convergent}\} = \{ \overline{\lim_{n\to\infty}}\, \overline{X}_n = \underline{\lim_{n\to\infty}}\, \overline{X}_n\} \, ,$$

which is a tail event and has probability equal to 0 or to 1. Therefore either the sequence $(\overline{X}_n)_n$ converges a.s. with probability 1 (and in this case the limit is a.s. constant) or it does not converge with probability 1.

A similar argument can be developed when investigating the convergence of a series $\sum_{n=1}^\infty X_n$ of independent r.v.'s. Also in this case the event $\{the\ series\ converges\}$ belongs to the tail σ-algebra, as the convergence of a series does not depend on its first terms. Hence either the series does not converge with probability 1 or is a.s. convergent.

In this case, however, the sum of the series depends also on its first terms. Hence the r.v. $\sum_{n=1}^{\infty} X_n$ *does not* necessarily belong to the tail σ-algebra and need not be constant.

2.3 Computation of Laws

Many problems in probability boil down to the computation of the law of an r.v., which is the topic of this section.

Recall that if X is an r.v. with values in a measurable space $(E, \mathscr{E})$, its law is a probability μ on $(E, \mathscr{E})$ such that (Proposition 1.27, integration with respect to an image measure)

$$E[\phi(X)] = \int_E \phi(x) \, d\mu(x) \tag{2.13}$$

for every bounded measurable function $\phi : E \to \mathbb{R}$. More precisely, if μ is a probability on $(E, \mathscr{E})$ such that (2.13) holds for every bounded measurable function $\phi : E \to \mathbb{R}$, then μ is necessarily the law of X.

Let now X be an r.v. with values in $(E, \mathscr{E})$ having law μ and let $\Phi : E \to G$ be a measurable map from E to some other measurable space $(G, \mathscr{G})$. How to determine the law, ν say, of $\Phi(X)$? We have, by the integration rule with respect to an image probability (Proposition 1.27),

$$E[\phi(\Phi(X))] = \int_E \phi(\Phi(x)) \, d\mu(x) \,,$$

but also

$$E[\phi(\Phi(X))] = \int_G \phi(y) \, d\nu(y) \,,$$

which takes us to the relation

$$\int_E \phi(\Phi(x)) \, d\mu(x) = \int_G \phi(y) \, d\nu(y) \tag{2.14}$$

and a probability ν satisfying (2.14) is necessarily the law of $\Phi(X)$. Hence a possible way to compute the law of $\Phi(X)$ is to solve "equation" (2.14) for every bounded measurable function ϕ, with ν as the unknown. This is the method of the "dumb function". A closer look at (2.14) allows us to foresee that the question boils down naturally to a change of variable.

Let us now see some examples of application of this method. Other tools toward the goal of computing the law of an r.v. will be introduced in §2.6 (characteristic functions), §2.7 (Laplace transforms) and §4.3 (conditional laws).

Example 2.16 Let X, Y be $\mathbb{R}^d$- and $\mathbb{R}^m$-valued respectively r.v.'s, having joint density $f : \mathbb{R}^{d+m} \to \mathbb{R}$ with respect to the Lebesgue measure of $\mathbb{R}^{d+m}$. Do X and Y also have a law with a density with respect to the Lebesgue measure (of $\mathbb{R}^d$ and $\mathbb{R}^m$ respectively)? What are these densities?

In other words, how can we compute the marginal densities from the joint density?

We have, for every real bounded measurable function ϕ,

$$E[\phi(X)] = \int_{\mathbb{R}^d \times \mathbb{R}^m} \phi(x) f(x, y) \, dx \, dy = \int_{\mathbb{R}^d} \phi(x) \, dx \int_{\mathbb{R}^m} f(x, y) \, dy \, ,$$

from which we conclude that the law of X is

$$d\mu(x) = f_X(x) \, dx \, ,$$

where

$$f_X(x) = \int_{\mathbb{R}^m} f(x, y) \, dy \, .$$

Note that the measurability of f_X follows from Fubini's Theorem.

Example 2.17 Let X, Y be d-dimensional r.v.'s having joint density : $\mathbb{R}^d \times \mathbb{R}^d \to \mathbb{R}$. Does their sum $X+Y$ also have a density with respect to the Lebesgue measure?

We have

$$E[\phi(X + Y)] = \int_{\mathbb{R}^d} dy \int_{\mathbb{R}^d} \phi(x + y) f(x, y) \, dx \, .$$

With the change of variable $z = x + y$ in the inner integral and changing the order of integration we find

$$E[\phi(X+Y)] = \int_{\mathbb{R}^d} dy \int_{\mathbb{R}^d} \phi(z) f(z-y, y) \, dz = \int_{\mathbb{R}^d} \phi(z) \, dz \underbrace{\int_{\mathbb{R}^d} f(z-y, y) \, dy}_{:=g(z)} \, .$$

Comparing with (2.14), $X + Y$ has density

$$h(z) = \int_{\mathbb{R}^d} f(z - y, y) \, dy$$

with respect to the Lebesgue measure. A change of variable gives that also

$$h(z) = \int_{\mathbb{R}^d} f(x, z - x)\, dx \ .$$

Given two probabilities μ, ν on $\mathbb{R}^d$, their *convolution* is the image of the product measure $\mu \otimes \nu$ under the "sum" map $\mathbb{R}^d \times \mathbb{R}^d \rightarrow \mathbb{R}^d$, $(x, y) \mapsto x + y$. The convolution is denoted $\mu * \nu$ (see also Exercise 1.19).

Equivalently, if X, Y are independent r.v.'s having laws μ and ν respectively, then $\mu * \nu$ is the law of $X + Y$.

Proposition 2.18 *If μ, ν are probabilities on $\mathbb{R}^d$ with densities f, g with respect to the Lebesgue measure respectively, then their convolution $\mu * \nu$ has density, still with respect to the Lebesgue measure,*

$$h(z) = \int_{\mathbb{R}^d} f(z - y)g(y)\, dy = \int_{\mathbb{R}^d} g(z - y)f(y)\, dy \ .$$

Proof Immediate consequence of Remark 2.17 with $f(x, y)$ replaced by $f(x)g(y)$. ∎

Example 2.19 Let W, T be independent r.v.'s having density respectively exponential of parameter $\frac{1}{2}$ and uniform on $[0, 2\pi]$. Let $R = \sqrt{W}$. What is the joint law of (X, Y) where $X = R \cos T$, $Y = R \sin T$? Are X and Y independent?

Going back to (2.14) we must find a density g such that, for every bounded measurable $\phi : \mathbb{R}^2 \rightarrow \mathbb{R}$,

$$E[\phi(X, Y)] = \int_{-\infty}^{+\infty} \int_{-\infty}^{+\infty} \phi(x, y)g(x, y)\, dx\, dy \ . \tag{2.15}$$

Let us compute first the law of R. For $r > 0$ we have, recalling the expression of the d.f. of an exponential law,

$$F_R(r) = P(\sqrt{W} \le r) = P(W \le r^2) = 1 - e^{-r^2/2}, \quad r \ge 0$$

and, taking the derivative, the law of $R = \sqrt{W}$ has a density with respect to the Lebesgue measure given by

$$f_R(r) = r\, e^{-r^2/2} \qquad \text{for } r > 0$$

and $f_R(r) = 0$ for $r \le 0$. The law of T has a density with respect to the Lebesgue measure that is equal to $\frac{1}{2\pi}$ on the interval $[0, 2\pi]$ and vanishes elsewhere. Hence (R, T) has joint density

$$f(r, t) = \frac{1}{2\pi}\, r\, e^{-r^2/2}, \qquad \text{for } r > 0,\ 0 \le t \le 2\pi,$$

and $f(r, t) = 0$ otherwise. By the integration formula with respect to an image law, Proposition 1.27,

$$E[\phi(X, Y)] = E[\phi(R \cos T, R \sin T)]$$

$$= \frac{1}{2\pi} \int_0^{2\pi} dt \int_0^{+\infty} \phi(r \cos t, r \sin t)\, r\, e^{-r^2/2}\, dr$$

and in cartesian coordinates

$$\cdots = \frac{1}{2\pi} \cdot \int_{-\infty}^{+\infty} \int_{-\infty}^{+\infty} \phi(x, y)\, e^{-\frac{1}{2}(x^2+y^2)}\, dx\, dy \ .$$

Comparing with (2.15) we conclude that the joint density g of (X, Y) is

$$g(x, y) = \frac{1}{2\pi}\, e^{-\frac{1}{2}(x^2+y^2)} \ .$$

As

$$g(x, y) = \frac{1}{\sqrt{2\pi}}\, e^{-x^2/2} \times \frac{1}{\sqrt{2\pi}}\, e^{-y^2/2} \ ,$$

g is the density of the product of two $N(0, 1)$ laws. Hence both X and Y are $N(0, 1)$-distributed and, as the their joint law is the product of the marginals, they are independent. Note that this is a bit unexpected, as both X and Y depend on R and T.

Example 2.20 Let X be an m-dimensional r.v. having density f with respect to the Lebesgue measure. Let A be an $m \times m$ *invertible* matrix and $b \in \mathbb{R}^m$.

Does the r.v. $Y = AX + b$ also have density with respect to the Lebesgue measure?

For every bounded measurable function ϕ we have

$$E[\phi(Y)] = E[\phi(AX + b)] = \int_{\mathbb{R}^m} \phi(Ax + b) f(x) \, dx .$$

With the change of variable $y = Ax + b$, $x = A^{-1}(y - b)$, we have

$$E[\phi(Y)] = \int_{\mathbb{R}^m} \phi(y) f(A^{-1}(y - b)) |\det A^{-1}| \, dy ,$$

so that Y has density, with respect to the Lebesgue measure,

$$f_Y(y) = \frac{1}{|\det A|} f(A^{-1}(y - b)) .$$

If $b = 0$ and $A = -I$ (I=identical matrix) then we have

$$f_{-Y}(y) = f_Y(-y) .$$

An r.v. Y such that $Y \sim -Y$ is said to be *symmetric*. Of course such an r.v., if integrable, is centered, as then $E(Y) = -E(Y)$.

One might wonder what happens if A is not invertible. See Exercise 2.27.

The next examples show instances of the application of the change of variable formula for multiple integrals in order to solve the dumb function "equation" (2.14).

Example 2.21 Let X, Y be r.v.'s defined on a same probability space, i.i.d. and with density, with respect to the Lebesgue measure,

$$f(x) = \frac{1}{x^2}, \qquad x \geq 1$$

and $f(x) = 0$ otherwise. What is the joint law of $U = XY$ and $V = \frac{X}{Y}$?

Let us surmise that this joint law has a density g: we should have then, for every bounded Borel function $\phi : \mathbb{R}^2 \to \mathbb{R}^2$,

$$E[\phi(U, V)] = \int_{\mathbb{R}^2} \phi(u, v) g(u, v) \, du \, dv .$$

But

$$
E[\phi(U, V)] = E[\phi(XY, \tfrac{X}{Y})] = \int_1^{+\infty} \int_1^{+\infty} \phi(xy, \tfrac{x}{y}) \frac{1}{x^2 y^2} \, dx \, dy \,.
$$

Let us make the change of variable $(u, v) = \Psi(x, y) = (xy, \tfrac{x}{y})$, whose inverse is

$$
\Psi^{-1}(u, v) = \left(\sqrt{uv}, \sqrt{\frac{u}{v}} \right) .
$$

Its differential is

$$
D\Psi^{-1}(u, v) = \frac{1}{2} \begin{pmatrix} \sqrt{\frac{v}{u}} & \sqrt{\frac{u}{v}} \\ \frac{1}{\sqrt{uv}} & -\sqrt{\frac{u}{v^3}} \end{pmatrix}
$$

and therefore

$$
| \det D\Psi^{-1}(u, v)| = \frac{1}{4} \left| -\frac{1}{v} - \frac{1}{v} \right| = \frac{1}{2v} \,.
$$

Moreover the condition $x > 1$, $y > 1$ becomes $u > 1$, $\frac{1}{u} \le v \le u$. Hence

$$
E[\phi(U, V)] = \int_1^{+\infty} du \int_{1/u}^{u} \phi(u, v) \frac{1}{2u^2 v} \, du \, dv
$$

and the density of (U, V) is

$$
g(u, v) = \frac{1}{2u^2 v} 1_{\{u > 1\}} 1_{\{\frac{1}{u} \le v \le u\}} \,.
$$

g is strictly positive in the shaded region of Fig. 2.1.

Sometimes, even in a multidimensional setting, it is not necessary to use the change of variables formula for multiple integrals, which requires some effort as in Example 2.21: the simpler formula for the one-dimensional integrals may be sufficient, as in the following example.

Example 2.22 Let X and Y be independent and exponential r.v.'s with parameter $\lambda = 1$. What is the joint law of X and $Z = \tfrac{X}{Y}$? And the law of $\tfrac{X}{Y}$?

Fig. 2.1 The joint density g is positive in the shaded region

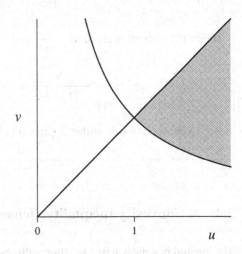

The joint law of X and Y has density

$$f(x, y) = f_X(x) f_Y(y) = e^{-x} e^{-y} = e^{-(x+y)}, \quad x > 0, y > 0.$$

Let $\phi : \mathbb{R}^2 \to \mathbb{R}$ be bounded and measurable, then

$$E\big[\phi(X, \tfrac{X}{Y})\big] = \int_0^{+\infty} dx \int_0^{+\infty} \phi(x, \tfrac{x}{y}) e^{-x} e^{-y} \, dy.$$

With the change of variable $\tfrac{x}{y} = z$, $dy = -\tfrac{x}{z^2} dz$, in the inner integral we have

$$E\big[\phi(X, \tfrac{X}{Y})\big] = \int_0^{+\infty} dx \int_0^{+\infty} \phi(x, z) \frac{x}{z^2} e^{-x} e^{-x/z} \, dz.$$

Hence the required joint law has density with respect to the Lebesgue measure

$$g(x, z) = \frac{x}{z^2} e^{-x(1 + \frac{1}{z})}, \quad x > 0, z > 0.$$

The density of $Z = \tfrac{X}{Y}$ is the second marginal of g:

$$g_Z(z) = \int g(x, z) \, dx = \frac{1}{z^2} \int_0^{+\infty} x e^{-x(1 + \frac{1}{z})} \, dx.$$

This integral can be computed easily by parts, keeping in mind that the integration variable is x and that here z is just a constant. More cleverly, just recognize in the integrand, but for the constant, a Gamma$(2, 1 + \frac{1}{z})$ density.

Hence the integral is equal to $\frac{1}{(1+\frac{1}{z})^2}$ and

$$g_Z(z) = \frac{1}{z^2(1 + \frac{1}{z})^2} = \frac{1}{(1 + z)^2} , \qquad z > 0 .$$

See Exercise 2.19 for another approach to the computation of the law of $\frac{X}{Y}$.

2.4 A Convexity Inequality: Jensen

The integral of a measurable function with respect to a probability (i.e. mathematical expectation) enjoys a convexity inequality. This property is typical of probabilities (see Exercise 2.23) and is related to the fact that, for probabilities, the integral takes the meaning of a *mean* or, for $\mathbb{R}^m$-valued r.v.'s, of a *barycenter*.

Recall that a function $\phi : \mathbb{R}^m \to \mathbb{R} \cup \{+\infty\}$ (the value $+\infty$ is also possible) is *convex* if and only if for every $0 \le \lambda \le 1$ and $x, y \in \mathbb{R}^m$ we have

$$\phi(\lambda x + (1 - \lambda)y) \le \lambda \phi(x) + (1 - \lambda)\phi(y) . \tag{2.16}$$

It is *concave* if $-\phi$ is convex, i.e. if in (2.16) $\le$ is replaced by $\ge$. ϕ is *strictly convex* if (2.16) holds with $<$ instead of $\le$ whenever $x \ne y$ and $0 < \lambda < 1$.

Note that an affine-linear function $f(x) = \langle \alpha, x \rangle + b$, $\alpha \in \mathbb{R}^m, b \in \mathbb{R}$, is continuous and convex: (2.16) actually becomes an equality so that such an f is also concave.

In the sequel we shall take advantage of the fact that if ϕ is convex and lower semi-continuous (l.s.c.) then

$$\phi(x) = \sup_f f(x) , \tag{2.17}$$

the supremum being taken among all affine-linear functions f such that $f \le \phi$. A similar result holds of course for concave and u.s.c functions (with inf).

Recall (p.14) that a function f is lower semi-integrable (l.s.i.) with respect to a measure μ if it is bounded from below by a μ-integrable function and that in this case the integral $\int f \, d\mu$ is defined (possibly $= +\infty$).

Theorem 2.23 (Jensen's Inequality) *Let X be an m-dimensional integrable r.v. and $\phi \colon \mathbb{R}^m \to \mathbb{R} \cup \{+\infty\}$ a convex l.s.c. function (resp. concave and u.s.c). Then $\phi(X)$ is l.s.i. (resp. u.s.i.) and*

$$E[\phi(X)] \geq \phi(E[X]) \qquad (resp. E[\phi(X)] \leq \phi(E[X])) .$$

Moreover, if ϕ is strictly convex and X is not a.s. constant, in the previous relation the inequality is strict.

Proof Let us assume first $\phi(E(X)) < +\infty$. A hyperplane crossing the graph of ϕ at $x = E(X)$ is an affine-linear function of the form

$$f(x) = \langle \alpha, x - E(X) \rangle + \phi(E(X))$$

for some $\alpha \in \mathbb{R}^m$. Note that f and ϕ take the same value at $x = E(X)$. As ϕ is convex, there exists such a hyperplane *minorizing* ϕ, i.e. such that

$$\phi(x) \geq \langle \alpha, x - E(X) \rangle + \phi(E(X)) \quad \text{for all } x \qquad (2.18)$$

and therefore

$$\phi(X) \geq \langle \alpha, X - E(X) \rangle + \phi(E(X)) . \qquad (2.19)$$

As the r.v. on the right-hand side is integrable, $\phi(X)$ is l.s.i. Taking the mathematical expectation in (2.19) we find

$$E[\phi(X)] \geq \langle \alpha, E(X) - E(X) \rangle + \phi(E(X)) = \phi(E(X)) . \qquad (2.20)$$

If ϕ is strictly convex, then in (2.18) the inequality is strict for $x \neq E(X)$. If X is not a.s. equal to its mean $E(X)$, then the inequality (2.19) is strict on an event of strictly positive probability and therefore in (2.20) a strict inequality holds.

If $\phi(E(X)) = +\infty$ instead, let f be an affine function minorizing ϕ; then $f(X)$ is integrable and $\phi(X) \geq f(X)$ so that $\phi(X)$ is l.s.i. Moreover,

$$E[\phi(X)] \geq E[f(X)] = f(E(X)) .$$

Taking the supremum on all affine functions f minorizing ϕ, thanks to (2.17) we find

$$E[\phi(X)] \geq \phi(E(X))$$

concluding the proof. ∎

By taking particular choices of ϕ, from Jensen's inequality we can derive the classical inequalities that we have already seen in Chap. 1 (see p. 27).

Hölder's Inequality: If p, q are positive numbers such that $\frac{1}{p} + \frac{1}{q} = 1$ then

$$E(|XY|) \leq E(|X|^p)^{1/p} E(|Y|^q)^{1/q} . \qquad (2.21)$$

If one among $|X|^p$ or $|Y|^q$ is not integrable there is nothing to prove. Otherwise note that the function

$$\phi(x, y) = \begin{cases} x^{1/p} y^{1/q} & x, y \geq 0 \\ -\infty & \text{otherwise} \end{cases}$$

is concave and u.s.c. so that

$$E(|XY|) = E[\phi(|X|^p, |Y|^p)] \leq \phi(E[|X|^p], E[|Y|^p]) = E(|X|^p)^{1/p} E(|Y|^q)^{1/q} .$$

Note that the condition $\frac{1}{p} + \frac{1}{q} = 1$ requires that both p and q are ≥ 1. Equivalently, if $0 \leq \alpha, \beta \leq 1$ with $\alpha + \beta = 1$, (2.21) becomes

$$E(X^\alpha Y^\beta) \leq E(X)^\alpha E(Y)^\beta \qquad (2.22)$$

for every pair of positive r.v.'s X, Y.

The particular case $p = q = 2$ (2.21) becomes the *Cauchy-Schwarz inequality*

$$E(|XY|) \leq E(|X|^2)^{1/2} E(|Y|^2)^{1/2} . \qquad (2.23)$$

Minkowski's Inequality: For every $p \geq 1$

$$E(|X + Y|^p)^{1/p} \leq E(|X|^p)^{1/p} + E(|Y|^p)^{1/p} . \qquad (2.24)$$

Again there is nothing to prove unless both X and Y belong to L^p. Otherwise (2.24) follows from Jensen's inequality applied to the concave u.s.c. function

$$\phi(x, y) = \begin{cases} (x^{1/p} + y^{1/p})^p & x, y \geq 0 \\ -\infty & \text{otherwise} \end{cases}$$

and to the r.v.'s $|X|^p, |Y|^p$: with this notation $\phi(|X|^p, |Y|^p) = (|X| + |Y|)^p$ and we have

$$E(|X + Y|^p) \leq E[(|X| + |Y|)^p] = E[\phi(|X|^p, |Y|^p)] \leq \phi(E[|X|^p], E[|Y|^p])$$
$$= (E[|X|^p]^{1/p} + E[|Y|^p]^{1/p})^p$$

and now just take the $\frac{1}{p}$-th power on both sides.

As we have seen in Chap. 1, the Hölder, Cauchy-Schwarz and Minkowski inequalities hold for every σ-finite measure. In the case of probabilities however they are particular instances of Jensen's inequality.

From Jensen's inequality we can deduce an inequality between L^p norms: if $p > q$, as $\phi(x) = |x|^{p/q}$ is a continuous convex function, we have

$$\|X\|_p^p = E(|X|^p) = E[\phi(|X|^q)] \geq \phi(E[|X|^q]) = E(|X|^q)^{p/q}$$

and, taking the p-th root,

$$\|X\|_p \geq \|X\|_q \tag{2.25}$$

i.e. the L^p norm is an increasing function of p. In particular, if $p \geq q$, $L^p \subset L^q$. This inclusion holds for all finite measures, as seen in Exercise 1.16 a), but inequality (2.25) only holds for L^p spaces with respect to probabilities.

2.5 Moments, Variance, Covariance

Given an m-dimensional r.v. X and $\alpha > 0$, its *absolute moment* of order α is the quantity $E(|X|^\alpha) = \|X\|_\alpha^\alpha$. Its *absolute centered moment* of order α is the quantity $E(|X - E(X)|^\alpha)$.

The *variance* of a real r.v. X is its second order centered moment, i.e.

$$\text{Var}(X) = E[(X - E(X))^2] . \tag{2.26}$$

Note that X has finite variance if and only if $X \in L^2$: if X has finite variance, then as $X = (X - E(X)) + E(X)$, X is in L^2, being the sum of square integrable r.v.'s. And if $X \in L^2$, also $X - E(X) \in L^2$ for the same reason.

Recalling that $E(X)$ is a constant we have

$$E\big[(X - E(X))^2\big] = E\big[X^2 - 2XE(X) + E(X)^2\big] = E(X^2) - 2E\big[XE(X)\big] + E(X)^2$$
$$= E(X^2) - E(X)^2 \,,$$

which provides an alternative expression for the variance:

$$\text{Var}(X) = E(X^2) - E(X)^2 \,. \tag{2.27}$$

This is the formula that is used in practice for the computation of the variance. As the variance is always positive, this relation also shows that we always have $E(X^2) \geq E(X)^2$, which we already know from Jensen's inequality.

The following properties are immediate from the definition of the variance.

$$\text{Var}(X + a) = \text{Var}(X) \,, \qquad a \in \mathbb{R}$$
$$\text{Var}(\lambda X) = \lambda^2 \text{Var}(X) \,, \qquad \lambda \in \mathbb{R} \,.$$

As for mathematical expectation, the moments of an r.v. X also only depend on the law μ of X: by Proposition 1.27, integration with respect to an image law,

$$E\big(|X|^\alpha\big) = \int_{\mathbb{R}^m} |x|^\alpha \mu(dx) \,,$$
$$E\big(|X - E(X)|^\alpha\big) = \int_{\mathbb{R}^m} |x - E(X)|^\alpha \mu(dx) \,.$$

The moments of X give information about the probability for X to take large values. The centered moments, similarly, give information about the probability for X to take values far from the mean. This aspect is made precise by the following two (very) important inequalities.

Markov's Inequality: For every $t > 0$, $\alpha > 0$,

$$P\big(|X| > t\big) \leq \frac{E\big(|X|^\alpha\big)}{t^\alpha} \tag{2.28}$$

which is immediate as

$$E\big(|X|^\alpha\big) \geq E\big(|X|^\alpha 1_{\{|X|>t\}}\big) \geq t^\alpha P\big(|X| > t\big) \,,$$

where we use the obvious fact that $|X|^\alpha \geq t^\alpha$ on the event $\{|X| > t\}$.

Applied to the r.v. $X - E(X)$ with $\alpha = 2$ Markov's inequality (2.28) becomes

Chebyshev's Inequality: For every $t > 0$

$$P(|X - E(X)| \geq t) \leq \frac{\text{Var}(X)}{t^2} \cdot \qquad (2.29)$$

Let us investigate now the variance of the sum of two r.v.'s:

$$\text{Var}(X + Y) = E\big[(X + Y - E(X) - E(Y))^2\big]$$
$$= E\big[(X - E(X))^2\big] + E\big[(Y - E(Y))^2\big] + 2E\big[(X - E(X))(Y - E(Y))\big]$$

and, if we set

$$\text{Cov}(X, Y) := E\big[(X - E(X))(Y - E(Y))\big] = E(XY) - E(X)E(Y) ,$$

then

$$\text{Var}(X + Y) = \text{Var}(X) + \text{Var}(Y) + 2\text{Cov}(X, Y) .$$

$\text{Cov}(X, Y)$ is the *covariance* of X and Y. Note that $\text{Cov}(X, Y)$ is nothing else than the scalar product in L^2 of $X - E(X)$ and $Y - E(Y)$. Hence it is well defined and finite if X and Y have finite variance and, by the Cauchy-Schwarz inequality,

$$|\text{Cov}(X, Y)| \leq E(|X - E(X)| \cdot |Y - E(Y)|)$$
$$\leq E\big[(X - E[X])^2\big]^{1/2} E\big[(Y - E[Y])^2\big]^{1/2} = \text{Var}(X)^{1/2}\text{Var}(Y)^{1/2} . \qquad (2.30)$$

If X and Y are independent, by Corollary 2.10,

$$\text{Cov}(X, Y) = E\big[(X - E(X))(Y - E(Y))\big] = E[X - E(X)] E[Y - E(Y)] = 0$$

hence if X and Y are independent

$$\text{Var}(X + Y) = \text{Var}(X) + \text{Var}(Y) .$$

The converse is not true: there are examples of r.v.'s having vanishing covariance, without being independent. This is hardly surprising: as remarked above $\text{Cov}(X, Y) = 0$ means that $E(XY) = E(X)E(Y)$, whereas independence requires $E[g(X)h(Y)] = E[g(X)]E[h(Y)]$ *for every* pair of bounded Borel functions g, h. Lack of correlation appears to be a much weaker condition.

If $\text{Cov}(X, Y) = 0$, X and Y are said to be *uncorrelated*. For an intuitive interpretation of the covariance, see Example 2.24 below.

If $X = (X_1, \ldots, X_m)$ is an m-dimensional r.v., its *covariance matrix* is the $m \times m$ matrix C whose elements are

$$c_{ij} = \text{Cov}(X_i, X_j) = \text{E}\big[(X_i - \text{E}(X_i))(X_j - \text{E}(X_j))\big].$$

C is a symmetric matrix having on the diagonal the variances of the components of X and outside the diagonal their covariances. Therefore if $X_1, \ldots, X_m$ are independent their covariance matrix is diagonal. The converse of course is not true.

An elegant way of manipulating the covariance matrix is to write

$$C = \text{E}\big[(X - \text{E}(X))(X - \text{E}(X))^*\big] \tag{2.31}$$

where $X - \text{E}(X)$ is a column vector. Indeed $(X - \text{E}(X))(X - \text{E}(X))^*$ is a matrix whose entries are the r.v.'s $(X_i - \text{E}(X_i))(X_j - \text{E}(X_j))$ whose expectation is c_{ij}. From (2.31) we easily see how a covariance matrix transforms under a linear map: if A is a $d \times m$ matrix, then the covariance matrix of the d-dimensional r.v. AX is

$$\begin{aligned}
C_{AX} &= \text{E}\big[(AX - \text{E}(AX))(AX - \text{E}(AX))^*\big] \\
&= \text{E}\big[A(X - \text{E}(X))(X - \text{E}(X))^* A^*\big] \\
&= A\,\text{E}\big[(X - \text{E}(X))(X - \text{E}(X))^*\big] A^* = ACA^*.
\end{aligned} \tag{2.32}$$

An important remark: the covariance matrix is always positive definite, i.e. for every $\xi \in \mathbb{R}^m$

$$\langle C\xi, \xi \rangle = \sum_{i,j=1}^m c_{ij}\xi_i\xi_j \geq 0.$$

Actually

$$\begin{aligned}
\langle C\xi, \xi \rangle &= \sum_{i,j=1}^m c_{ij}\xi_i\xi_j = \sum_{i,j=1}^m \text{E}\big[\xi_i(X_i - \text{E}(X_i))\xi_j(X_j - \text{E}(X_j))\big] \\
&= \text{E}\Big[\sum_{i,j=1}^m \xi_i(X_i - \text{E}(X_i))\xi_j(X_j - \text{E}(X_j))\Big] \\
&= \text{E}\Big[\Big(\sum_{i=1}^m \xi_i(X_i - \text{E}(X_i))\Big)^2\Big] = \text{E}\big[\langle \xi, X - \text{E}(X)\rangle^2\big] \geq 0.
\end{aligned} \tag{2.33}$$

Recall that a matrix is positive definite if and only if (it is symmetric) and all its eigenvalues are ≥ 0.

Example 2.24 (The Regression "Line") Let us consider a real r.v. Y and an m-dimensional r.v. X, defined on the same probability space $(\Omega, \mathcal{F}, P)$, both of them square integrable. What is the affine-linear function of X that best approximates Y? This is, we need to find a number $b \in \mathbb{R}$ and a vector $a \in \mathbb{R}^m$ such that the difference $\langle a, X \rangle + b - Y$ is "smallest". The simplest way (not the only one) to measure this discrepancy is to use the L^2 norm of this difference, which leads us to search for the values a and b that minimize the quantity $E[(\langle a, X \rangle + b - Y)^2]$.

We shall assume that the covariance matrix C of X is strictly positive definite, hence invertible. Let us add and subtract the expectations so that

$$E[(\langle a, X \rangle + b - Y)^2] = E[(\langle a, X - E(X) \rangle + \tilde{b} - (Y - E(Y)))^2],$$

where $\tilde{b} = b + \langle a, E(X) \rangle - E(Y)$. Let us find the minimizer of

$$a \mapsto S(a) := E[((\langle a, X - E(X) \rangle) + \tilde{b} - (Y - E(Y)))^2].$$

Expanding the square we have

$$\begin{aligned} S(a) = E[\langle a, X - E(X) \rangle^2] + \tilde{b}^2 + E[(Y - E(Y))^2] \\ -2E[\langle a, X - E(X) \rangle (Y - E(Y))] \end{aligned} \tag{2.34}$$

(the contribution of the other two double products vanishes because $X - E(X)$ and $Y - E(Y)$ have expectation equal to 0 and $\tilde{b}$ is a number). Thanks to (2.33) (read from right to left) $E[\langle a, X - E(X) \rangle^2] = \langle Ca, a \rangle$ and also

$$E[\langle a, X - E(X) \rangle (Y - E(Y))] = \sum_{i=1}^{m} a_i E((X_i - E(X_i))(Y - E(Y)))$$

$$= \sum_{i=1}^{m} a_i \mathrm{Cov}(X_i, Y) = \langle a, R \rangle,$$

where by R we denote the vector of the covariances, $R_i = \mathrm{Cov}(X_i, Y)$. Hence (2.34) can be written

$$S(a) = \langle Ca, a \rangle + \tilde{b}^2 + \mathrm{Var}(Y) - 2\langle a, R \rangle.$$

Let us look for the critical points of S: its differential is

$$DS(a) = 2Ca - 2R$$

and in order for $DS(a)$ to vanish we must have

$$a = C^{-1}R .$$

We see easily that this critical value is also a minimizer (S is a polynomial of the second degree that tends to infinity as $|a| \to \infty$). Now just choose for b the value such that $\tilde{b} = 0$, i.e.

$$b = \mathrm{E}(Y) - \langle a, \mathrm{E}(X) \rangle = \mathrm{E}(Y) - \langle C^{-1}R, \mathrm{E}(X) \rangle .$$

If $m = 1$ (i.e. X is a real r.v.) C is the scalar $\mathrm{Var}(X)$ and $R = \mathrm{Cov}(X, Y)$, so that

$$a = \frac{\mathrm{Cov}(X, Y)}{\mathrm{Var}(X)}$$

$$b = \mathrm{E}(Y) - a\mathrm{E}(X) = \mathrm{E}(Y) - \frac{\mathrm{Cov}(X, Y)}{\mathrm{Var}(X)} \mathrm{E}(X) . \tag{2.35}$$

The function $x \mapsto ax + b$ for the values a, b determined in (2.35) is the *regression line* of Y on X. Note that the angular coefficient a has the same sign as the covariance. The regression line is therefore increasing or decreasing as a function of x according as $\mathrm{Cov}(X, Y)$ is positive or negative.

The covariance therefore has an intuitive interpretation: r.v.'s having positive covariance will be associated to quantities that, essentially, take small or large values at the unison, while r.v.'s having a negative covariance will instead be associated to quantities that take small or large values in countertrend.

Independent (and therefore uncorrelated) r.v.'s have a horizontal regression line, in agreement with intuition: the knowledge of the values of one of them does not give information concerning the values of the other one.

A much more interesting problem will be to find the function ϕ, not necessarily affine-linear, such that $\mathrm{E}[(\phi(X) - Y)^2]$ is minimum. This problem will be treated in Chap. 4.

Let Y be an m-dimensional square integrable r.v. What is the vector $b \in \mathbb{R}^m$ such that the quantity $b \mapsto \mathrm{E}(|Y - b|^2)$ is minimum? Similar arguments as in Example 2.24 can be used: let us consider the map

$$b \mapsto \mathrm{E}(|Y - b|^2) = \mathrm{E}\big(|Y|^2 - 2\langle b, Y \rangle + |b|^2\big) = \mathrm{E}(|Y|^2) - 2\langle b, \mathrm{E}(Y) \rangle + |b|^2 .$$

Its differential is $b \mapsto -2E(Y) + 2b$ and vanishes at $b = E(Y)$, which is the minimizer we were looking for. Hence its mathematical expectation is also the constant r.v. that best approximates Y (still in the sense of L^2).

Even if, as noted above, there are many examples of pairs X, Y of non-independent r.v.'s whose covariance vanishes, the covariance is often used to measure "how independent X and Y are". However, when used to this purpose, the covariance has a drawback, as it is sensitive to changes of scale: if $\alpha, \beta > 0$, $\mathrm{Cov}(\alpha X, \beta Y) = \alpha\beta\mathrm{Cov}(X, Y)$ whereas, intuitively, the dependence between X and Y should be the same as the dependence between αX and βY (think of a change of unit of measure, for instance). More useful in this sense is the *correlation coefficient* $\rho_{X,Y}$ of X and Y, which is defined as

$$\rho_{X,Y} := \frac{\mathrm{Cov}(X, Y)}{\sqrt{\mathrm{Var}(X)\mathrm{Var}(Y)}}$$

and is invariant under scale changes. Thanks to (2.30) we have $-1 \le \rho_{X,Y} \le 1$.

In some sense, values of $\rho_{X,Y}$ close to 0 indicate "almost independence" whereas values close to 1 or -1 indicate a "strong dependence", at the unison or in countertrend respectively.

2.6 Characteristic Functions

Let X be an m-dimensional r.v. Its *characteristic function* is the function $\phi : \mathbb{R}^m \to \mathbb{C}$ defined as

$$\phi(\theta) = E(e^{i\langle\theta,X\rangle}) = E(\cos\langle\theta, X\rangle) + i\,E(\sin\langle\theta, X\rangle)\,. \tag{2.36}$$

The characteristic function is defined for every m-dimensional r.v. X because, for every $\theta \in \mathbb{R}^m$, $|e^{i\langle\theta,X\rangle}| = 1$, so that the complex r.v. $e^{i\langle\theta,X\rangle}$ is always integrable. Moreover, thanks to (1.7) (the integral of the modulus is larger than the modulus of the integral), $|E(e^{i\langle\theta,X\rangle})| \le E(|e^{i\langle\theta,X\rangle}|) = 1$, so that

$$|\phi(\theta)| \le 1 \qquad \text{for every } \theta \in \mathbb{R}^m$$

and obviously $\phi(0) = 1$. Proposition 1.27, integration with respect to an image law, gives

$$\phi(\theta) = \int_{\mathbb{R}^m} e^{i\langle\theta,x\rangle}\,d\mu(x)\,, \tag{2.37}$$

where μ denotes the law of X. The characteristic function therefore depends only on the law of X and we can speak equally of the characteristic function of an r.v. or of a probability law.

Whenever there is a danger of ambiguity we shall write ϕ_X or ϕ_μ in order to specify the characteristic function of the r.v. X or of its law μ. Sometimes we shall write $\widehat{\mu}(\theta)$ instead of $\phi_\mu(\theta)$, which stresses the close ties between characteristic functions and Fourier transforms.

Characteristic functions enjoy many properties that make them a very useful computation tool.

If μ and ν are probabilities on $\mathbb{R}^m$ we have

$$\phi_{\mu*\nu}(\theta) = \phi_\mu(\theta)\phi_\nu(\theta) . \tag{2.38}$$

Indeed, if X and Y are independent with laws μ and ν respectively, then

$$\phi_{\mu*\nu}(\theta) = \phi_{X+Y}(\theta) = \mathrm{E}(e^{i\langle\theta,X+Y\rangle}) = \mathrm{E}(e^{i\langle\theta,X\rangle}e^{i\langle\theta,Y\rangle})$$
$$= \mathrm{E}(e^{i\langle\theta,X\rangle})\mathrm{E}(e^{i\langle\theta,Y\rangle}) = \phi_\mu(\theta)\phi_\nu(\theta) .$$

Moreover

$$\phi_{-X}(\theta) = \mathrm{E}(e^{-i\langle\theta,X\rangle}) = \mathrm{E}(\overline{e^{i\langle\theta,X\rangle}}) = \overline{\phi_X(\theta)} . \tag{2.39}$$

Therefore if X is symmetric (i.e. such that $X \sim -X$) then ϕ_X is real-valued. What about the converse? If ϕ_X is real-valued is it true that X is symmetric? See below. It is easy to see how characteristic functions transform under affine-linear maps: if $Y = AX + b$, with A a $d \times m$ matrix and $b \in \mathbb{R}^d$, Y is $\mathbb{R}^d$-valued and for $\theta \in \mathbb{R}^d$

$$\phi_Y(\theta) = \mathrm{E}(e^{i\langle\theta,\,AX+b\rangle}) = e^{i\langle\theta,\,b\rangle}\mathrm{E}(e^{i\langle A^*\theta,\,X\rangle}) = \phi_X(A^*\theta)e^{i\langle\theta,\,b\rangle} . \tag{2.40}$$

Example 2.25 In the following examples $m = 1$ and therefore $\theta \in \mathbb{R}$. For the computations we shall always take advantage of (2.37).

(a) Binomial $B(n, p)$: thanks to the binomial rule

$$\phi(\theta) = \sum_{k=0}^{n}\binom{n}{k}p^k(1-p)^{n-k}e^{i\theta k} = \sum_{k=0}^{n}\binom{n}{k}(pe^{i\theta})^k(1-p)^{n-k}$$
$$= (1 - p + pe^{i\theta})^n .$$

(b) Geometric

$$\phi(\theta) = \sum_{k=0}^{\infty} p(1-p)^k e^{i\theta k} = p\sum_{k=0}^{\infty}((1-p)e^{i\theta})^k = \frac{p}{1-(1-p)e^{i\theta}} .$$

(c) Poisson

$$\phi(\theta) = e^{-\lambda} \sum_{k=0}^{\infty} \frac{\lambda^k}{k!} e^{i\theta k} = e^{-\lambda} \sum_{k=0}^{\infty} \frac{(\lambda e^{i\theta})^k}{k!} = e^{-\lambda} e^{\lambda e^{i\theta}} = e^{\lambda(e^{i\theta} - 1)} .$$

(d) Exponential

$$\phi(\theta) = \lambda \int_0^{+\infty} e^{-\lambda x} e^{i\theta x} \, dx = \lambda \int_0^{+\infty} e^{x(i\theta - \lambda)} \, dx = \frac{\lambda}{i\theta - \lambda} e^{x(i\theta - \lambda)} \Big|_{x=0}^{x=+\infty}$$

$$= \frac{\lambda}{i\theta - \lambda} \Big(\lim_{x \to +\infty} e^{x(i\theta - \lambda)} - 1 \Big) .$$

As the complex number $e^{x(i\theta - \lambda)}$ has modulus $|e^{x(i\theta - \lambda)}| = e^{-\lambda x}$ vanishing as $x \to +\infty$, we have $\lim_{x \to +\infty} e^{x(i\theta - \lambda)} = 0$ and

$$\phi(\theta) = \frac{\lambda}{\lambda - i\theta} .$$

Let us now investigate regularity of characteristic functions. Looking at (2.37), $\widehat{\mu}$ appears to be an integral depending on a parameter. Let us begin with continuity. We have

$$|\widehat{\mu}(\theta) - \widehat{\mu}(\theta_0)| = \big| E(e^{i\langle \theta, X \rangle}) - E(e^{i\langle \theta_0, X \rangle}) \big| \le E\big(|e^{i\langle \theta, X \rangle} - e^{i\langle \theta_0, X \rangle}|\big) .$$

If $\theta \to \theta_0$, then $|e^{i\langle \theta, X \rangle} - e^{i\langle \theta_0, X \rangle}| \to 0$. As also $|e^{i\langle \theta, X \rangle} - e^{i\langle \theta_0, X \rangle}| \le 2$, by Lebesgue's Theorem

$$\lim_{\theta \to \theta_0} |\widehat{\mu}(\theta) - \widehat{\mu}(\theta_0)| = 0 ,$$

so that $\widehat{\mu}$ is continuous. $\widehat{\mu}$ is actually always uniformly continuous (Exercise 2.41).

In order to investigate differentiability, let us assume first $m = 1$ (i.e. μ is a probability on $\mathbb{R}$). Proposition 1.21 (differentiability of integrals depending on a parameter) states that in order for

$$\theta \mapsto E[f(\theta, X)] = \int f(\theta, x) \, d\mu(x)$$

to be differentiable it is sufficient that the derivative $\frac{\partial f}{\partial \theta}(\theta, x)$ exists for μ a.s. every x and that the bound

$$\sup_{\theta \in \mathbb{R}} \Big| \frac{\partial}{\partial \theta} f(\theta, x) \Big| \le g(x)$$

holds for some function g such that $g(X)$ is integrable. In our case

$$\left| \frac{\partial}{\partial \theta} e^{i\theta x} \right| = |ixe^{i\theta x}| = |x| \,.$$

Hence

$$\sup_{\theta \in \mathbb{R}} \left| \frac{\partial}{\partial \theta} e^{i\theta X} \right| \le |X|$$

and if X is integrable $\widehat{\mu}$ is differentiable and we can take the derivative under the integral sign, i.e.

$$\widehat{\mu}'(\theta) = \int_{-\infty}^{+\infty} ixe^{i\theta x}\, \mu(dx) = E(iXe^{i\theta X}) \,. \tag{2.41}$$

A repetition of the same argument for the integrand $f(\theta, x) = ixe^{i\theta x}$ gives

$$\left| \frac{\partial}{\partial \theta} ixe^{i\theta x} \right| = |-x^2 e^{i\theta x}| = |x|^2 \,,$$

hence, if X has a finite second order moment, $\widehat{\mu}$ is twice differentiable and

$$\widehat{\mu}''(\theta) = -\int_{-\infty}^{+\infty} x^2 e^{i\theta x}\, \mu(dx) \,. \tag{2.42}$$

Repeating the argument above we see, by induction, that if μ has a finite absolute moment of order k, then $\widehat{\mu}$ is k times differentiable and

$$\widehat{\mu}^{(k)}(\theta) = \int_{-\infty}^{+\infty} (ix)^k e^{i\theta x}\, \mu(dx) \,. \tag{2.43}$$

We have the following much more precise result.

Proposition 2.26 *If μ has a finite moment of order k then $\widehat{\mu}$ is k times differentiable and (2.43) holds. Conversely if $\widehat{\mu}$ is k times differentiable and k is even then μ has a finite moment of order k and (therefore) (2.43) holds.*

Proof The first part of the statement has already been proved. Assume, first, that $k = 2$. As $\widehat{\mu}$ is twice differentiable we know that

$$\lim_{\theta \to 0} \frac{\widehat{\mu}(\theta) + \widehat{\mu}(-\theta) - 2\widehat{\mu}(0)}{\theta^2} = \widehat{\mu}''(0)$$

(just replace $\widehat{\mu}$ by its order two Taylor polynomial). But

$$\frac{2\widehat{\mu}(0) - \widehat{\mu}(\theta) - \widehat{\mu}(-\theta)}{\theta^2} = \int_{-\infty}^{+\infty} \frac{2 - e^{i\theta x} - e^{-i\theta x}}{\theta^2} \mu(dx)$$

$$= \int_{-\infty}^{+\infty} 2 \frac{1 - \cos(\theta x)}{x^2 \theta^2} x^2 \mu(dx) .$$

The last integrand *is positive* and converges to x^2 as $\theta \to 0$. Hence taking the limit as $\theta \to 0$, by Fatou's Lemma,

$$- \widehat{\mu}''(0) \geq \int_{-\infty}^{+\infty} x^2 \mu(dx) ,$$

which proves that μ has a finite moment of the second order and, thanks to the first part of the statement, for every $\theta \in \mathbb{R}$,

$$\widehat{\mu}''(\theta) = - \int_{-\infty}^{+\infty} x^2 e^{i\theta x} \mu(dx) .$$

The proof is completed by induction: let us assume that it has already been proved that if $\widehat{\mu}$ is k times differentiable (k even) then μ has a finite moment of order k and

$$\widehat{\mu}^{(k)}(\theta) = \int_{-\infty}^{+\infty} (ix)^k e^{i\theta x} \mu(dx) . \tag{2.44}$$

If $\widehat{\mu}$ is $k + 2$ times differentiable then

$$\lim_{\theta \to 0} \frac{\widehat{\mu}^{(k)}(\theta) + \widehat{\mu}^{(k)}(-\theta) - 2\widehat{\mu}^{(k)}(0)}{\theta^2} = \widehat{\mu}^{(k+2)}(0)$$

and, multiplying (2.44) by i^k and noting that $i^{2k} = 1$ (recall that k is even),

$$i^k \frac{2\widehat{\mu}^{(k)}(0) - \widehat{\mu}^{(k)}(\theta) - \widehat{\mu}^{(k)}(-\theta)}{\theta^2}$$

$$= i^k \int_{-\infty}^{+\infty} (ix)^k \frac{2 - e^{i\theta x} - e^{-i\theta x}}{\theta^2} \mu(dx)$$

$$= \int_{-\infty}^{+\infty} \frac{2 - e^{i\theta x} - e^{-i\theta x}}{\theta^2} x^k \mu(dx) \tag{2.45}$$

$$= \int 2 \frac{1 - \cos(\theta x)}{x^2 \theta^2} x^{k+2} \mu(dx) ,$$

so that the left-hand side above is real and positive and, as $\theta \to 0$, by Fatou's Lemma as above,

$$i^k \widehat{\mu}^{(k+2)}(0) \geq \int_{-\infty}^{+\infty} x^{k+2} \mu(dx) ,$$

hence μ has a finite $(k+2)$-th order moment (note that (2.45) ensures that the quantity $i^k \widehat{\mu}^{(k+2)}(0)$ is real). ∎

Remark 2.27 A closer look at the previous proof allows us to say something more: if k is even it is sufficient for $\widehat{\mu}$ to be differentiable k times *at the origin* in order to ensure that the moment of order k of μ is finite: if $\widehat{\mu}$ is differentiable k times at 0 and k is even, then $\widehat{\mu}$ is differentiable k times everywhere.

For $\theta = 0$ (2.43) becomes

$$\widehat{\mu}^{(k)}(0) = i^k \int_{-\infty}^{+\infty} x^k \mu(dx) , \tag{2.46}$$

which allows us to compute the moments of μ simply by taking the derivatives of $\widehat{\mu}$ at 0. Beware however: examples are known where $\widehat{\mu}$ is differentiable but does not have a finite mathematical expectation. If, instead, $\widehat{\mu}$ is twice differentiable, thanks to Proposition 2.26 (2 is even) X has a moment of order 2 which is finite (and therefore also a finite mathematical expectation). In order to find a necessary and sufficient condition for the characteristic function to be differentiable the curious reader can look at Brancovan and Jeulin book [3], Proposition 8.6, p. 154.

Similar arguments (only more complicated to express) give analogous results for probabilities on $\mathbb{R}^m$. More precisely, let $\alpha = (\alpha_1, \dots, \alpha_m)$ be a multiindex and let us denote as usual

$$|\alpha| = \alpha_1 + \cdots + \alpha_m, \qquad x^\alpha = x_1^{\alpha_1} \cdots x_m^{\alpha_m}$$
$$\frac{\partial^\alpha}{\partial \theta^\alpha} = \frac{\partial^{\alpha_1}}{\partial \theta^{\alpha_1}} \cdots \frac{\partial^{\alpha_m}}{\partial \theta^{\alpha_m}} .$$

Then if

$$\int_{\mathbb{R}^m} |x|^{|\alpha|} \mu(dx) < +\infty$$

$\widehat{\mu}$ is $|\alpha|$ times differentiable and

$$\frac{\partial^\alpha}{\partial \theta^\alpha} \widehat{\mu}(\theta) = \int_{\mathbb{R}^m} (ix)^\alpha e^{i\langle \theta, x \rangle} \mu(dx) .$$

In particular,

$$\frac{\partial \widehat{\mu}}{\partial \theta_k}(0) = i \int_{\mathbb{R}^m} x_k \, \mu(dx) \,,$$

$$\frac{\partial^2 \widehat{\mu}}{\partial \theta_k \partial \theta_h}(0) = - \int_{\mathbb{R}^m} x_h x_k \, \mu(dx) \,,$$

i.e. the gradient of μ at the origin is equal to i times the expectation and, if μ is centered, the Hessian of $\widehat{\mu}$ at the origin is equal to minus the covariance matrix.

Example 2.28 (Characteristic Function of Gaussian Laws, First Method)
If $\mu = N(0, 1)$ then

$$\widehat{\mu}(\theta) = \frac{1}{\sqrt{2\pi}} \int_{-\infty}^{+\infty} e^{i\theta x} e^{-x^2/2} \, dx \,. \tag{2.47}$$

This integral can be computed by the following argument (which is also valid for other characteristic functions). As μ has finite mean, by (2.41) and integrating by parts,

$$\widehat{\mu}'(\theta) = \frac{1}{\sqrt{2\pi}} \int_{-\infty}^{+\infty} i x \, e^{i\theta x} e^{-x^2/2} \, dx$$

$$= -\frac{1}{\sqrt{2\pi}} i e^{i\theta x} e^{-x^2/2} \Big|_{-\infty}^{+\infty} + \frac{1}{\sqrt{2\pi}} \int_{-\infty}^{+\infty} i \cdot i\theta e^{i\theta x} e^{-x^2/2} \, dx = -\theta \widehat{\mu}(\theta) \,,$$

i.e. $\widehat{\mu}$ solves the linear differential equation

$$u'(\theta) = -\theta u(\theta)$$

with the initial condition $u(0) = 1$. Its solution is

$$\widehat{\mu}(\theta) = e^{-\theta^2/2} \,.$$

If $Y \sim N(b, \sigma^2)$, as $Y = \sigma X + b$ with $X \sim N(0, 1)$ and thanks to (2.40),

$$\widehat{\mu}_Y(\theta) = e^{-\frac{1}{2}\sigma^2 \theta^2} e^{i\theta b} \,.$$

We shall soon see another method of computation (Example 2.37 b)) of the characteristic function of Gaussian laws.

The computation of the characteristic function of the $N(0, 1)$ law of the previous example allows us to derive a relation that is important in view of the next statement.

Let $X_1, \ldots, X_m$ be i.i.d. $N(0, \sigma^2)$-distributed r.v.'s. Then X has a density with respect to the Lebesgue measure of $\mathbb{R}^m$ given by

$$
\begin{aligned}
f_\sigma(x) &= \frac{1}{\sqrt{2\pi}\,\sigma} e^{-\frac{1}{2\sigma^2} x_1^2} \cdots \frac{1}{\sqrt{2\pi}\,\sigma} e^{-\frac{1}{2\sigma^2} x_m^2} \\
&= \frac{1}{(2\pi)^{m/2}\,\sigma^m} e^{-\frac{1}{2\sigma^2}|x|^2} .
\end{aligned}
\tag{2.48}
$$

Its characteristic function, for $\theta = (\theta_1, \ldots, \theta_m)$, is

$$
\begin{aligned}
\phi_\sigma(\theta) &= \mathrm{E}(e^{i\langle\theta,X\rangle}) = \mathrm{E}(e^{i\theta_1 X_1 + \cdots + i\theta_m X_m}) = \mathrm{E}(e^{i\theta_1 X_1}) \cdots \mathrm{E}(e^{i\theta_m X_m}) \\
&= e^{-\frac{1}{2}\sigma^2\theta_1^2} \cdots e^{-\frac{1}{2}\sigma^2\theta_m^2} = e^{-\frac{1}{2}\sigma^2|\theta|^2} .
\end{aligned}
\tag{2.49}
$$

We have therefore

$$
e^{-\frac{1}{2}\sigma^2|\theta|^2} = \frac{1}{(2\pi)^{m/2}\,\sigma^m} \int_{\mathbb{R}^m} e^{-\frac{1}{2\sigma^2}|x|^2} e^{i\langle\theta,x\rangle}\, dx
$$

and exchanging the roles of x and θ, replacing σ by $\frac{1}{\sigma}$ we obtain the relation

$$
e^{-\frac{1}{2\sigma^2}|x|^2} = \frac{\sigma^m}{(2\pi)^{m/2}} \int_{\mathbb{R}^m} e^{-\frac{1}{2}\sigma^2|\theta|^2} e^{i\langle\theta,x\rangle}\, d\theta ,
$$

which finally gives that

$$
f_\sigma(x) = \frac{1}{(2\pi)^m} \int_{\mathbb{R}^m} e^{-\frac{1}{2}\sigma^2|\theta|^2} e^{i\langle\theta,x\rangle}\, d\theta .
\tag{2.50}
$$

Given a function $\psi \in C_0(\mathbb{R}^m)$, let

$$
\psi_\sigma(x) = \int_{\mathbb{R}^m} f_\sigma(x - y)\psi(y)\, dy .
\tag{2.51}
$$

Lemma 2.29 *For every $\psi \in C_0(\mathbb{R}^m)$ we have*

$$
\psi_\sigma \underset{\sigma\to 0+}{\to} \psi
$$

uniformly.

Proof We have, for every $\delta > 0$,

$$|\psi(x) - \psi_\sigma(x)| = \left| \int_{\mathbb{R}^m} f_\sigma(x - y)(\psi(x) - \psi(y)) \, dy \right|$$

$$\leq \int_{\mathbb{R}^m} f_\sigma(x - y)|\psi(x) - \psi(y)| \, dy$$

$$= \int_{\{|y-x| \leq \delta\}} f_\sigma(x - y)|\psi(x) - \psi(y)| \, dy + \int_{\{|y-x| > \delta\}} f_\sigma(x - y)|\psi(x) - \psi(y)| \, dy$$

$$:= I_1 + I_2 .$$

First, let $\delta > 0$ be such that $|\psi(x) - \psi(y)| \leq \varepsilon$ whenever $|x - y| \leq \delta$ (ψ is uniformly continuous), so that $I_1 \leq \varepsilon$. Moreover,

$$I_2 \leq 2\|\psi\|_\infty \int_{\{|y-x| > \delta\}} f_\sigma(x - y) \, dy$$

and, if $X = (X_1, \ldots, X_m)$ denotes an r.v. with density f_σ, by Markov's inequality,

$$\int_{\{|y-x| > \delta\}} f_\sigma(x - y) \, dy = \int_{\{|z| > \delta\}} f_\sigma(z) \, dz = P(|X| \geq \delta) \leq \frac{1}{\delta^2} E(|X|^2)$$

$$\leq \frac{1}{\delta^2} E(|X_1|^2 + \cdots + |X_m|^2) = \frac{m\sigma^2}{\delta^2} .$$

Then just choose σ small enough so that $2\|\psi\|_\infty \frac{m\sigma^2}{\delta^2} \leq \varepsilon$, which gives

$$|\psi(x) - \psi_\sigma(x)| \leq 2\varepsilon \qquad \text{for every } x \in \mathbb{R}^m .$$

■

Note, in addition, that $\psi_\sigma \in C_0(\mathbb{R}^m)$ (Exercise 2.6).

Theorem 2.30 *Let μ, ν be probabilities on $\mathbb{R}^m$ such that*

$$\widehat{\mu}(\theta) = \widehat{\nu}(\theta) \qquad \text{for every } \theta \in \mathbb{R}^m .$$

Then $\mu = \nu$.

Proof Note that the relation $\widehat{\mu}(\theta) = \widehat{\nu}(\theta)$ for every $\theta \in \mathbb{R}^m$ means that

$$\int_{\mathbb{R}^m} f \, d\mu = \int_{\mathbb{R}^m} f \, d\nu \tag{2.52}$$

for every function of the form $f(x) = e^{i\langle\theta,x\rangle}$. Theorem 2.30 will follow as soon as we prove that (2.52) holds for every function $\psi \in C_K(\mathbb{R}^m)$ (Lemma 1.25). Let $\psi \in C_K(\mathbb{R}^m)$ and ψ_σ as in (2.51). We have

$$\int_{\mathbb{R}^m} \psi_\sigma(x)\,d\mu(x) = \int_{\mathbb{R}^m} d\mu(x) \int_{\mathbb{R}^m} \psi(y) f_\sigma(x-y)\,dy$$

and thanks to (2.50) and then to Fubini's Theorem

$$\cdots = \frac{1}{(2\pi)^m} \int_{\mathbb{R}^m} d\mu(x) \int_{\mathbb{R}^m} \psi(y)\,dy \int_{\mathbb{R}^m} e^{-\frac{1}{2}\sigma^2|\theta|^2} e^{i\langle\theta,x-y\rangle}\,d\theta$$

$$= \frac{1}{(2\pi)^m} \int_{\mathbb{R}^m} \psi(y)\,dy \int_{\mathbb{R}^m} e^{-\frac{1}{2}\sigma^2|\theta|^2} e^{-i\langle\theta,y\rangle}\,d\theta \int_{\mathbb{R}^m} e^{i\langle\theta,x\rangle}\,d\mu(x) \qquad (2.53)$$

$$= \frac{1}{(2\pi)^m} \int_{\mathbb{R}^m} \psi(y)\,dy \int_{\mathbb{R}^m} e^{-\frac{1}{2}\sigma^2|\theta|^2} e^{-i\langle\theta,y\rangle} \widehat{\mu}(\theta)\,d\theta\ .$$

Of course we have previously checked that, as $\psi \in C_K$, the function

$$(y,x,\theta) \mapsto \left|\psi(y)e^{-\frac{1}{2}\sigma^2|\theta|^2} e^{i\langle\theta,x-y\rangle}\right| = |\psi(y)|e^{-\frac{1}{2}\sigma^2|\theta|^2}$$

is integrable with respect to $\lambda_m(dy)\otimes\lambda_m(d\theta)\otimes\mu(dx)$ (λ_m = the Lebesgue measure of $\mathbb{R}^m$), which authorizes the application of Fubini's Theorem. As the integral only depends on $\widehat{\mu}$ and $\widehat{\mu} = \widehat{\nu}$ we obtain

$$\int_{\mathbb{R}^m} \psi_\sigma(x)\,d\mu(x) = \int_{\mathbb{R}^m} \psi_\sigma(x)\,d\nu(x)$$

and now, thanks to Lemma 2.29,

$$\int_{\mathbb{R}^m} \psi(x)\,d\mu(x) = \lim_{\sigma\to 0+} \int_{\mathbb{R}^m} \psi_\sigma(x)\,d\mu(x) = \lim_{\sigma\to 0+} \int_{\mathbb{R}^m} \psi_\sigma(x)\,d\nu(x)$$

$$= \int_{\mathbb{R}^m} \psi(x)\,d\nu(x)\ .$$

∎

Example 2.31 Let $\mu \sim N(a,\sigma^2)$ and $\nu \sim N(b,\tau^2)$. What is the law $\mu * \nu$?
Note that

$$\phi_{\mu*\nu}(\theta) = \widehat{\mu}(\theta)\widehat{\nu}(\theta) = e^{ia\theta}e^{-\frac{1}{2}\sigma^2\theta^2}e^{ib\theta}e^{-\frac{1}{2}\tau^2\theta^2} = e^{i(a+b)\theta}e^{-\frac{1}{2}(\sigma^2+\tau^2)\theta^2}\ .$$

Therefore $\mu * \nu$ has the same characteristic function as an $N(a + b, \sigma^2 + \tau^2)$ law, hence $\mu * \nu = N(a+b, \sigma^2 + \tau^2)$. The same result can also be obtained by computing the convolution integral of Proposition 2.18, but the computation, although elementary, is neither short nor amusing.

Example 2.32 Let X be an r.v. whose characteristic function is real-valued. Then X is symmetric.

Indeed $\phi_{-X}(\theta) = \phi_X(-\theta) = \overline{\phi_X(\theta)} = \phi_X(\theta)$: X and $-X$ have the same characteristic function, hence the same law.

Theorem 2.30 is of great importance from a theoretical point of view but unfortunately it is not constructive, i.e. it does not give any indication about how, knowing the characteristic function $\widehat{\mu}$, it is possible to obtain, for instance, the distribution function of μ or its density, with respect to the Lebesgue measure or the counting measure of $\mathbb{Z}$, if it exists.

This question has a certain importance also because, as in Example 2.31, characteristic functions provide a simple method of computation of the law of the sum of independent r.v.'s: just compute their characteristic functions, then the characteristic function of their sum (easy, it is the product). At this point, what can we do in order to derive from this characteristic function some information on the law?

The following theorem gives an element of an answer in this sense. Example 2.34 and Exercises 2.40 and 2.32 are also concerned with this question of "inverting" the characteristic function.

Theorem 2.33 (Inversion) *Let μ be a probability on $\mathbb{R}^m$. If $\widehat{\mu}$ is integrable then μ is absolutely continuous and has a density with respect to the Lebesgue measure given by*

$$f(x) = \frac{1}{(2\pi)^m} \int_{-\infty}^{+\infty} e^{-i\langle\theta,x\rangle} \widehat{\mu}(\theta)\,d\theta\,. \qquad (2.54)$$

A proof and more general inversion results (giving answers also when μ does not have a density) can be found in almost all books listed in the references section.

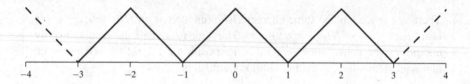

Fig. 2.2 Graph of the characteristic function ϕ of Example 2.34

Example 2.34 Let ϕ be the function $\phi(\theta) = 1 - |\theta|$ for $-1 \leq \theta \leq 1$ and then extended periodically on the whole of $\mathbb{R}$ as in Fig. 2.2.

Let us prove that ϕ is a characteristic function and determine the corresponding law.

As ϕ is periodic, we can consider its Fourier series

$$\phi(\theta) = \frac{1}{2} a_0 + \sum_{k=1}^{\infty} a_k \cos(k\pi\theta) = \sum_{k=-\infty}^{\infty} b_k \cos(k\pi\theta)$$

$$= \sum_{k=-\infty}^{\infty} b_k e^{i k\pi\theta} \tag{2.55}$$

where $b_k = \frac{1}{2} a_{|k|}$ for $k \neq 0$, $b_0 = \frac{1}{2} a_0$. The series converges uniformly, ϕ being continuous. In the series only the cosines appear as ϕ is even.

A closer look at (2.55) indicates that ϕ is the characteristic function of an r.v. X taking the values $k\pi$, $k \in \mathbb{Z}$, with probability b_k, provided we can prove that the numbers a_k are positive. Note that we know already that the sum of the b_k's is equal to 1, as $1 = \phi(0) = \sum_{k=-\infty}^{\infty} b_k$. Let us compute these Fourier coefficients: we have

$$a_0 = \int_{-1}^{1} \phi(\theta) \, d\theta = \int_{-1}^{1} (1 - |\theta|) \, d\theta = 2 \int_{0}^{1} (1 - \theta) \, d\theta = 1$$

and, for $k > 0$,

$$a_k = \int_{-1}^{1} (1 - |\theta|) \cos(k\pi\theta) \, d\theta = - \int_{-1}^{1} |\theta| \cos(k\pi\theta) \, d\theta$$

$$= -2 \int_{0}^{1} \theta \cos(k\pi\theta) \, d\theta = -2 \left(\frac{1}{k\pi} \theta \sin(k\pi\theta) \Big|_{0}^{1} + \frac{1}{k\pi} \int_{0}^{1} \sin(k\pi\theta) \, d\theta \right)$$

$$= -\frac{2}{(k\pi)^2} \cos(k\pi\theta) \Big|_{0}^{1} = \frac{2}{(k\pi)^2} (1 - \cos(k\pi))$$

i.e. $\frac{1}{2} a_0 = \frac{1}{2}$ and

$$a_k = \begin{cases} \dfrac{4}{(k\pi)^2} & k \text{ odd} \\ 0 & k \text{ even} . \end{cases}$$

Therefore ϕ is the characteristic function of a $\mathbb{Z}$-valued r.v. X such that, for $m \neq 0$,

$$P\big(X = \pm(2m + 1)\pi\big) = \frac{1}{2} a_{|2m+1|} = \frac{2}{\pi^2} \cdot \frac{1}{(2m + 1)^2}$$

and $P(X = 0) = \frac{1}{2}$. Note that X does not have a finite mathematical expectation, but this we already knew, as ϕ is not differentiable.

This example shows, on one hand, the link between characteristic functions and Fourier series, in the case of $\mathbb{Z}$-valued r.v.'s.

On the other hand, together with Exercise 2.32, it provides an example of a pair of characteristic functions that coincide in a neighborhood of the origin but that correspond to very different laws (the one of Exercise 2.32 is absolutely continuous with respect to the Lebesgue measure, whereas ϕ is the characteristic function of a discrete law).

Let $X_1, \ldots, X_m$ be r.v.'s with values in $\mathbb{R}^{n_1}, \ldots, \mathbb{R}^{n_m}$ respectively and let us consider, for $n = n_1 + \cdots + n_m$, the $\mathbb{R}^n$-valued r.v. $X = (X_1, \ldots, X_m)$. Let us denote by ϕ its characteristic function. Then it is easy to obtain the characteristic function ϕ_{X_k} of the k-th marginal of X. Indeed, recalling that ϕ is defined on $\mathbb{R}^n$ whereas ϕ_{X_k} is defined on $\mathbb{R}^{n_k}$,

$$\phi_{X_k}(\theta) = \mathrm{E}(e^{i\langle \theta, X_k \rangle}) = \mathrm{E}(e^{i\langle \tilde{\theta}, X \rangle}) = \phi(\tilde{\theta}), \qquad \theta \in \mathbb{R}^{n_k} ,$$

where $\tilde{\theta} = (0, \ldots, 0, \theta, 0, \ldots, 0)$ is the vector of $\mathbb{R}^n$ all of whose components vanish except for those in the $(n_1 + \cdots + n_{k-1} + 1)$-th to the $(n_1 + \cdots + n_k)$-th position.

Assume the r.v.'s $X_1, \ldots, X_m$ to be independent; if $\theta_1 \in \mathbb{R}^{n_1}, \ldots, \theta_m \in \mathbb{R}^{n_m}$ and $\theta = (\theta_1, \ldots, \theta_m) \in \mathbb{R}^n$ then

$$\phi_X(\theta) = \mathrm{E}(e^{i\langle \theta, X \rangle}) = \mathrm{E}(e^{i\langle \theta_1, X_1 \rangle} \cdots e^{i\langle \theta_m, X_m \rangle}) = \phi_{X_1}(\theta_1) \cdots \phi_{X_m}(\theta_m) . \qquad (2.56)$$

(2.56) can also be expressed in terms of laws: if $\mu_1, \ldots, \mu_m$ are probabilities respectively on $\mathbb{R}^{n_1}, \ldots, \mathbb{R}^{n_m}$ and $\mu = \mu_1 \otimes \cdots \otimes \mu_m$ then

$$\widehat{\mu}(\theta) = \widehat{\mu}_1(\theta_1) \ldots \widehat{\mu}_m(\theta_m) . \tag{2.57}$$

Actually we have the following result which provides a characterization of independence in terms of characteristic functions.

Proposition 2.35 *Let* $X_1, \ldots, X_m$ *be r.v.'s with values in* $\mathbb{R}^{n_1}, \ldots, \mathbb{R}^{n_m}$ *respectively and* $X = (X_1, \ldots, X_m)$. *Then* $X_1, \ldots, X_m$ *are independent if and only if, for every* $\theta_1 \in \mathbb{R}^{n_1}, \ldots, \theta_m \in \mathbb{R}^{n_m}$, *and* $\theta = (\theta_1, \ldots, \theta_m)$, *we have*

$$\phi_X(\theta) = \phi_{X_1}(\theta_1) \cdots \phi_{X_m}(\theta_m) . \tag{2.58}$$

Proof If the X_i's are independent we have already seen that (2.58) holds. Conversely, if (2.58) holds, then X has the same characteristic function as the product of the laws of the X_i's. Therefore by Theorem 2.30 the law of X is the product law and the X_i's are independent. ∎

2.7 The Laplace Transform

Let X be an m-dimensional r.v., μ its law and $z \in \mathbb{C}^m$. The *complex Laplace transform* (CLT) of X (or of μ) is the function

$$L(z) = \mathrm{E}(\mathrm{e}^{\langle z, X \rangle}) = \int_{\mathbb{R}^m} \mathrm{e}^{\langle z, x \rangle} \, d\mu(x) \tag{2.59}$$

defined for those values $z \in \mathbb{C}^m$ such that $\mathrm{e}^{\langle z, X \rangle}$ is integrable. Obviously L is always defined on the imaginary axes, as on them $|\mathrm{e}^{\langle z, X \rangle}| = 1$, and actually between the CLT L and the characteristic function ϕ we have the relation

$$L(i\theta) = \phi(\theta) \qquad \text{for every } \theta \in \mathbb{R}^m .$$

Hence the knowledge of the CLT L implies the knowledge of the characteristic function ϕ, which is the restriction of L to the imaginary axes. The *domain* of the CLT is the set of complex vectors $z \in \mathbb{C}^m$ such that $\mathrm{e}^{\langle z, X \rangle}$ is integrable. Recalling that $e^{\langle z,x \rangle} = e^{\Re\langle z,x \rangle}(\cos\langle z,x \rangle + i\sin\langle z,x \rangle)$, the domain of L is the set of the $z \in \mathbb{C}^m$ such that

$$\int_{\mathbb{R}^m} |e^{\langle z,x \rangle}| \, d\mu(x) = \int_{\mathbb{R}^m} e^{\Re\langle z,x \rangle} \, d\mu(x) < +\infty.$$

The domain of the CLT of μ will be denoted $\mathscr{D}_\mu$. We shall restrict ourselves to the case $m = 1$ from now on. We have

$$\int_{-\infty}^{+\infty} e^{\Re z\, x} \, d\mu(x) = \int_{-\infty}^0 e^{\Re z\, x} \, d\mu(x) + \int_0^{+\infty} e^{\Re z\, x} \, d\mu(x) := I_1 + I_2.$$

Clearly if $\Re z \leq 0$ then $I_2 < +\infty$, as the integrand is then smaller than 1. Moreover the function $t \mapsto \int_0^{+\infty} e^{tx} \, d\mu(x)$ is increasing. Therefore if

$$x_2 := \sup\left\{ t;\ \int_0^{+\infty} e^{tx} \, d\mu(x) < +\infty \right\}$$

(possibly $x_2 = +\infty$), then $x_2 \geq 0$ and $I_2 < +\infty$ for $\Re z < x_2$, whereas $I_2 = +\infty$ if $\Re z > x_2$.

Similarly, on the negative side, by the same argument there exists a number $x_1 \leq 0$ such that $I_1(z) < +\infty$ if $x_1 < \Re z$ and $I_1(z) = +\infty$ if $\Re z < x_1$.

Putting things together the domain $\mathscr{D}_\mu$ contains the open strip $S = \{z;\ x_1 < \Re z < x_2\}$, and it does not contain the complex numbers z outside the closure of S, i.e. such that $\Re z > x_2$ or $\Re z < x_1$.

Actually we have the following result.

> **Theorem 2.36** *Let μ be a probability on $\mathbb{R}$. Then there exist $x_1, x_2 \in \overline{\mathbb{R}}$ (the convergence abscissas) with $x_1 \leq 0 \leq x_2$ (possibly $x_1 = 0 = x_2$) such that the Laplace transform, L, of μ is defined in the strip $S = \{z;\ x_1 < \Re z < x_2\}$, whereas it is not defined for $\Re z > x_2$ or $\Re z < x_1$. Moreover L is holomorphic in S.*

Proof We need only prove that the CLT is holomorphic in S and this will follow as soon as we check that in S the Cauchy-Riemann equations are satisfied, i.e., if $z = x + iy$ and $L = L_1 + iL_2$,

$$\frac{\partial L_1}{\partial x} = \frac{\partial L_2}{\partial y}, \qquad \frac{\partial L_1}{\partial y} = -\frac{\partial L_2}{\partial x}.$$

The idea is simple: if $t \in \mathbb{R}$, then $z \mapsto e^{zt}$ is holomorphic, hence satisfies the Cauchy-Riemann equations and

$$L(z) = \int_{-\infty}^{+\infty} e^{zt} \, d\mu(t) , \qquad (2.60)$$

so that we must just verify that in (2.60) we can take the derivatives under the integral sign. Let us check that the conditions of Proposition 1.21 (derivation under the integral sign) are satisfied. We have

$$L_1(x, y) = \int_{-\infty}^{+\infty} e^{xt} \cos(yt) \, d\mu(t), \qquad L_2(x, y) = \int_{-\infty}^{+\infty} e^{xt} \sin(yt) \, d\mu(t) .$$

As we assume $x + iy \in S$, there exists an $\varepsilon > 0$ such that $x_1 + \varepsilon < x < x_2 - \varepsilon$ (x_1, x_2 are the convergence abscissas). For L_1, the derivative of the integrand with respect to x is $t \mapsto te^{xt} \cos(yt)$. Now the map $t \mapsto te^{xt} e^{-(x_2 - \varepsilon)t}$ is bounded on $\mathbb{R}^+$ (a global maximum is attained at $\bar{t} = (x_2 - \varepsilon - x)^{-1}$). Hence for some constant c_2 we have

$$|t| e^{xt} \leq c_2 e^{(x_2 - \varepsilon)t} \qquad \text{for } t \geq 0 .$$

Similarly there exists a constant c_1 such that

$$|t| e^{xt} \leq c_1 e^{(x_1 + \varepsilon)t} \qquad \text{for } t \leq 0 .$$

Hence the condition of Proposition 1.21 (derivation under the integral sign) is satisfied with $g(t) = c_2 e^{(x_2 - \varepsilon)t} + c_1 e^{(x_1 + \varepsilon)t}$, which is integrable with respect to μ, as $x_2 - \varepsilon$ and $x_1 + \varepsilon$ both belong to the convergence strip S. The same argument allows us to prove that also for L_2 we can take the derivative under the integral sign, and the first Cauchy-Riemann equation is satisfied:

$$\frac{\partial L_1}{\partial x}(x, y) = \int_{-\infty}^{+\infty} \frac{\partial}{\partial x} e^{xt} \cos(yt) \, d\mu(t) = \int_{-\infty}^{+\infty} te^{xt} \cos(yt) \, d\mu(t)$$

$$= \int_{-\infty}^{+\infty} \frac{\partial}{\partial y} e^{xt} \sin(yt) \, d\mu(t) = \frac{\partial L_2}{\partial y}(x, y) .$$

We can argue in the same way for the second Cauchy-Riemann equation.   ■

Recall that a holomorphic function is identified as soon as its value is known on a set having at least one cluster point (uniqueness of analytic continuation). Typically, therefore, the knowledge of the Laplace transform on the real axis (or on a nonvoid open interval) determines its value on the whole of the convergence strip (which, recall, is an open set). This also provides a method of computation for characteristic functions, as shown in the next example.

Example 2.37 (a) Let X be a Cauchy-distributed r.v., i.e. with density with respect to the Lebesgue measure

$$f(x) = \frac{1}{\pi} \frac{1}{1+x^2} .$$

Then

$$L(t) = \frac{1}{\pi} \int_{-\infty}^{+\infty} \frac{e^{tx}}{1+x^2} dx$$

and therefore $L(t) = +\infty$ for every $t \neq 0$. In this case the domain is the imaginary axis $\Re z = 0$ only and the convergence strip is empty.

(b) Assume $X \sim N(0, 1)$. Then, for $t \in \mathbb{R}$,

$$L(t) = \frac{1}{\sqrt{2\pi}} \int_{-\infty}^{+\infty} e^{tx} e^{-x^2/2} dx = \frac{e^{t^2/2}}{\sqrt{2\pi}} \int_{-\infty}^{+\infty} e^{-\frac{1}{2}(x-t)^2} dx = e^{t^2/2}$$

and the convergence strip is the whole of $\mathbb{C}$. Moreover, by analytic continuation, the Laplace transform of X is $L(z) = e^{z^2/2}$ for all $z \in \mathbb{C}$. In particular, for $z = it$, on the imaginary axis we have $L(it) = e^{-t^2/2}$ which gives, in a different way, the characteristic function of an $N(0, 1)$ law.

(c) If $X \sim \text{Gamma}(\alpha, \lambda)$ then, for $t \in \mathbb{R}$,

$$L(t) = \frac{\lambda^\alpha}{\Gamma(\alpha)} \int_0^{+\infty} x^{\alpha-1} e^{tx} e^{-\lambda x} dx .$$

This integral converges if and only if $t < \lambda$, hence the convergence strip is $S = \{\Re z < \lambda\}$ and does not depend on α. If $t < \lambda$, recalling the integrals of the Gamma distributions,

$$L(t) = \frac{\lambda^\alpha}{(\lambda - t)^\alpha} .$$

Thanks to the uniqueness of the analytic continuation we have, for $\Re z < \lambda$,

$$L(z) = \left(\frac{\lambda}{\lambda - z}\right)^\alpha \tag{2.61}$$

from which we obtain the characteristic function

$$\phi(t) = L(it) = \left(\frac{\lambda}{\lambda - it}\right)^\alpha .$$

(d) If X is Poisson distributed with parameter λ, then, again for $z \in \mathbb{R}$,

$$L(z) = e^{-\lambda} \sum_{k=0}^{\infty} \frac{\lambda^k}{k!} e^{zk} = e^{-\lambda} \sum_{k=0}^{\infty} \frac{(e^z \lambda)^k}{k!} = e^{-\lambda} e^{e^z \lambda} = e^{\lambda(e^z - 1)} \qquad (2.62)$$

and the convergence abscissas are infinite.

The Laplace transform of the sum of independent r.v.'s is easy to compute, in a similar way to the case of characteristic functions: if X and Y are independent, then

$$L_{X+Y}(z) = E(e^{\langle z, X+Y \rangle}) = E(e^{\langle z, X \rangle} e^{\langle z, Y \rangle}) = E(e^{\langle z, X \rangle})E(e^{\langle z, Y \rangle}) = L_X(z)L_Y(z) .$$

Note however that as, in general, the Laplace transform is not everywhere defined, the domain of L_{X+Y} is the intersection of the domains of L_X and L_Y.

If the abscissas of convergence are both different from 0, then the CLT is analytic at 0, thanks to Theorem 2.36. Hence the characteristic function $\phi_X(t) = L_X(it)$ is infinitely many times differentiable and (Theorem 2.26) *the moments of all orders are finite*. Moreover, as

$$i L'_X(0) = \phi'_X(0) = i \, E(X)$$

we have $L'(0) = E(X)$. Also the higher order moments of X can be obtained by taking the derivatives of the CLT: it is easy to see that

$$L_X^{(k)}(0) = E(X^k) . \qquad (2.63)$$

More information on the law of X can be gathered from the Laplace transform, see e.g. Exercises 2.44 and 2.47.

2.8 Multivariate Gaussian Laws

Let $X_1, \ldots, X_m$ be i.i.d. $N(0, 1)$-distributed r.v.'s; we have seen in (2.48) and (2.49) that the vector $X = (X_1, \ldots, X_m)$ has density

$$f(x) = \frac{1}{(2\pi)^{m/2}} e^{-\frac{1}{2}|x|^2}$$

with respect to the Lebesgue measure and characteristic function

$$\phi(\theta) = e^{-\frac{1}{2}|\theta|^2} . \qquad (2.64)$$

This law is the prototype of a particularly important family of multidimensional laws. If $Y = AX + b$ for an $m \times m$ matrix A and $b \in \mathbb{R}^m$, then, by (2.40),

$$\phi_Y(\theta) = e^{i\langle\theta,b\rangle}\phi_X(A^*\theta) = e^{i\langle\theta,b\rangle}e^{-\frac{1}{2}|A^*\theta|^2} = e^{i\langle\theta,b\rangle}e^{-\frac{1}{2}\langle A^*\theta,A^*\theta\rangle}$$
$$= e^{i\langle\theta,b\rangle}e^{-\frac{1}{2}\langle AA^*\theta,\theta\rangle} . \tag{2.65}$$

Recall that throughout this book "positive" means ≥ 0.

Theorem 2.38 *Given a vector $b \in \mathbb{R}^m$ and an $m \times m$ positive definite matrix C, there exists a probability μ on $\mathbb{R}^m$ such that*

$$\widehat{\mu}(\theta) = e^{i\langle\theta,b\rangle}e^{-\frac{1}{2}\langle C\theta,\theta\rangle} .$$

We shall say that such a μ is an $N(b, C)$ law (normal, or Gaussian, with mean b and covariance matrix C).

Proof Taking into account (2.65), it suffices to prove that a matrix A exists such that $AA^* = C$. It is a classical result of linear algebra that such a matrix always exists, provided C is positive definite, and even that A can be chosen symmetric (and therefore such that $A^2 = C$); in this case we say that A is the square root of C. Actually if C is diagonal

$$C = \begin{pmatrix} \lambda_1 & & 0 \\ & \ddots & \\ 0 & & \lambda_m \end{pmatrix}$$

as all the eigenvalues λ_i are ≥ 0 (C is positive definite) we can just choose

$$A = \begin{pmatrix} \sqrt{\lambda_1} & & 0 \\ & \ddots & \\ 0 & & \sqrt{\lambda_m} \end{pmatrix} .$$

Otherwise (i.e. if C is not diagonal) there exists an orthogonal matrix O such that OCO^{-1} is diagonal. It is immediate that OCO^{-1} is also positive definite so that there exists a matrix B such that $B^2 = OCO^{-1}$. Then if $A := O^{-1}BO$, A is symmetric (as $O^{-1} = O^*$) and is the matrix we were looking for as

$$A^2 = O^{-1}BO \cdot O^{-1}BO = O^{-1}B^2O = C .$$

The r.v. X introduced at the beginning of this section is therefore $N(0, I)$-distributed (I = the identity matrix). In the remainder of this chapter we draw attention to the many important properties of this class of distributions.

Note that, according to the definition, an r.v. having characteristic function $\theta \mapsto e^{i\langle\theta, b\rangle}$ is Gaussian. Hence Dirac masses are Gaussian and a Gaussian r.v. need not have a density with respect to the Lebesgue measure. See also below.

- A remark that simplifies the manipulation of the $N(b, C)$ laws consists in recalling (2.65), i.e. that it is the law of an r.v. of the form $AX + b$ with $X \sim N(0, I)$ and A a square root of C. Hence an r.v. $Y \sim N(b, C)$ can always be written $Y = AX + b$, with $X \sim N(0, I)$.
- If $Y \sim N(b, C)$, then b is indeed the mean of Y and C its covariance matrix, as anticipated in the statement of Theorem 2.38. This is obvious if $b = 0$ and $C = I$, recalling the way we defined the $N(0, I)$ laws. In general, as $Y = AX + b$, where A is the square root of C and $X \sim N(0, I)$, we have immediately $E(Y) = E(AX + b) = AE(X) + b = b$. Moreover, the covariance matrix of Y is $AIA^* = AA^* = C$, thanks to the transformation rule of covariance matrices under linear maps (2.32).
- If C is invertible then the $N(b, C)$ law has a density with respect to the Lebesgue measure. Indeed in this case the square root A of C is also invertible (the eigenvalues of A are the square roots of those of C, which are all > 0). If $Y \sim N(b, C)$, hence of the form $Y = AX + b$ with $X \sim N(0, I)$, then Y has density (see the computation of a density under a linear-affine transformation, Example 2.20)

$$f_Y(y) = \frac{1}{|\det A|} f\left(A^{-1}(y - b)\right) = \frac{1}{(2\pi)^{m/2}|\det A|} e^{-\frac{1}{2}\langle A^{-1}(y-b), A^{-1}(y-b)\rangle}$$

$$= \frac{1}{(2\pi)^{m/2}(\det C)^{1/2}} e^{-\frac{1}{2}\langle C^{-1}(y-b), y-b\rangle}.$$

If C is not invertible, then the $N(b, C)$ law cannot have a density with respect to the Lebesgue measure. In this case the image of the linear map associated to A is a proper hyperplane of $\mathbb{R}^m$, hence Y also takes its values in a proper hyperplane with probability 1 and cannot have a density, as such a hyperplane has Lebesgue measure 0.

This is actually a general fact: any r.v. having a covariance matrix that is not invertible cannot have a density with respect to the Lebesgue measure (Exercise 2.27).

- If $X \sim N(b, C)$ is m-dimensional and R is a $d \times m$ matrix and $\tilde{b} \in \mathbb{R}^d$, then the d-dimensional r.v. $Y = RX + \tilde{b}$ has characteristic function (see (2.40) again)

$$\phi_Y(\theta) = e^{i\langle\theta, \tilde{b}\rangle}\phi_X(R^*\theta) = e^{i\langle\theta, \tilde{b}\rangle}e^{i\langle R^*\theta, b\rangle}e^{-\frac{1}{2}\langle CR^*\theta, R^*\theta\rangle}$$

$$= e^{i\langle\theta, \tilde{b}+Rb\rangle}e^{-\frac{1}{2}\langle RCR^*\theta, \theta\rangle} \tag{2.66}$$

and therefore $Y \sim N(\tilde{b} + Rb, RCR^*)$. Therefore

affine-linear maps transform Gaussian laws into Gaussian laws.

This is one of the most important properties of Gaussian laws and we shall use it throughout.

In particular, for instance, if $X = (X_1, \ldots, X_m) \sim N(b, C)$, then also its components $X_1, \ldots, X_m$ are necessarily Gaussian (real of course), as the component X_i is a linear function of X.

Hence *the marginals of a multivariate Gaussian law are also Gaussian*. Moreover, taking into account that X_i has mean b_i and covariance c_{ii}, X_i is $N(b_i, c_{ii})$-distributed.

- If X is $N(0, I)$ and O is an orthogonal matrix then the "rotated" r.v. OX is itself Gaussian, being a linear function of a Gaussian r.v. It is moreover obviously centered and, recalling how covariance matrices transform under linear transformations (see (2.32)), it has covariance matrix $C = OIO^* = OO^* = I$. Hence $OX \sim N(0, I)$.

- Let $X \sim N(b, C)$ and assume C to be diagonal. Then we have

$$\phi_X(\theta) = e^{i\langle \theta, b \rangle} e^{-\frac{1}{2}\langle C\theta, \theta \rangle} = e^{i\langle \theta, b \rangle} \exp\left(-\frac{1}{2} \sum_{h=1}^{m} c_{hh}\theta_h^2 \right)$$

$$= e^{i\theta_1 b_1} e^{-\frac{1}{2}c_{11}\theta_1^2} \cdots e^{i\theta_m b_m} e^{-\frac{1}{2}c_{mm}\theta_m^2} = \phi_{X_1}(\theta_1) \cdots \phi_{X_m}(\theta_m) .$$

Thanks to Proposition 2.35 therefore the r.v.'s $X_1, \ldots, X_m$ are independent. Recalling that C is the covariance matrix of X, we have that *uncorrelated r.v.'s are also independent if their joint distribution is Gaussian*.

Note that, in order for this property to hold, the r.v.'s $X_1, \ldots, X_m$ must be *jointly* Gaussian. It is possible for them each to have a Gaussian law without having a Gaussian joint law (see Exercise 2.56). Individually but non-jointly Gaussian r.v.'s are however a rare occurrence.

More generally, let X, Y be r.v.'s with values in $\mathbb{R}^m$, $\mathbb{R}^d$ respectively and jointly Gaussian, i.e such that the pair (X, Y) (with values in $\mathbb{R}^n$, $n = m + d$) has Gaussian distribution. Then if

$$\text{Cov}(X_i, Y_j) = 0 \qquad \text{for every } 1 \le i \le m, 1 \le j \le d , \qquad (2.67)$$

i.e. the components of X are uncorrelated with the components of Y, then X and Y are independent.

Actually (2.67) is equivalent to the assumption that the covariance matrix C of (X, Y) is block diagonal

$$C = \begin{pmatrix} & & 0 \ldots 0 \\ & C_X & \vdots \ddots \vdots \\ & & 0 \ldots 0 \\ 0 \ldots 0 & & \\ \vdots \ddots \vdots & & C_Y \\ 0 \ldots 0 & & \end{pmatrix}$$

so that, if $\theta_1 \in \mathbb{R}^m$, $\theta_2 \in \mathbb{R}^d$ and $\theta := (\theta_1, \theta_2) \in \mathbb{R}^n$, and denoting by b_1, b_2 respectively the expectations of X and Y and $b = (b_1, b_2)$,

$$e^{i\langle\theta,b\rangle}e^{-\frac{1}{2}\langle C\theta,\theta\rangle} = e^{i\langle\theta_1,b_1\rangle}e^{-\frac{1}{2}\langle C_X\theta_1,\theta_1\rangle} \; e^{i\langle\theta_2,b_2\rangle}e^{-\frac{1}{2}\langle C_Y\theta_2,\theta_2\rangle} \,,$$

i.e.

$$\phi_{(X,Y)}(\theta) = \phi_X(\theta_1)\phi_Y(\theta_2) \,,$$

and again X and Y are independent thanks to the criterion of Proposition 2.35.

The argument above of course also works in the case of m r.v.'s: if $X_1, \ldots, X_m$ are jointly Gaussian with values in $\mathbb{R}^{n_1}, \ldots, \mathbb{R}^{n_m}$ respectively and the covariances of the components of X_k and of $X_j, k \neq j$, are uncorrelated, then again the covariance matrix of the vector $X = (X_1, \ldots, X_m)$ is block diagonal and by Proposition 2.35 $X_1, \ldots, X_m$ are independent.

2.9 Quadratic Functionals of Gaussian r.v.'s, a Bit of Statistics

Recall that if $X \sim N(0, I)$ is an m-dimensional r.v. then $|X|^2 = X_1^2 + \cdots + X_m^2 \sim \chi^2(m)$. In this section we go further into the investigation of quadratic functionals of Gaussian r.v.'s. Exercises 2.7, 2.51, 2.52 and 2.53 also go in this direction. The key tool is Cochran's theorem below. Let us however first recall some notions concerning orthogonal projections.

If V is a subspace of a Hilbert space H let us denote by $V^\perp$ its orthogonal, i.e. the set of vectors $x \in H$ such that $\langle x, z \rangle = 0$ for every $z \in V$. The orthogonal $V^\perp$ is always a closed subspace.

The following statements introduce the notion of projector on a subspace.

> **Lemma 2.39** *Let H be a Hilbert space, $F \subset H$ a closed convex set and $x \in F^c$ a point not belonging to F. Then there exists a unique $y_0 \in F$ such that*
>
> $$|x - y_0| = \min_{y \in F} |x - y|.$$

Proof By subtraction, possibly replacing F by $F - x$, we can assume $x = 0$ and $0 \notin F$. Let $\eta = \min_{y \in F} |y|$. It is immediate that, for every $z, y \in H$,

$$\left| \frac{1}{2}(z - y) \right|^2 + \left| \frac{1}{2}(z + y) \right|^2 = \frac{1}{2}|z|^2 + \frac{1}{2}|y|^2 \qquad (2.68)$$

and therefore

$$\left| \frac{1}{2}(z - y) \right|^2 = \frac{1}{2}|z|^2 + \frac{1}{2}|y|^2 - \left| \frac{1}{2}(z + y) \right|^2.$$

If $z, y \in F$, as also $\frac{1}{2}(z + y) \in F$ (F is convex), we obtain

$$\left| \frac{1}{2}(z - y) \right|^2 \le \frac{1}{2}|z|^2 + \frac{1}{2}|y|^2 - \eta^2. \qquad (2.69)$$

Let now $(y_n)_n \subset F$ be a minimizing sequence, i.e. such that $|y_n| \to_{n \to \infty} \eta$. Then (2.69) gives

$$|y_n - y_m|^2 \le 2|y_n|^2 + 2|y_m|^2 - 4\eta^2.$$

As $|y_n|^2 \to_{n \to \infty} \eta^2$ this relation proves that $(y_n)_n$ is a Cauchy sequence, hence converges to some $y_0 \in F$ that is the required minimizer. The fact that every minimizing sequence is a Cauchy sequence implies uniqueness. ∎

Let $V \subset H$ be a closed subspace, hence also a closed convex set. Lemma 2.39 allows us to define, for $x \in H$,

$$Px := \operatorname*{argmin}_{v \in V} |x - v| \qquad (2.70)$$

i.e. Px is the (unique) element of V that is closest to x.

Let us investigate the properties of the operator P. It is immediate that $Px = x$ if and only if already $x \in V$ and that $P(Px) = Px$.

Proposition 2.40 *Let P be as in (2.70) and $Qx = x - Px$. Then $Qx \in V^\perp$, so that Px and Qx are orthogonal. Moreover P and Q are linear operators.*

Proof Let us prove that, for every $v \in V$,

$$\langle Qx, v \rangle = \langle x - Px, v \rangle = 0 . \tag{2.71}$$

By the definition of P, as $Px + tv \in V$ for every $t \in \mathbb{R}$, for all $v \in V$ the function

$$t \mapsto |x - (Px + tv)|^2$$

is minimum at $t = 0$. But

$$|x - (Px + tv)|^2 = |x - Px|^2 - 2t\langle x - Px, v \rangle + t^2|v|^2 .$$

The derivative with respect to t at $t = 0$ must therefore vanish, which gives (2.71). For every $x, y \in H, \alpha, \beta \in \mathbb{R}$ we have, thanks to the relation $x = Px + Qx$,

$$\alpha x + \beta y = P(\alpha x + \beta y) + Q(\alpha x + \beta y) \tag{2.72}$$

but also $\alpha x = \alpha(Px + Qx), \beta y = \beta(Py + Qy)$ and by (2.72)

$$\alpha(Px + Qx) + \beta(Py + Qy) = P(\alpha x + \beta y) + Q(\alpha x + \beta y)$$

i.e.

$$\alpha Px + \beta Py - P(\alpha x + \beta y) = Q(\alpha x + \beta y) - \alpha Qx - \beta Qy .$$

As in the previous relation the left-hand side is a vector of V whereas the right-hand side belongs to $V^\perp$, both are necessarily equal to 0, which proves linearity. ∎

P is the *orthogonal projector on V.*

We shall need Proposition 2.40 in this generality later. In this section we shall be confronted with orthogonal projectors only in the simpler case $H = \mathbb{R}^m$.

Example 2.41 Let $V \subset \mathbb{R}^m$ be the subspace of the vectors of the form

$$v = (v_1, \dots, v_k, 0, \dots, 0) \qquad v_1, \dots, v_k \in \mathbb{R} \, .$$

In this case, if $x = (x_1, \dots, x_m)$,

$$Px = \operatorname*{argmin}_{v \in V} |x - v|^2 = \operatorname*{argmin}_{v_1, \dots, v_k \in \mathbb{R}} \sum_{i=1}^{k} (x_i - v_i)^2 + \sum_{i=k+1}^{m} x_i^2$$

i.e. $Px = (x_1, \dots, x_k, 0, \dots, 0)$. Here, of course, $V^\perp$ is formed by the vectors of the form

$$v = (0, \dots, 0, v_{k+1}, \dots, v_m) \, .$$

Theorem 2.42 (Cochran) *Let X be an m-dimensional $N(0, I)$-distributed r.v. and $V_1, \dots, V_k$ pairwise orthogonal vector subspaces of $\mathbb{R}^m$. For $i = 1, \dots, k$ let n_i denote the dimension of V_i and P_i the orthogonal projector onto V_i. Then the r.v.'s $P_i X, i = 1, \dots, k$, are independent and $|P_i X|^2$ is $\chi^2(n_i)$-distributed.*

Proof Assume for simplicity $k = 2$. Except for a rotation we can assume that V_1 is the subspace of the first n_1 coordinates and V_2 the subspace of the subsequent n_2 as in Example 2.41 (recall that the $N(0, I)$ laws are invariant with respect to orthogonal transformations). Hence

$$P_1 X = (X_1, \dots, X_{n_1}, 0, \dots, 0) \, ,$$
$$P_2 X = (0, \dots, 0, X_{n_1+1}, \dots, X_{n_1+n_2}, 0, \dots, 0) \, .$$

$P_1 X$ and $P_2 X$ are jointly Gaussian (the vector $(P_1 X, P_2 X)$ is a linear function of X) and it is clear that (2.67) (orthogonality of the components of $P_1 X$ and $P_2 X$) holds; therefore $P_1 X$ and $P_2 X$ are independent. Moreover

$$|P_1 X|^2 = (X_1^2 + \cdots + X_{n_1}^2) \sim \chi^2(n_1) \, ,$$
$$|P_2 X|^2 = (X_{n_1+1}^2 + \cdots + X_{n_1+n_2}^2) \sim \chi^2(n_2) \, .$$

∎

A first important application of Cochran's Theorem is the following.

Let $V_0 \subset \mathbb{R}^m$ be the subspace generated by the vector $e = (1, 1, \ldots, 1)$ (i.e. the subspace of the vectors whose components are equal); let us show that the orthogonal projector on V_0 is $P_{V_0} x = (\bar{x}, \ldots, \bar{x})$, where

$$\bar{x} = \frac{1}{m}(x_1 + \cdots + x_m) .$$

In order to determine $P_{V_0} x$ we must find the number $\lambda_0 \in \mathbb{R}$ such that the function $\lambda \mapsto |x - \lambda e|$ is minimum at $\lambda = \lambda_0$. That is we must find the minimizer of

$$\lambda \mapsto \sum_{i=1}^{m} (x_i - \lambda)^2 .$$

Taking the derivative we find for the critical value the relation $2 \sum_{i=1}^{m}(x_i - \lambda) = 0$, i.e. $\sum_{i=1}^{m} x_i = m\lambda$. Hence $\lambda_0 = \bar{x}$.

If $X \sim N(0, I)$ and $\overline{X} = \frac{1}{m}(X_1 + \cdots + X_m)$, then $\overline{X}e$ is the orthogonal projection of X on V_0 and therefore $X - \overline{X}e$ is the orthogonal projection of X on the orthogonal subspace $V_0^{\perp}$. By Cochran's Theorem $\overline{X}e$ and $X - \overline{X}e$ are independent (which is not completely obvious as both these r.v.'s depend on $\overline{X}$). Moreover, as $V_0^{\perp}$ has dimension $m - 1$, Cochran's Theorem again gives

$$\sum_{i=1}^{m} (X_i - \overline{X})^2 = |X - \overline{X}e|^2 \sim \chi^2(m - 1) . \tag{2.73}$$

Let us introduce a new probability law: the *Student t* with n degrees of freedom is the law of an r.v. of the form

$$Z = \frac{X}{\sqrt{Y}} \sqrt{n} , \tag{2.74}$$

where X and Y are independent and $N(0, 1)$- and $\chi^2(n)$-distributed respectively. This law is usually denoted $t(n)$.

Student laws are symmetric, i.e. Z and $-Z$ have the same law. This follows immediately from their definition: the r.v.'s X, Y and $-X, Y$ in (2.74) have the same joint law, as their components have the same distribution and are independent. Hence the laws of $\frac{X}{Y} \sqrt{n}$ and $-\frac{X}{Y} \sqrt{n}$ are the images of the same joint law under the same map and therefore coincide.

It is not difficult to compute the density of a $t(n)$ law (see Example 4.17 p. 192) but we shall skip this computation for now. Actually it will be apparent that the important thing about Student laws are the distribution functions and quantiles, which are provided by appropriate software (tables in ancient times...).

Example 2.43 (Quantiles) Let F be the d.f. of some r.v. X. The *quantile of order* α, $0 < \alpha < 1$, of F is the infimum, q_α say, of the numbers x such that $F(x) = P(X \le x) \ge \alpha$, i.e.

$$q_\alpha = \inf\{x; \, F(x) \ge \alpha\}$$

(actually this is a minimum as F is right continuous). If F is continuous then, by the intermediate value theorem, the equation

$$F(x) = \alpha \qquad (2.75)$$

has (at least) one solution for every $0 < \alpha < 1$. If moreover F is strictly increasing (which is the case for instance if X has a strictly positive density) then the solution of equation (2.75) is unique. In this case q_α is therefore the unique real number x such that

$$F(x) = P(X \le x) = \alpha \, .$$

If X is symmetric (i.e. X and $-X$ have the same law), as is the case for $N(0, 1)$ and Student laws, we have the relations

$$1 - \alpha = P(X \ge q_\alpha) = P(-X \ge q_\alpha) = P(X \le -q_\alpha) \, ,$$

from which we obtain that $q_{1-\alpha} = -q_\alpha$. Moreover, we have the relation (see Fig. 2.3)

$$P(|X| \le q_{1-\alpha/2}) = P(-q_{1-\alpha/2} \le X \le q_{1-\alpha/2})$$
$$= P(X \le q_{1-\alpha/2}) - P(X \le -q_{1-\alpha/2}) = 1 - \frac{\alpha}{2} - \frac{\alpha}{2} = 1 - \alpha \, . \qquad (2.76)$$

Going back to the case $X \sim N(0, I)$, we have seen that, as a consequence of Cochran's theorem, the r.v.'s $\overline{X}$ and $\sum_{i=1}^m (X_i - \overline{X})^2$ are independent and that $\sum_{i=1}^m (X_i - \overline{X})^2 \sim \chi^2(m-1)$. As $\overline{X} = \frac{1}{m}(X_1 + \cdots + X_m)$ is $N(0, \frac{1}{m})$-distributed, $\sqrt{m}\,\overline{X} \sim N(0, 1)$ and

$$T := \frac{\sqrt{m}\,\overline{X}}{\sqrt{\frac{1}{m-1} \sum_{i=1}^m (X_i - \overline{X})^2}} \sim t(m-1) \, . \qquad (2.77)$$

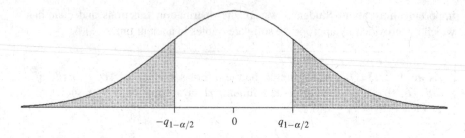

Fig. 2.3 Each of the two shaded regions has an area equal to $\frac{\alpha}{2}$. Hence the probability of a value between $-q_{1-\alpha/2}$ and $q_{1-\alpha/2}$ is equal to $1 - \alpha$

Corollary 2.44 *Let* $Z_1, \ldots, Z_m$ *be i.i.d.* $N(b, \sigma^2)$*-distributed r.v.'s. Let*

$$\overline{Z} = \frac{1}{m}(Z_1 + \cdots + Z_m) \,,$$

$$S^2 = \frac{1}{m-1}\sum_{i=1}^{m}(Z_i - \overline{Z})^2 \,.$$

Then $\overline{Z}$ *and* S^2 *are independent. Moreover,*

$$\frac{m-1}{\sigma^2}S^2 \sim \chi^2(m-1) \,, \tag{2.78}$$

$$\frac{\sqrt{m}\,(\overline{Z}-b)}{S} \sim t(m-1) \,. \tag{2.79}$$

Proof Let us trace back to the case of $N(0, I)$-distributed r.v.'s that we have already seen. If $X_i = \frac{1}{\sigma}(Z_i - b)$, then $X = (X_1, \ldots, X_m) \sim N(0, I)$ and we know already that $\overline{X}$ and $\sum_i (X_i - \overline{X})^2$ are independent. Moreover,

$$\overline{Z} = \sigma\overline{X} + b \,,$$

$$\frac{m-1}{\sigma^2}S^2 = \frac{1}{\sigma^2}\sum_{i=1}^{m}(Z_i - \overline{Z})^2 = \sum_{i=1}^{m}(X_i - \overline{X})^2 \tag{2.80}$$

so that $\overline{Z}$ and S^2 are also independent, being functions of independent r.v.'s. Finally $\frac{m-1}{\sigma^2}S^2 \sim \chi^2(m-1)$ thanks to (2.73) and the second of the formulas (2.80), and as

$$\frac{\sqrt{m}\,(\overline{Z}-b)}{S} = \frac{\sqrt{m}\,\overline{X}}{\sqrt{\frac{1}{m-1}\sum_{i=1}^{m}(X_i - \overline{X})^2}} \,,$$

(2.79) follows by (2.77). ∎

Example 2.45 (A Bit of Statistics...) Let $X_1, \ldots, X_n$ be i.i.d. $N(b, \sigma^2)$-distributed r.v.'s, where both b and σ^2 are unknown. How can we, from the observed values $X_1, \ldots, X_n$, estimate the two unknown parameters b and σ^2?

If

$$\overline{X} = \frac{1}{n}(X_1 + \cdots + X_n),$$

$$S^2 = \frac{1}{n-1}\sum_{i=1}^{n}(X_i - \overline{X})^2$$

then by Corollary 2.44

$$\frac{n-1}{\sigma^2}S^2 \sim \chi^2(n-1)$$

and

$$T := \frac{\sqrt{n}\,(\overline{X} - b)}{S} \sim t(n-1).$$

If we denote by $t_\alpha(n-1)$ the quantile of order α of a $t(n-1)$ law, then

$$P\big(|T| > t_{1-\alpha/2}(n-1)\big) = \alpha$$

(this is (2.76), as Student laws are symmetric). On the other hand,

$$\big\{|T| > t_{1-\alpha/2}(n-1)\big\} = \Big\{|\overline{X} - b| > t_{1-\alpha/2}(n-1)\frac{S}{\sqrt{n}}\Big\}.$$

Therefore the probability for the empirical mean $\overline{X}$ to differ from the expectation b by more than $t_{1-\alpha/2}(n-1)\frac{S}{\sqrt{n}}$ is $\leq \alpha$. Or, in other words, the unknown mean b lies in the interval

$$I = \Big[\overline{X} - t_{1-\alpha/2}(n-1)\frac{S}{\sqrt{n}}\,,\ \overline{X} + t_{1-\alpha/2}(n-1)\frac{S}{\sqrt{n}}\Big] \qquad (2.81)$$

with probability $1-\alpha$. We say that I is a *confidence interval* for b *of level* $1-\alpha$.

The same idea allows us to estimate the variance σ^2, but with some changes as the χ^2 laws are not symmetric. If we denote by $\chi^2_\alpha(n-1)$ the quantile of order α of a $\chi^2(n-1)$ law, we have

$$P\Big(\frac{n-1}{\sigma^2}S^2 < \chi^2_{\alpha/2}(n-1)\Big) = \frac{\alpha}{2}, \qquad P\Big(\frac{n-1}{\sigma^2}S^2 > \chi^2_{1-\alpha/2}(n-1)\Big) = \frac{\alpha}{2}$$

and therefore

$$1 - \alpha = P\left(\chi^2_{\alpha/2}(n-1) \leq \frac{n-1}{\sigma^2} S^2 \leq \chi^2_{1-\alpha/2}(n-1)\right)$$

$$= P\left(\frac{n-1}{\chi^2_{1-\alpha/2}(n-1)} S^2 \leq \sigma^2 \leq \frac{n-1}{\chi^2_{\alpha/2}(n-1)} S^2\right).$$

In other words

$$\left[\frac{n-1}{\chi^2_{1-\alpha/2}(n-1)} S^2, \frac{n-1}{\chi^2_{\alpha/2}(n-1)} S^2\right]$$

is a confidence interval for σ^2 of level $1 - \alpha$.

Example 2.46 In 1879 the physicist A. A. Michelson made $n = 100$ measurements of the speed of the light, obtaining the value

$$\overline{X} = 299\,852.4$$

with $S = 79.0$. If we assume that these values are equal to the true value of the speed of light with the addition of a Gaussian measurement error, (2.81) gives, for the confidence interval (2.81), intending 299,000 plus the indicated value,

$$[836.72, 868.08].$$

The latest measurements of the speed of the light give the value 792.4574 with a confidence interval ensuring precision up to the third decimal place. It appears that the 1879 measurements were biased. Michelson obtained much more precise results later on.

Exercises

2.1 (p. 270) Let $(\Omega, \mathcal{F}, P)$ be a probability space and $(A_n)_n$ a sequence of events, each having probability 1. Prove that their intersection $\bigcap_n A_n$ also has probability 1.

2.2 (p. 271) Let $(\Omega, \mathcal{F}, P)$ be a probability space and $\mathcal{G} \subset \mathcal{F}$ a P-trivial σ-algebra, i.e. such that, for every $A \in \mathcal{G}$, either $P(A) = 0$ or $P(A) = 1$. In this exercise we prove that a $\mathcal{G}$-measurable r.v. X with values in a separable metric space E is a.s. constant. This fact has already been established in Theorem 2.15 in the case $E = \mathbb{R}^m$. Let X be an E-valued $\mathcal{G}$-measurable r.v.

(a) Prove that for every $n \in \mathbb{N}$ there exists a ball $B_{x_n}(\frac{1}{n})$ centered at some $x_n \in E$ and with radius $\frac{1}{n}$ such that $P(X \in B_{x_n}(\frac{1}{n})) = 1$.
(b) Prove that there exists a decreasing sequence $(A_n)_n$ of Borel sets of E such that $P(X \in A_n) = 1$ for every n and such that the diameter of A_n is $\leq \frac{2}{n}$.
(c) Prove that there exists an $x_0 \in E$ such that $P(X = x_0) = 1$.

2.3 (p. 271)

(a) Let $(X_n)_n$ be a sequence of real independent r.v.'s and let

$$Z = \sup_{n \geq 1} X_n .$$

Assume that, for some $a \in \mathbb{R}$, $P(Z \leq a) > 0$. Prove that $Z < +\infty$ a.s.
(b) Let $(X_n)_n$ be a sequence of real independent r.v.'s with X_n exponential of parameter λ_n.
(b1) Assume that $\lambda_n = \log n$. Prove that

$$Z := \sup_{n \geq 1} X_n < +\infty \qquad \text{a.s.}$$

(b2) Assume that $\lambda_n \equiv c > 0$. Prove that $Z = +\infty$ a.s.

2.4 (p. 272) Let X and Y be real independent r.v.'s such that $X + Y$ has finite mathematical expectation. Prove that both X and Y have finite mathematical expectation.

2.5 (p. 272) Let X, Y be d-dimensional independent r.v.'s μ- and ν-distributed respectively. Assume that μ has density f with respect to the Lebesgue measure of $\mathbb{R}^d$ (no assumption is made concerning the law of Y).

(a) Prove that $X + Y$ also has density, g say, with respect to the Lebesgue measure and compute it.
(b) Prove that if f is k times differentiable with bounded derivatives up to the order k, then g is also k times differentiable (again whatever the law of Y).

2.6 (p. 273) Let μ be a probability on $\mathbb{R}^d$.

(a) Prove that, for every $\varepsilon > 0$, there exists an $M_1 > 0$ such that $\mu(|x| \geq M_1) < \varepsilon$.
(b) Let $f \in C_0(\mathbb{R}^d)$, i.e. continuous and such that for every $\varepsilon > 0$ there exists an $M_2 > 0$ such $|f(x)| \leq \varepsilon$ for $|x| > M_2$. Prove that if

$$g(x) = \mu * f(x) := \int_{\mathbb{R}^d} f(x - y)\, \mu(dy) \tag{2.82}$$

then also $g \in C_0(\mathbb{R}^d)$. In particular, as obviously $\|\mu * f\|_\infty \le \|f\|_\infty$, the map $f \mapsto \mu * f$ is continuous from $C_0(\mathbb{R}^d)$ to itself.

2.7 (p. 273) Let $X \sim N(0, \sigma^2)$. Compute $E(e^{tX^2})$ for $t \in \mathbb{R}$.

2.8 (p. 274) Let X be an $N(0, 1)$-distributed r.v., σ, b real numbers and $x, K > 0$. Show that

$$E\big[(xe^{b+\sigma X} - K)^+\big] = xe^{b+\frac{1}{2}\sigma^2}\Phi(-\zeta + |\sigma|) - K\Phi(-\zeta), \qquad (2.83)$$

where $\zeta = \frac{1}{|\sigma|}(\log\frac{K}{x} - b)$ and Φ denotes the distribution function of an $N(0, 1)$ law. This quantity appears naturally in mathematical finance.

2.9 (p. 274) (Weibull Laws) Let, for $\alpha > 0, \lambda > 0$,

$$f(t) = \begin{cases} \lambda \alpha t^{\alpha-1} e^{-\lambda t^\alpha} & \text{for } t > 0 \\ 0 & \text{for } t \le 0 . \end{cases}$$

(a) Prove that f is a probability density with respect to the Lebesgue measure and compute its d.f.

(b1) Let X be an exponential r.v. with parameter λ and let $\beta > 0$. Compute $E(X^\beta)$. What is the law of X^β?

(b2) Compute the expectation and the variance of an r.v. that is Weibull-distributed with parameters α, λ.

(b3) Deduce that for the Gamma function we have $\Gamma(1 + 2t) \ge \Gamma(1 + t)^2$ holds for every $t \ge 0$.

2.10 (p. 275) A pair of r.v.'s X, Y has joint density

$$f(x, y) = (\theta + 1)\frac{e^{\theta x}e^{\theta y}}{(e^{\theta x} + e^{\theta y} - 1)^{2+\frac{1}{\theta}}}, \qquad x > 0, y > 0$$

and $f(x, y) = 0$ otherwise, where $\theta > 0$. Compute the densities of X and of Y.

2.11 (p. 276) Let X, Y, Z be independent r.v.'s uniform on $[0, 1]$.

(a1) Compute the laws of $-\log X$ and of $-\log Y$.

(a2) Compute the law of $-\log X - \log Y$ and then of XY.

(b) Prove that $P(XY < Z^2) = \frac{5}{9}$.

2.12 (p. 277) Let Z be an exponential r.v. with parameter λ and let $Z_1 = \lfloor Z \rfloor$, $Z_2 = Z - \lfloor Z \rfloor$, respectively the integer and fractional parts of Z.

(a) Compute the laws of Z_1 and of Z_2.

(b1) Compute, for $0 \le a < b \le 1$ and $k \in \mathbb{N}$, the probability $P(Z_1 = k, Z_2 \in [a, b])$.

(b2) Prove that Z_1 and Z_2 are independent.

2.13 (p. 277) (Recall first Remark 2.1) Let F be the d.f. of a *positive* r.v. X having finite mean $b > 0$ and let $\overline{F}(t) = 1 - F(t)$. Let

$$g(t) = \frac{1}{b} \overline{F}(t) .$$

(a) Prove that g is a probability density.
(b) Determine g when X is
(b1)   exponential with parameter λ;
(b2)   uniform on $[0, 1]$;
(b3)   Pareto with parameters $\alpha > 1$ and $\theta > 0$, i.e. with density

$$f(t) = \begin{cases} \dfrac{\alpha \theta^\alpha}{(\theta + t)^{\alpha+1}} & \text{if } t > 0 \\ 0 & \text{otherwise .} \end{cases}$$

(c) Let $X \sim \text{Gamma}(n, \lambda)$, with n an integer ≥ 1. Prove that g is a linear combination of $\text{Gamma}(k, \lambda)$ densities for $1 \leq k \leq n$.
(d) Assume that X has finite variance σ^2. Compute the mean of the law having density g with respect to the Lebesgue measure.

2.14 (p. 279) In this exercise we determine the image law of the uniform distribution on the sphere under the projection on the north-south diameter (or, indeed, on any diameter). Recall that in polar coordinates the parametrization of the sphere $\mathbb{S}_2$ of $\mathbb{R}^3$ is

$$z = \cos\theta ,$$
$$y = \sin\theta \cos\phi ,$$
$$x = \sin\theta \sin\phi$$

where $(\theta, \phi) \in [0, \pi] \times [0, 2\pi]$. θ is the *colatitude* (i.e. the latitude but with values in $[0, \pi]$ instead of $[-\frac{\pi}{2}, \frac{\pi}{2}]$) and ϕ the *longitude*. The Lebesgue measure of the sphere, normalized so that the total measure is equal to 1, is $f(\theta, \phi) \, d\theta \, d\phi$, where

$$f(\theta, \phi) = \frac{1}{4\pi} \sin\theta \qquad (\theta, \phi) \in [0, \pi] \times [0, 2\pi] . \tag{2.84}$$

Let us consider the map $\mathbb{S}_2 \to [-1, 1]$ defined as

$$(x, y, z) \mapsto z ,$$

i.e. the projection of $\mathbb{S}_2$ on the north-south diameter.

What is the image of the normalized Lebesgue measure of the sphere under this map? Are the points at the center of the interval $[-1, 1]$ (corresponding to the equator) the most likely? Or those near the endpoints (the poles)?

2.15 (p. 279) Let Z be an r.v. uniform on $[0, \pi]$. Determine the law of $W = \cos Z$.

2.16 (p. 280) Let X, Y be r.v.'s whose joint law has density, with respect to the Lebesgue measure of $\mathbb{R}^2$, of the form

$$f(x, y) = g(x^2 + y^2),\tag{2.85}$$

where $g : \mathbb{R}^+ \to \mathbb{R}^+$ is a Borel function.

(a) Prove that necessarily

$$\int_0^{+\infty} g(t)\, dt = \frac{1}{\pi}\,.$$

(b1) Prove that X and Y have the same law.

(b2) Assume that X and (hence) Y are integrable. Compute $E(X)$ and $E(Y)$.

(b3) Assume that X and (hence) Y are square integrable. Prove that X and Y are uncorrelated. Give an example with X and Y independent and an example with X and Y non-independent.

(c1) Prove that $Z := \frac{X}{Y}$ has a Cauchy law, i.e. with density with respect to the Lebesgue measure

$$z \mapsto \frac{1}{\pi(1 + z^2)}\,.$$

In particular, the law of $\frac{X}{Y}$ does not depend on g.

(c2) Let X, Y be independent $N(0, 1)$-distributed r.v.'s. What is the law of $\frac{X}{Y}$?

(c3) Let Z be a Cauchy-distributed r.v. Prove that $\frac{1}{Z}$ also has a Cauchy law.

2.17 (p. 281) Let $(\Omega, \mathcal{F}, P)$ be a probability space and X a positive r.v. such that $E(X) = 1$. Let us define a new measure Q on $(\Omega, \mathcal{F})$ by

$$\frac{dQ}{dP} = X\,,$$

i.e. $Q(A) = E(X 1_A)$ for all $A \in \mathcal{F}$.

(a) Prove that Q is a probability and that $Q \ll P$.

(b) We now address the question of whether also $P \ll Q$.

(b1) Prove that the event $\{X = 0\}$ has probability 0 with respect to Q.

(b2) Let $\widetilde{P}$ be the measure on $(\Omega, \mathscr{F})$ defined as

$$\frac{d\widetilde{P}}{dQ} = \frac{1}{X}$$

(which is well defined as $X > 0$ Q-a.s.). Prove that $\widetilde{P} = P$ if and only if $\{X = 0\}$ has probability 0 also with respect to P and that in this case $P \ll Q$.

(c) Let μ be the law of X with respect to P. What is the law of X with respect to Q? If $X \sim$ Gamma(λ, λ) under P, what is its law under Q?

(d) Let Z be an r.v. independent of X (under P).

(d1) Prove that if Z is integrable under P then it is also integrable with respect to Q and that $E^Q(Z) = E(Z)$.

(d2) Prove that Z has the same law with respect to Q as with respect to P.

(d3) Prove that Z is also independent of X under Q.

2.18 (p. 282) Let $(\Omega, \mathscr{F}, P)$ be a probability space, and X and Z independent exponential r.v.'s of parameter λ. Let us define on $(\Omega, \mathscr{F})$ the new measure

$$\frac{dQ}{dP} = \frac{\lambda}{2}(X + Z)$$

i.e. $Q(A) = \frac{\lambda}{2} E[(X + Z)1_A]$.

(a) Prove that Q is a probability and that $Q \ll P$.

(b) Compute $E^Q(XZ)$.

(c1) Compute the joint law of X and Z with respect to Q. Are X and Z also independent with respect to Q?

(c2) What are the laws of X and of Z under Q?

2.19 (p. 283)

(a) Let X, Y be real r.v.'s having joint density f with respect to the Lebesgue measure. Prove that both XY and $\frac{X}{Y}$ have a density with respect to the Lebesgue measure and compute it.

(b) Let X, Y be independent r.v.'s Gamma(α, λ)- and Gamma(β, λ)-distributed respectively.

(b1) Compute the law of $W = \frac{X}{Y}$.

(b2) This law turns out not to depend on λ. Was this to be expected?

(b3) For which values of p does W have a finite moment of order p? Compute these moments.

(c1) Let X, Y, Z be $N(0, 1)$-distributed independent r.v.'s. Compute the laws of

$$W_1 = \frac{X^2}{Z^2 + Y^2}$$

and of

$$W_2 = \frac{|X|}{\sqrt{Z^2 + Y^2}} \, .$$

(c2) Compute the law of $\frac{X}{Y}$.

2.20 (p. 286) Let X and Y be independent r.v.'s, $\Gamma(\alpha, 1)$- and $\Gamma(\beta, 1)$-distributed respectively with $\alpha, \beta > 0$.

(a) Prove that $U = X + Y$ and $V = \frac{1}{X}(X + Y)$ are independent.
(b) Determine the laws of V and of $\frac{1}{V}$.

2.21 (p. 287) Let T be a *positive* r.v. having density f with respect to the Lebesgue measure and X an r.v. uniform on $[0, 1]$, independent of T. Let $Z = XT$, $W = (1 - X)T$.

(a) Determine the joint law of Z and W.
(b) Explicitly compute this joint law when f is Gamma$(2, \lambda)$. Prove that in this case Z and W are independent.

2.22 (p. 288)

(a) Let X, Y be real r.v.'s, having joint density f with respect to the Lebesgue measure and such that $X \leq Y$ a.s. Let, for every $x, y, x \leq y$,

$$G(x, y) := P(x \leq X \leq Y \leq y) \, .$$

Deduce from G the density f.
(b1) Let Z, W be i.i.d. real r.v.'s having density h with respect to the Lebesgue measure. Determine the joint density of $X = \min(Z, W)$ and $Y = \max(Z, W)$.
(b2) Explicitly compute this joint density when Z, W are uniform on $[0, 1]$ and deduce the value of $E[|Z - W|]$.

2.23 (p. 289) Let $(E, \mathscr{E}, \mu)$ be a σ-finite measure space. Assume that, for every integrable function $f : E \to \mathbb{R}$ and for every convex function ϕ,

$$\int_E \phi(f(x)) \, d\mu(x) \geq \phi\left(\int_E f(x) \, d\mu(x)\right) \tag{2.86}$$

(note that, as in the proof of Jensen's inequality, $\phi \circ f$ is lower semi-integrable, so that the l.h.s above is always well defined).

(a) Prove that for every $A \in \mathscr{E}$ such that $\mu(A) < +\infty$ necessarily $\mu(A) \leq 1$. Deduce that μ is finite.
(b) Prove that μ is a probability.

In other words, Jensen's inequality only holds for probabilities.

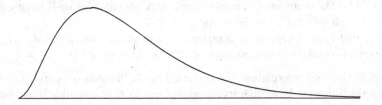

Fig. 2.4 A typical example of a density with a positive skewness

2.24 (p. 289) Given two probabilities μ, ν on a measurable space $(E, \mathscr{E})$, the *relative entropy* (or Kullback-Leibler divergence) of ν with respect to μ is defined as

$$H(\nu; \mu) := \int_E \log \frac{d\nu}{d\mu}\, \nu(dx) = \int_E \frac{d\nu}{d\mu} \log \frac{d\nu}{d\mu}\, \mu(dx) \tag{2.87}$$

if $\nu \ll \mu$ and $H(\nu; \mu) = +\infty$ otherwise.

(a1) Prove that $H(\nu; \mu) \geq 0$ and that $H(\nu; \mu) > 0$ unless $\nu = \mu$. Moreover, H is a convex function of ν.

(a2) Let $A \in \mathscr{E}$ be a set such that $0 < \mu(A) < 1$ and $d\nu = \frac{1}{\mu(A)} 1_A\, d\mu$. Compute $H(\nu; \mu)$ and $H(\mu; \nu)$ and note that $H(\nu; \mu) \neq H(\mu; \nu)$.

(b1) Let $\mu = B(n, p)$ and $\nu = B(n, q)$ with $0 < p, q < 1$. Compute $H(\nu; \mu)$.

(b2) Compute $H(\nu; \mu)$ when ν and μ are exponential of parameters ρ and λ respectively.

(c) Let $\nu_i, \mu_i, i = 1, \ldots, n$, be probabilities on the measurable spaces $(E_i, \mathscr{E}_i)$. Prove that, if $\nu = \nu_1 \otimes \cdots \otimes \nu_n, \mu = \mu_1 \otimes \cdots \otimes \mu_n$, then

$$H(\nu; \mu) = \sum_{i=1}^n H(\nu_i; \mu_i). \tag{2.88}$$

2.25 (p. 291) The *skewness* (or asymmetry) index of an r.v. X is the quantity

$$\gamma = \frac{E[(X - b)^3]}{\sigma^3}, \tag{2.89}$$

where $b = E(X)$ and $\sigma^2 = \text{Var}(X)$ (provided X has a finite moment of order 3). The index γ, intuitively, measures the asymmetry of the law of X: values of γ that are positive indicate the presence of a "longish tail" on the right (as in Fig. 2.4), whereas negative values indicate the same thing on the left.

(a) What is the skewness of an $N(b, \sigma^2)$ law?

(b) And of an exponential law? Of a Gamma(α, λ)? How does the skewness depend on α and λ?

Recall the binomial expansion of third degree: $(a + b)^3 = a^3 + 3a^2b + 3ab^2 + b^3$.

2.26 (p. 292) (The problem of moments) Let μ, ν be probabilities on $\mathbb{R}$ having equal moments of all orders. Can we infer that $\mu = \nu$?

Prove that if their support is contained in a bounded interval $[-M, M]$, then $\mu = \nu$ (this is not the weakest assumption, see e.g. Exercise 2.45).

2.27 (p. 293) (Some information that is carried by the covariance matrix) Let X be an m-dimensional r.v. Prove that its covariance matrix C is invertible if and only if the support of the law of X *is not* contained in a proper hyperplane of $\mathbb{R}^d$. Deduce that if C is not invertible, then the law of X cannot have a density with respect to the Lebesgue measure.

Recall Eq. (2.33). Proper hyperplanes have Lebesgue measure 0...

2.28 (p. 293) Let X, Y be real square integrable r.v.'s and $x \mapsto ax + b$ the regression line of Y on X.

(a) Prove that $Y - (aX + b)$ is centered and that the r.v.'s $Y - (aX + b)$ and $aX + b$ are orthogonal in L^2.
(b) Prove that the squared discrepancy $E[(Y - (aX + b))^2]$ is equal to $E(Y^2) - E[(aX + b)^2]$.

2.29 (p. 294)

(a) Let Y, W be independent r.v.'s $N(0, 1)$- and $N(0, \sigma^2)$-distributed respectively and let $X = Y + W$. What is the regression line $x \mapsto ax + b$ of Y with respect to X? What is the value of the quadratic error

$$E[(Y - aX - b)^2] \, ?$$

(b) Assume, instead, the availability of two measurements of the same quantity Y, $X_1 = Y + W_1$ and $X_2 = Y + W_2$, where the r.v.'s Y, W_1 and W_2 are independent and $W_1, W_2 \sim N(0, \sigma^2)$. What is now the best estimate of Y by an affine-linear function of the two observations X_1 and X_2? What is the value of the quadratic error now?

2.30 (p. 295) Let Y, W be exponential r.v.'s with parameters respectively λ and ρ. Determine the regression line of Y with respect to $X = Y + W$.

2.31 (p. 295) Let ϕ be a characteristic function. Show that $\overline{\phi}$, ϕ^2, $|\phi|^2$ are also characteristic functions.

2.32 (p. 296) (a) Let X_1, X_2 be independent r.v.'s uniform on $[-\frac{1}{2}, \frac{1}{2}]$.

(a) Compute the characteristic function of $X_1 + X_2$.
(b) Compute the characteristic function, ϕ say, of the probability with density, with respect to the Lebesgue measure, $f(x) = 1 - |x|$, $|x| \le 1$ and $f(x) = 0$ for $|x| > 1$ and deduce the law of $X_1 + X_2$.

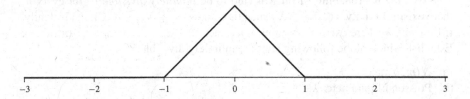

Fig. 2.5 The graph of f of Exercise 2.32 (and of ψ as well)

(c) Prove that the function (Fig. 2.5)

$$\kappa(\theta) = \begin{cases} 1 - |\theta| & \text{if } -1 \le \theta \le 1 \\ 0 & \text{otherwise} \end{cases}$$

is a characteristic function and determine the corresponding law.

Recall the trigonometric relation $1 - \cos x = 2 \sin^2 \frac{x}{2}$.

2.33 (p. 296) (Characteristic functions are positive definite) A function $f : \mathbb{R}^d \to \mathbb{C}$ is said to be *positive definite* if, for every choice of $n \in \mathbb{N}$ and $x_1, \ldots, x_n \in \mathbb{R}^d$, the complex matrix $(f(x_h - x_k))_{h,k}$ is positive definite, i.e. Hermitian and such that

$$\sum_{h,k=1}^{n} f(x_h - x_k)\xi_h \overline{\xi_k} \ge 0 \qquad \text{for every } \xi_1, \ldots, \xi_n \in \mathbb{C}.$$

Prove that characteristic functions are positive definite.

2.34 (p. 297)

(a) Let ν be a Laplace law with parameter $\lambda = 1$, i.e. having density $h(x) = \frac{1}{2} e^{-|x|}$ with respect to the Lebesgue measure. Prove that

$$\widehat{\nu}(\theta) = \frac{1}{1 + \theta^2}. \qquad (2.90)$$

(b1) Let μ be a Cauchy law, i.e. the probability having density

$$f(x) = \frac{1}{\pi(1 + x^2)}$$

with respect to the Lebesgue measure. Prove that $\widehat{\mu}(\theta) = e^{-|\theta|}$.

(b2) Let X, Y be independent Cauchy r.v.'s. Prove that $\frac{1}{2}(X + Y)$ is also Cauchy distributed.

2.35 (p. 298) A probability μ on $\mathbb{R}$ is said to be *infinitely divisible* if, for every n, there exist n i.i.d. r.v.'s $X_1, \ldots, X_n$ such that $X_1 + \cdots + X_n \sim \mu$. Or, equivalently, if for every n, there exists a probability μ_n such that $\mu_n * \cdots * \mu_n = \mu$ (n times). Establish which of the following laws are infinitely divisible.

(a) $N(m, \sigma^2)$.
(b) Poisson of parameter λ.
(c) Exponential of parameter λ.
(d) Cauchy.

2.36 (p. 299) Let μ, ν be probabilities on $\mathbb{R}^d$ such that

$$\mu(H_{\theta,a}) = \nu(H_{\theta,a}) \tag{2.91}$$

for every half-space $H_{\theta,a} = \{x; \langle \theta, x \rangle \le a\}, \theta \in \mathbb{R}^d, a \in \mathbb{R}$.

(a) Let μ_θ, ν_θ denote the images of μ and ν respectively through the map ψ_θ : $\mathbb{R}^d \to \mathbb{R}$ defined by $\psi_\theta(x) = \langle \theta, x \rangle$. Prove that $\mu_\theta = \nu_\theta$.
(b) Deduce that $\mu = \nu$.

2.37 (p. 299) Let $(\Omega, \mathscr{F}, \mathrm{P})$ be a probability space and X a positive integrable r.v. on it, such that $\mathrm{E}(X) = 1$. Let us denote by μ and ϕ respectively the law and the characteristic function of X. Let Q be the probability on $(\Omega, \mathscr{F})$ having density X with respect to P.

(a1) Compute the characteristic function of X under Q and deduce that $-i\phi'$ also is a characteristic function.
(a2) Compute the law of X under Q and determine the law having characteristic function $-i\phi'$.
(a3) Determine the probability corresponding to $-i\phi'$ when $X \sim \mathrm{Gamma}(\lambda, \lambda)$ and when X is geometric of parameter $p = 1$.
 (b) Prove that if X is a positive integrable r.v. but $\mathrm{E}(X) \ne 1$, then $-i\phi'$ cannot be a characteristic function.

2.38 (p. 299) A professor says: "let us consider a real r.v. X with characteristic function $\phi(\theta) = \mathrm{e}^{-\theta^4} \ldots$". What can we say about the values of mean and variance of such an X? Comments?

2.39 (p. 300) (Stein's characterization of the Gaussian law)

(a) Let $Z \sim N(0, 1)$. Prove that

$$\mathrm{E}[Zf(Z)] = \mathrm{E}[f'(Z)] \qquad \text{for every } f \in C_b^1 \tag{2.92}$$

where C_b^1 denotes the vector space of bounded continuous functions $\mathbb{R} \to \mathbb{C}$ with bounded derivative.
(b) Let Z be a real r.v. satisfying (2.92).

(b1) Prove that Z is integrable.

(b2) What is its characteristic function? Prove that necessarily $Z \sim N(0, 1)$.

2.40 (p. 301) Let X be a $\mathbb{Z}$-valued r.v. and ϕ its characteristic function.

(a) Prove that

$$P(X = 0) = \frac{1}{2\pi} \int_0^{2\pi} \phi(\theta) \, d\theta . \qquad (2.93)$$

(b) Are you able to find a similar formula in order to obtain from ϕ the probabilities $P(X = m)$, $m \in \mathbb{Z}$?

(c) What about the integrability of ϕ on the whole of $\mathbb{R}$?

2.41 (p. 302) (Characteristic functions are *uniformly* continuous) Let μ be a probability on $\mathbb{R}^d$.

(a) Prove that for every $\eta > 0$ there exist $R = R_\eta > 0$ such that $\mu(B_R^c) \le \eta$, where B_R denotes the ball centered at 0 and with radius R.

(b) Prove that, for every $\theta_1, \theta_2 \in \mathbb{R}^d$,

$$\left| e^{i \langle \theta_1, x \rangle} - e^{i \langle \theta_2, x \rangle} \right| \le |x| |\theta_1 - \theta_2| . \qquad (2.94)$$

In particular the functions $\theta \mapsto e^{i \langle \theta, x \rangle}$ are uniformly continuous as x ranges over a bounded set.

(c) Prove that $\widehat{\mu}$ is uniformly continuous.

2.42 (p. 303) Let X be an r.v. and let us denote by L its Laplace transform.

(a) Prove that, for every λ, $0 \le \lambda \le 1$, and $s, t \in \mathbb{R}$,

$$L\big(\lambda s + (1 - \lambda)t\big) \le L(s)^\lambda L(t)^{1-\lambda} .$$

(b) Prove that L restricted to the real axis and its logarithm are both convex functions.

2.43 (p. 303) Let X be an r.v. with a Laplace law of parameter λ, i.e. of density

$$f(x) = \frac{\lambda}{2} e^{-\lambda |x|}$$

with respect to the Lebesgue measure.

(a) Compute the Laplace transform and the characteristic function of X.

(b) Let Y and W be independent r.v.'s, both exponential of parameter λ. Compute the Laplace transform of $Y - W$. What is the law of $Y - W$?

(c1) Prove that the Laplace law is infinitely divisible (see Exercise 2.35 for the definition).

(c2) Prove that

$$\phi(\theta) = \frac{1}{(1+\theta^2)^{1/n}} \tag{2.95}$$

is a characteristic function.

2.44 (p. 304) (Some information about the tail of a distribution that is carried by its Laplace transform) Let X be an r.v. and x_2 the right convergence abscissa of its Laplace transform L.

(a) Prove that if $x_2 > 0$ then for every $\lambda < x_2$ we have for some constant $c > 0$

$$P(X \geq t) \leq c e^{-\lambda t} .$$

(b) Prove that if there exists a $t_0 > 0$ such that $P(X \geq t) \leq c e^{-\lambda t}$ for $t > t_0$, then $x_2 \geq \lambda$.

2.45 (p. 304) Let μ, ν be probabilities on $\mathbb{R}$ such that all their moments coincide:

$$\int_{-\infty}^{+\infty} x^k \, d\mu(x) = \int_{-\infty}^{+\infty} x^k \, d\nu(x) \qquad k = 1, 2, \ldots$$

and assume, in addition, that their Laplace transform is finite in a neighborhood of 0.

Then $\mu = \nu$.

2.46 (p. 305) (Exponential families) Let μ be a probability on $\mathbb{R}$ whose Laplace transform L is finite in an interval $]a, b[$, $a < 0 < b$ (hence containing the origin in its interior). Let, for $t \in \mathbb{R}$,

$$\psi(t) = \log L(t) . \tag{2.96}$$

As mentioned in Sect. 2.7, L, hence also its logarithm ψ, are infinitely many times differentiable in $]a, b[$.

(a) Express the mean and variance of μ using the derivatives of ψ.

(b) Let, for $\gamma \in]a, b[$,

$$d\mu_\gamma(x) = \frac{e^{\gamma x}}{L(\gamma)} d\mu(x) .$$

(b1) Prove that μ_γ is a probability and that its Laplace transform is

$$L_\gamma(t) := \frac{L(t+\gamma)}{L(\gamma)} .$$

(b2) Express the mean and variance of μ_γ using the derivatives of ψ.

(b3) Prove that ψ is a convex function and deduce that the mean of μ_γ is an increasing function of γ.

(c) Determine μ_γ when

(c1) $\mu \sim N(0, \sigma^2)$;

(c2) $\mu \sim \Gamma(\alpha, \lambda)$;

(c3) μ has a Laplace law of parameter θ, i.e. having density $f(x) = \frac{\lambda}{2} e^{-\lambda|x|}$ with respect to the Lebesgue measure;

(c4) $\mu \sim B(n, p)$;

(c5) μ is geometric of parameter p.

2.47 (p. 308) Let μ, ν be probabilities on $\mathbb{R}$ and denote by L_μ and L_ν respectively their Laplace transforms. Assume that $L_\mu = L_\nu$ on an open interval $]a, b[, a < b$.

(a) Assume $a < 0 < b$. Prove that $\mu = \nu$.

(b1) Let $a < \gamma < b$ and

$$d\mu_\gamma(x) = \frac{e^{\gamma x}}{L_\mu(\gamma)} d\mu(x), \qquad d\nu_\gamma(x) = \frac{e^{\gamma x}}{L_\nu(\gamma)} d\nu(x) .$$

Compute the Laplace transforms L_{μ_γ} and L_{ν_γ} and prove that $\mu_\gamma = \nu_\gamma$.

(b2) Prove that $\mu = \nu$ also if $0 \notin]a, b[$.

2.48 (p. 308) Let $X_1, \ldots, X_n$ be independent r.v.'s having an exponential law of parameter λ and let

$$Z_n = \max(X_1, \ldots, X_n) .$$

The aim of this exercise is to compute the expectation of Z_n.

(a) Prove that Z_n has a law having a density with respect to the Lebesgue measure and compute it. What is the value of the mean of Z_2? And of Z_3?

(b) Prove that the Laplace transform of Z_n is

$$L_n(z) = n\Gamma(n) \frac{\Gamma(1 - \frac{z}{\lambda})}{\Gamma(n + 1 - \frac{z}{\lambda})} \tag{2.97}$$

and determine its domain.

(c) Prove that for the derivative of $\log \Gamma$ we have the relation

$$\frac{\Gamma'(\alpha+1)}{\Gamma(\alpha+1)} = \frac{1}{\alpha} + \frac{\Gamma'(\alpha)}{\Gamma(\alpha)} \tag{2.98}$$

and deduce that $E(Z_n) = \frac{1}{\lambda}(1 + \cdots + \frac{1}{n})$.

Recall the Beta integral $\int_0^1 t^{\alpha-1}(1-t)^{\beta-1}\,dt = \frac{\Gamma(\alpha)\Gamma(\beta)}{\Gamma(\alpha+\beta)}$.

2.49 (p. 310) Let X be a d-dimensional r.v.

(a) Prove that if X is Gaussian then, for every $\xi \in \mathbb{R}^d$, the real r.v. $\langle \xi, X \rangle$ is Gaussian.
(b) Assume that, for every $\xi \in \mathbb{R}^d$, the real r.v. $\langle \xi, X \rangle$ is Gaussian.
(b1) Prove that X is square integrable.
(b2) Prove that X is Gaussian.

• This is a useful criterion.

2.50 (p. 311) Let X, Y be independent $N(0, 1)$-distributed r.v.'s.

(a) Prove that

$$U = \frac{X}{\sqrt{X^2 + Y^2}} \quad \text{and} \quad V = X^2 + Y^2$$

are independent and deduce the laws of U and of V.
(b) Prove that, for $\theta \in \mathbb{R}$, the r.v.'s

$$U' = \frac{X \cos\theta + Y \sin\theta}{\sqrt{X^2 + Y^2}} \quad \text{and} \quad V' = X^2 + Y^2$$

are independent and deduce the law of U'.

2.51 (p. 312) (Quadratic functions of Gaussian r.v.'s) Let X be an m-dimensional $N(0, I)$-distributed r.v. and A an $m \times m$ symmetric matrix.

(a) Compute

$$E(e^{\langle AX, X \rangle}) \tag{2.99}$$

under the assumption that all eigenvalues of A are $< \frac{1}{2}$.
(b) Prove that if A has an eigenvalue which is $\geq \frac{1}{2}$ then $E(e^{\langle AX, X \rangle}) = +\infty$.
(c) Compute the expectation in (2.99) if A is not symmetric.

Compare with Exercises 2.7 and 2.53.

2.52 (p. 313) (Non-central chi-square distributions)

(a) Let $X \sim N(\rho, 1)$. Compute the Laplace transform L of X^2 (with specification of the domain).

(b) Let $X_1, \ldots, X_m$ be independent r.v.'s with $X_i \sim N(b_i, 1)$, let $X = (X_1, \ldots, X_m)$ and $W = |X|^2$.

(b1) Prove that the law of W depends only on $\lambda = b_1^2 + \cdots + b_m^2$. Compute $E(W)$.

(b2) Prove that the Laplace transform of W is, for $\Re z < \frac{1}{2}$,

$$L(z) = \frac{1}{(1 - 2z)^{m/2}} \exp\left(\frac{z\lambda}{1 - 2z}\right).$$

• The law of W is the *non-central chi-square with m degrees of freedom*; λ is the *parameter of non-centrality*.

2.53 (p. 314)

(a) Let X be an m-dimensional Gaussian $N(0, I)$-distributed r.v. What is the law of $|X|^2$?

(b) Let C be an $m \times m$ positive definite matrix and X an m-dimensional Gaussian $N(0, C)$-distributed r.v. Prove that $|X|^2$ has the same law as an r.v. of the form

$$\sum_{k=1}^{m} \lambda_k Z_k \tag{2.100}$$

where $Z_1, \ldots, Z_m$ are independent $\chi^2(1)$-distributed r.v.'s and $\lambda_1, \ldots, \lambda_m$ are the eigenvalues of C. Prove that $E(|X|^2) = \operatorname{tr} C$.

2.54 (p. 315) Let $X = (X_1, \ldots, X_n)$ be an $N(0, I)$-distributed Gaussian vector. Let, for $k = 1, \ldots, n$, $Y_k = X_1 + \cdots + X_k - k X_{k+1}$ (with the understanding $X_{n+1} = 0$). Are $Y_1, \ldots, Y_n$ independent?

2.55 (p. 315)

(a) Let A and B be $d \times d$ real positive definite matrices. Let G be the matrix whose elements are obtained by multiplying A and B entrywise, i.e. $g_{ij} = a_{ij} b_{ij}$. Prove that G is itself positive definite (where is probability here?).

(b) A function $f : \mathbb{R}^d \to \mathbb{R}$ is said to be *positive definite* if $f(x) = f(-x)$ and if for every choice of $n \in \mathbb{N}$, of $x_1, \ldots, x_n \in \mathbb{R}^d$ and of $\xi_1, \ldots, \xi_n \in \mathbb{R}$, we have

$$\sum_{h,k=1}^{n} f(x_h - x_k) \xi_h \xi_k \geq 0.$$

Prove that the product of two positive definite functions is also positive definite.

Let X, Y be d-dimensional independent r.v.'s having covariance matrices A and B respectively...

2.56 (p. 315) Let X and Y be independent r.v.'s, where $X \sim N(0, 1)$ and Y is such that $P(Y = \pm 1) = \frac{1}{2}$. Let $Z = XY$.

(a) What is the law of Z?
(b) Are Z and X correlated? Independent?
(c) Compute the characteristic function of $X + Z$. Prove that X and Z are not jointly Gaussian.

2.57 (p. 316) Let $X_1, \dots, X_n$ be independent $N(0, 1)$-distributed r.v.'s and let

$$\overline{X} = \frac{1}{n} \sum_{k=1}^{n} X_k \ .$$

(a) Prove that, for every $i = 1, \dots, n$, $\overline{X}$ and $X_i - \overline{X}$ are independent.
(b) Prove that $\overline{X}$ is independent of

$$Y = \max_{i=1,\dots,n} X_i - \min_{i=1,\dots,n} X_i \ .$$

2.58 (p. 316) Let $X = (X_1, \dots, X_m)$ be an $N(0, I)$-distributed r.v. and $a \in \mathbb{R}^m$ a vector of modulus 1.

(a) Prove that the real r.v. $\langle a, X \rangle$ is independent of the m-dimensional r.v. $X - \langle a, X \rangle a$.
(b) What is the law of $|X - \langle a, X \rangle a|^2$?

Chapter 3
Convergence

Convergence is an important aspect of the computation of probabilities. It can be defined in many ways, each type of convergence having its own interest and its specific field of application. Note that the notions of convergence and approximation are very close.

As usual we shall assume an underlying probability space $(\Omega, \mathcal{F}, P)$.

3.1 Convergence of r.v.'s

Definition 3.1 Let $X, X_n, n \geq 1$, be r.v.'s on the same probability space $(\Omega, \mathcal{F}, P)$.

(a) If $X, X_n, n \geq 1$, take their values in a metric space (E, d), we say that the sequence $(X_n)_n$ converges to X *in probability* (written $\lim_{n\to\infty} X_n \overset{P}{=} X$) if for every $\delta > 0$

$$\lim_{n\to\infty} P\big(d(X_n, X) > \delta\big) = 0 .$$

(b) If $X, X_n, n \geq 1$, take their values in a topological space E, we say that $(X_n)_n$ converges to X *almost surely* (a.s.) if there exists a negligible event $N \in \mathcal{F}$ such that for every $\omega \in N^c$

$$\lim_{n\to\infty} X_n(\omega) = X(\omega) .$$

(continued)

© The Author(s), under exclusive license to Springer Nature Switzerland AG 2023
P. Baldi, *Probability*, Universitext, https://doi.org/10.1007/978-3-031-38492-9_3

Definition 3.1 (continued)

(c) If $X, X_n, n \geq 1$, are $\mathbb{R}^m$-valued, we say that $(X_n)_n$ *converges to X in L^p* if $X_n \in L^p$ for every n and

$$\lim_{n \to \infty} \mathrm{E}\big(|X_n - X|^p\big)^{1/p} = \lim_{n \to \infty} \|X_n - X\|_p = 0 .$$

Remark 3.2

(a) Recalling that for probabilities the L^p norm is an increasing function of p (see p. 63), L^p convergence implies L^q convergence for every $q \leq p$.

(b) Indeed L^p convergence can be defined for r.v.'s with values in a normed space. We shall restrict ourselves to the Euclidean case, but all the properties that we shall see also hold for r.v.'s with values in a general complete normed space. In this case L^p is a Banach space.

(c) Recall (see Remark 1.30) the inequality

$$\big| \|X\|_p - \|Y\|_p \big| \leq \|X - Y\|_p .$$

Therefore L^p convergence entails convergence of the L^p norms.

Let us compare these different types of convergence. Assume the r.v.'s $(X_n)_n$ to be $\mathbb{R}^m$-valued: by Markov's inequality we have, for every $p > 0$,

$$P\big(|X_n - X| > \delta\big) \leq \frac{1}{\delta^p} \, \mathrm{E}\big(|X_n - X|^p\big) ,$$

hence

L^p convergence, $p > 0$, implies convergence in probability.

If the sequence $(X_n)_n$, with values in a metric space (E, d), converges a.s. to an r.v. X, then $d(X_n, X) \to_{n \to \infty} 0$ a.s., i.e., for every $\delta > 0$, $1_{\{d(X_n, X) > \delta\}} \to_{n \to \infty} 0$ a.s. and by Lebesgue's Theorem

$$\lim_{n \to \infty} P\big(d(X_n, X) > \delta\big) = \lim_{n \to \infty} \mathrm{E}(1_{\{d(X_n, X) > \delta\}}) = 0 ,$$

i.e.

a.s. convergence implies convergence in probability.

The converse is not true, as shown in Example 3.5 below. Note that convergence in probability only depends on the joint laws of X and each of the X_n, whereas a.s. convergence depends in a deeper way on the joint distributions of the X_n's and X.

It is easy to construct examples of sequences converging a.s. but not in L^p: these two modes of convergence are not comparable, even if a.s. convergence is usually considered to be stronger.

The investigation of a.s. convergence requires an important tool that is introduced in the next section.

3.2 Almost Sure Convergence and the Borel-Cantelli Lemma

Let $(A_n)_n \subset \mathcal{F}$ be a sequence of events and let

$$A = \overline{\lim_{n \to \infty}} A_n := \bigcap_{n=1}^{\infty} \bigcup_{k \geq n} A_k .$$

A is the *superior limit* of the events $(A_n)_n$.

A closer look at this definition shows that $\omega \in A$ if and only if

$$\omega \in \bigcup_{k \geq n} A_k \qquad \text{for every } n ,$$

that is if and only if $\omega \in A_k$ for infinitely many indices k, i.e.

$$\overline{\lim_{n \to \infty}} A_n = \{\omega; \omega \in A_k \text{ for infinitely many indices } k\} . \tag{3.1}$$

The name "superior limit" comes from the fact that

$$1_A = \overline{\lim_{n \to \infty}} 1_{A_n} .$$

Clearly the superior limit of a sequence $(A_n)_n$ does not depend on the "first" events $A_1, \ldots, A_k$. Hence it belongs to the tail σ-algebra

$$\mathscr{B}^{\infty} = \bigcap_{i=1}^{\infty} \sigma(1_{A_i}, 1_{A_{i+1}}, \ldots)$$

and, if the events $A_1, A_2, \ldots$ are independent, by Kolmogorov's Theorem 2.15 their superior limit can only have probability 0 or 1. The following result provides a simple and powerful tool to establish which one of these contingencies holds.

Theorem 3.3 (The Borel-Cantelli Lemma) *Let $(A_n)_n \subset \mathscr{F}$ be a sequence of events.*

(a) *If $\sum_{n=1}^{\infty} P(A_n) < +\infty$ then $P(\overline{\lim}_{n\to\infty} A_n) = 0$.*
(b) *If $\sum_{n=1}^{\infty} P(A_n) = +\infty$ and the events A_n are independent then $P(\overline{\lim}_{n\to\infty} A_n) = 1$.*

Proof (a) We have

$$\sum_{n=1}^{\infty} P(A_n) = E\left(\sum_{n=1}^{\infty} 1_{A_n}\right)$$

but $\overline{\lim}_{n\to\infty} A_n$ is exactly the event $\{\sum_{n=1}^{\infty} 1_{A_n} = +\infty\}$: if $\omega \in \overline{\lim}_{n\to\infty} A_n$ then $\omega \in A_n$ for infinitely many indices and therefore in the series on the right-hand side there are infinitely many terms that are equal to 1. Hence if $\sum_{n=1}^{\infty} P(A_n) < +\infty$, then $\sum_{n=1}^{\infty} 1_{A_n}$ is integrable and the event $\overline{\lim}_{n\to\infty} A_n$ is negligible (the set of ω's on which an integrable function takes the value $+\infty$ is negligible, Exercise 1.9).

(b) By definition the sequence of events

$$\left(\bigcup_{k\geq n} A_k\right)_n$$

decreases to $\overline{\lim}_{n\to\infty} A_n$. Hence

$$P\left(\overline{\lim}_{n\to\infty} A_n\right) = \lim_{n\to\infty} P\left(\bigcup_{k\geq n} A_k\right). \tag{3.2}$$

Let us prove that, for every n, $P\left(\bigcup_{k\geq n} A_k\right) = 1$ or, what is the same, that

$$P\left(\bigcap_{k\geq n} A_k^c\right) = 0 .$$

We have, the A_n being independent,

$$P\left(\bigcap_{k\geq n} A_k^c\right) = \lim_{N\to\infty} P\left(\bigcap_{k=n}^{N} A_k^c\right) = \lim_{N\to\infty} \prod_{k=n}^{N} P(A_k^c)$$

$$= \lim_{N\to\infty} \prod_{k=n}^{N} \left(1 - P(A_k)\right) = \prod_{k=n}^{\infty} \left(1 - P(A_k)\right) .$$

As we assume $\sum_{n=1}^{\infty} P(A_n) = +\infty$, the infinite product above vanishes by a well-known convergence result for infinite products (recalled in the next proposition). Therefore $P\left(\bigcup_{k\geq n} A_k\right) = 1$ for every n and the limit in (3.2) is equal to 1. ∎

Proposition 3.4 *Let $(u_k)_k$ be a sequence of numbers with $0 \leq u_k \leq 1$ and let*

$$a := \prod_{k=1}^{\infty}(1 - u_k) .$$

Then

(a) If $\sum_{k=1}^{\infty} u_k = +\infty$ then $a = 0$.
(b) If $u_k < 1$ for every k and $\sum_{k=1}^{\infty} u_k < +\infty$ then $a > 0$.

Proof

(a) The inequality $1 - x \leq e^{-x}$ gives

$$a = \lim_{n\to\infty} \prod_{k=1}^{n}(1 - u_k) \leq \lim_{n\to\infty} \prod_{k=1}^{n} e^{-u_k} = \lim_{n\to\infty} \exp\left(-\sum_{k=1}^{n} u_k\right) = 0 .$$

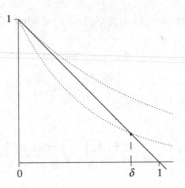

Fig. 3.1 The graphs of $x \mapsto 1 - x$ together with $x \mapsto e^{-x}$ (dots, the upper one) and $x \mapsto e^{-2x}$

(b) We have $1 - x \geq e^{-2x}$ for $0 \leq x \leq \delta$ for some $\delta > 0$ (see Fig. 3.1). As $\sum_{k=1}^{\infty} u_k < +\infty$, we have $u_k \to_{k \to \infty} 0$, so that $u_k \leq \delta$ for $k \geq n_0$. Hence

$$\prod_{k=1}^{n}(1 - u_k) = \prod_{k=1}^{n_0}(1 - u_k) \prod_{k=n_0+1}^{n}(1 - u_k) \geq \prod_{k=1}^{n_0}(1 - u_k) \prod_{k=n_0+1}^{n} e^{-2u_k}$$

$$= \prod_{k=1}^{n_0}(1 - u_k) \times \exp\left(-2 \sum_{k=n_0+1}^{n} u_k\right)$$

and as $n \to \infty$ this converges to $\prod_{k=1}^{n_0}(1 - u_k) \times \exp\left(-2 \sum_{k=n_0+1}^{\infty} u_k\right) > 0$. ∎

Example 3.5 Let $(X_n)_n$ be a sequence of i.i.d. r.v.'s having an exponential law of parameter λ and let $c > 0$. What is the probability of the event

$$\overline{\lim_{n \to \infty}} \{X_n \geq c \log n\} ? \qquad (3.3)$$

Note that the events $\{X_n \geq c \log n\}$ have a probability that decreases to 0, as the X_n have the same law. But, at least if the constant c is small enough, might it be true that $X_n \geq c \log n$ for infinitely many indices n a.s.?

The Borel-Cantelli lemma allows us to face this question in a simple way: as these events are independent, it suffices to determine the nature of the series

$$\sum_{n=1}^{\infty} P(X_n \geq c \log n).$$

Recalling the d.f. of the exponential laws,

$$P(X_n \geq c \log n) = e^{-\lambda c \log n} = \frac{1}{n^{\lambda c}} ,$$

which is the general term of a convergent series if and only if $c > \frac{1}{\lambda}$. Hence the superior limit (3.3) has probability 0 if $c > \frac{1}{\lambda}$ and probability 1 if $c \leq \frac{1}{\lambda}$.

The computation above provides an example of a sequence converging in probability but not a.s.: the sequence $(\frac{1}{\log n} X_n)_n$ tends to zero in L^p, and therefore also in probability as, for every $p > 0$,

$$\lim_{n \to \infty} E\left[\left(\frac{X_n}{\log n}\right)^p\right] = \lim_{n \to \infty} \frac{1}{(\log n)^p} E(X_1^p) = 0$$

(an exponential r.v. has finite moments of all orders). A.s. convergence however does not take place: as seen above, with probability 1

$$\frac{X_n}{\log n} \geq \varepsilon$$

infinitely many times as soon as $\varepsilon \leq \frac{1}{\lambda}$ so that a.s. convergence cannot take place.

We can now give practical conditions ensuring a.s. convergence.

Proposition 3.6 *Let $(X_n)_n$ be a sequence of r.v.'s with values in a metric space (E, d). Then $\lim_{n \to \infty} X_n = X$ a.s. if and only if*

$$P\left(\overline{\lim_{n \to \infty}} \{d(X_n, X) > \delta\} \right) = 0 \quad \text{for every } \delta > 0 . \tag{3.4}$$

Proof If $\lim_{n \to \infty} X_n = X$ a.s. then, with probability 1, $d(X_n, X)$ can be larger than $\delta > 0$ for a finite number of indices at most, hence (3.4). Conversely if (3.4) holds, then with probability 1 $d(X_n, X) > \delta$ only for a finite number of indices, so that $\overline{\lim}_{n \to \infty} d(X_n, X) \leq \delta$ and the result follows thanks to the arbitrariness of δ. ∎

Together with Proposition 3.6, the Borel-Cantelli Lemma provides a criterion for a.s. convergence:

Remark 3.7 If for every $\delta > 0$ the series $\sum_{n=1}^{\infty} P(d(X_n, X) > \delta)$ converges (no assumptions of independence), then (3.4) holds and $X_n \xrightarrow[n \to \infty]{a.s.} X$.

Note that, in comparison, only $\lim_{n \to \infty} P(d(X_n, X) > \delta) = 0$ for every $\delta > 0$ is required in order to have convergence in probability.

In the sequel we shall use often the following very useful elementary fact.

Criterion 3.8 (The Sub-Sub-Sequence Criterion) Let $(x_n)_n$ be a sequence in the metric space (E, d). Then $\lim_{n \to \infty} x_n = x$ if and only if from every subsequence $(x_{n_k})_k$ a further subsequence converging to x can be extracted.

Proposition 3.9

(a) *If $(X_n)_n$ converges to X in probability, then there exists a subsequence $(X_{n_k})_k$ such that $X_{n_k} \xrightarrow{k \to \infty} X$ a.s.*
(b) *$(X_n)_n$ converges to X in probability if and only if every subsequence $(X_{n_k})_k$ admits a further subsequence converging to X a.s.*

Proof (a) By the definition of convergence in probability we have, for every positive integer k,

$$\lim_{n \to \infty} P(d(X_n, X) > 2^{-k}) = 0 .$$

Let, for every k, n_k be an integer such that $P(d(X_n, X) > 2^{-k}) \leq 2^{-k}$ for every $n \geq n_k$. We can assume the sequence $(n_k)_k$ to be increasing. For $\delta > 0$ let k_0 be an integer such that $2^{-k} \leq \delta$ for $k > k_0$. Then, for $k > k_0$,

$$P(d(X_{n_k}, X) > \delta) \leq P(d(X_{n_k}, X) > 2^{-k}) \leq 2^{-k}$$

and the series $\sum_{k=1}^{\infty} P(d(X_{n_k}, X) > \delta)$ is summable as $P(d(X_{n_k}, X) > \delta) < 2^{-k}$ eventually. By the Borel-Cantelli lemma $P(\overline{\lim}_{k \to \infty} \{d(X_{n_k}, X) > \delta\}) = 0$ and $X_{n_k} \xrightarrow{k \to \infty} X$ a.s. by Proposition 3.6.

(b) The only if part follows from (a). Conversely, let us take advantage of Criterion 3.8: let us prove that from every subsequence of $(P(d(X_n, X) \geq \delta))_n$ we can extract a further subsequence converging to 0. But by assumption from every subsequence $(X_{n_k})_k$ we can extract a further subsequence $(X_{n_{k_h}})_h$ such that $X_{n_{k_h}} \to_{h \to \infty}^{a.s.} X$, hence also $\lim_{h \to 0} P(d(X_{n_{k_h}}, X) \geq \delta) = 0$ as a.s. convergence implies convergence in probability. ∎

Proposition 3.9, together with Criterion 3.8, allows us to obtain some valuable insights about convergence in probability.

• For convergence in probability many properties hold that are obvious for a.s. convergence. In particular, if $X_n \to_{n \to \infty}^{P} X$ and $\Phi : E \to G$ is a continuous function, G denoting another metric space, then also $\Phi(X_n) \to_{n \to \infty}^{P} \Phi(X)$. Actually from every subsequence of $(X_n)_n$ a further subsequence, $(X_{n_k})_k$ say, can be extracted converging to X a.s. and of course $\Phi(X_{n_k}) \to_{n \to \infty}^{a.s.} \Phi(X)$. Hence for every subsequence of $(\Phi(X_n))_n$ a further subsequence can be extracted converging a.s. to $\Phi(X)$ and the statement follows from Proposition 3.9.

In quite a similar way other useful properties of convergence in probability can be obtained. For instance, if $X_n \to_{n \to \infty}^{P} X$ and $Y_n \to_{n \to \infty}^{P} Y$, then also $X_n + Y_n \to_{n \to \infty}^{P} X + Y$.

• The a.s. limit is obviously unique: if Y and Z are two a.s. limits of the same sequence $(X_n)_n$, then $Y = Z$ a.s. Let us prove uniqueness also for the limit in probability, which is less immediate.

Let us assume that $X_n \to_{n \to \infty}^{P} Y$ and $X_n \to_{n \to \infty}^{P} Z$. By Proposition 3.9(a) we can find a subsequence of $(X_n)_n$ converging a.s. to Y. This subsequence obviously still converges to Z in probability and from it we can extract a further subsequence converging a.s. to Z. This sub-sub-sequence converges a.s. to both Y and Z and therefore $Y = Z$ a.s.

• The limits a.s. and in probability coincide: if $X_n \to_{n \to \infty}^{a.s.} Y$ and $X_n \to_{n \to \infty}^{P} Z$ then $Y = Z$ a.s.

• L^p convergence implies a.s. convergence for a subsequence.

Proposition 3.10 (Cauchy Sequences in Probability) *Let $(X_n)_n$ be a sequence of r.v.'s with values in the complete metric space E and such that for every $\delta, \varepsilon > 0$ there exists an n_0 such that*

$$P\big(d(X_n, X_m) > \delta\big) \leq \varepsilon \qquad \text{for every } n, m \geq n_0. \tag{3.5}$$

Then $(X_n)_n$ converges in probability to some E-valued r.v. X.

Proof For every $k > 0$ let n_k be an index such that, for every $m \geq n_k$,

$$P\big(d(X_{n_k}, X_m) \geq 2^{-k}\big) \leq 2^{-k} .$$

The sequence $(n_k)_k$ of course can be chosen to be increasing, therefore

$$\sum_{k=1}^{\infty} P\big(d(X_{n_k}, X_{n_{k+1}}) \geq 2^{-k}\big) < +\infty ,$$

and, by the Borel-Cantelli Lemma, the event $N := \overline{\lim}_{k \to \infty}\{d(X_{n_k}, X_{n_{k+1}}) \geq 2^{-k}\}$ has probability 0. Outside N we have $d(X_{n_k}, X_{n_{k+1}}) < 2^{-k}$ for every k larger than some k_0 and, for $\omega \in N^c$, $k \geq k_0$ and $m > k$,

$$d(X_{n_k}, X_{n_m}) \leq \sum_{i=k}^{m} 2^{-i} \leq 2 \cdot 2^{-k} .$$

Therefore, for $\omega \in N^c$, $(X_{n_k}(\omega))_k$ is a Cauchy sequence in E and converges to some limit $X(\omega) \in E$. Hence the sequence $(X_{n_k})_k$ converges a.s. to some r.v. X. Let us deduce that $X_n \to_{n \to \infty}^{P} X$: choose first an index n_k as above and large enough so that

$$P\big(d(X_n, X_{n_k}) \geq \tfrac{\delta}{2}\big) \leq \frac{\varepsilon}{2} \qquad \text{for every } n \geq n_k \tag{3.6}$$

$$P\big(d(X, X_{n_k}) \geq \tfrac{\delta}{2}\big) \leq \frac{\varepsilon}{2} . \tag{3.7}$$

An index n_k with these properties exists thanks to (3.5) and as $X_{n_k} \to_{n \to \infty}^{P} X$. Thus, for every $n \geq n_k$,

$$P\big(d(X_n, X) \geq \delta\big) \leq P\big(d(X_n, X_{n_k}) \geq \tfrac{\delta}{2}\big) + P\big(d(X, X_{n_k}) \geq \tfrac{\delta}{2}\big) \leq \varepsilon .$$

∎

In the previous proof we have been a bit careless: the limit X is only defined on N^c and we should prove that it can be defined on the whole of Ω in a measurable way. This recurring question is treated in Remark 1.15.

3.3   Strong Laws of Large Numbers

In this section we see that, under rather weak assumptions, if $(X_n)_n$ is a sequence of independent r.v.'s (or at least uncorrelated) and having finite mathematical expectation b, then their empirical means

$$\overline{X}_n := \frac{1}{n}(X_1 + \cdots + X_n)$$

converge a.s. to b. This type of result is a *strong law* of Large Numbers, as opposed to the *weak laws*, which are concerned with L^p convergence or in probability.

Note that we can assume $b = 0$: otherwise if $Y_n = X_n - b$ the r.v.'s Y_n have mean 0 and, as $\overline{Y}_n = \overline{X}_n - b$, to prove that $\overline{X}_n \to_{n\to\infty}^{\text{a.s.}} b$ or that $\overline{Y}_n \to_{n\to\infty}^{\text{a.s.}} 0$ is the same thing.

Theorem 3.11 (Rajchman's Strong Law) *Let $(X_n)_n$ be a sequence of pairwise uncorrelated r.v.'s having a common mean b and finite variance and assume that*

$$\sup_{n\geq 1} \text{Var}(X_n) := M < +\infty . \tag{3.8}$$

Then $\overline{X}_n \to_{n\to\infty} b$ a.s.

Proof Let $S_n := X_1 + \cdots + X_n$ and assume $b = 0$. For every $\delta > 0$ by Chebyshev's inequality

$$\text{P}(|\overline{X}_{n^2}| > \delta) \leq \frac{1}{\delta^2} \text{Var}(\overline{X}_{n^2}) = \frac{1}{\delta^2 n^4} \sum_{k=1}^{n^2} \text{Var}(X_k) \leq \frac{M}{\delta^2} \frac{1}{n^2} .$$

As the series $\sum_{n=1}^{\infty} \frac{1}{n^2}$ is summable, by Remark 3.7 the subsequence $(\overline{X}_{n^2})_n$ converges to 0 a.s. Now we need to investigate the behavior of $\overline{X}_n$ between two consecutive integers of the form n^2. With this goal let

$$D_n := \sup_{n^2 \leq k < (n+1)^2} |S_k - S_{n^2}|$$

(recall that $S_k = X_1 + \cdots + X_k$) so that if $n^2 \leq k < (n+1)^2$

$$|\overline{X}_k| = \frac{|S_k|}{k} \leq \frac{|S_{n^2}| + D_n}{k} \leq \frac{1}{n^2} (|S_{n^2}| + D_n) = |\overline{X}_{n^2}| + \frac{1}{n^2} D_n .$$

We are left to prove that $\frac{1}{n^2} D_n \to_{n\to\infty} 0$ a.s. This will follow as soon as we show that the term $n \mapsto \text{P}(|\frac{1}{n^2} D_n| > \delta)$ is summable and in order to do this, thinking of Markov's inequality, we shall look for estimates of the second order moment of D_n.

We have $D_n^2 \leq \sum_{n^2 \leq k < (n+1)^2} (S_k - S_{n^2})^2$, therefore

$$E(D_n^2) \leq \sum_{n^2 \leq k < (n+1)^2} E\big[(S_k - S_{n^2})^2\big]. \tag{3.9}$$

As the X_n are centered and uncorrelated, for $n^2 \leq k < (n+1)^2$,

$$E\big[(S_k - S_{n^2})^2\big] = E\big[(X_{n^2+1} + \cdots + X_k)^2\big] = \mathrm{Var}(X_{n^2+1} + \cdots + X_k)$$

$$= \sum_{i=n^2+1}^{k} \mathrm{Var}(X_i) \leq \big[(n+1)^2 - n^2 - 1\big] \cdot M = 2nM$$

and together with (3.9)

$$E(D_n^2) \leq \big[(n+1)^2 - n^2 - 1\big] \cdot 2nM = 4n^2 M$$

so that, for every $\delta > 0$, by Markov's inequality (2.28)

$$P\Big(\Big|\frac{1}{n^2} D_n\Big| > \delta\Big) \leq \frac{1}{\delta^2 n^4} E(D_n^2) \leq \frac{4M}{\delta^2} \frac{1}{n^2},$$

which is summable, completing the proof.   ∎

Note that, under the assumptions of Rajchman's Theorem 3.11, by Chebyshev's inequality,

$$P\big(|\overline{X}_n - b| \geq \delta\big) \leq \frac{1}{\delta^2} \mathrm{Var}(\overline{X}_n)$$

$$= \frac{1}{\delta^2 n^2}\big(\mathrm{Var}(X_1) + \cdots + \mathrm{Var}(X_n)\big) \leq \frac{M}{\delta^2 n} \xrightarrow[n \to \infty]{} 0,$$

so that the weak law, $\overline{X}_n \xrightarrow[n \to \infty]{P} b$, is immediate and much easier to prove than the strong law.

We state finally, without proof, the most celebrated Law of Large Numbers. It requires the r.v.'s to be independent and identically distributed, but the assumptions of existence of moments are weaker (the variances might be infinite) and the statement is much more precise. See [3, Theorem 10.42, p. 231], for a proof.

Theorem 3.12 (Kolmogorov's Strong Law)   *Let $(X_n)_n$ be a sequence of real i.i.d. r.v.'s. Then*

(a) *if the X_n are integrable, then $\overline{X}_n \xrightarrow[n \to \infty]{} b = E(X)$ a.s.;*

(continued)

Theorem 3.12 (continued)

(b) *if* $E(|X_n|) = +\infty$, *then at least one of the two terminal r.v.'s*

$$\varliminf_{n\to\infty} \overline{X}_n \quad \text{and} \quad \varlimsup_{n\to\infty} \overline{X}_n$$

is a.s. infinite (i.e. one of them at least takes the values $+\infty$ *or* $-\infty$ *a.s.).*

Example 3.13 Let $(X_n)_n$ be a sequence of i.i.d. Cauchy-distributed r.v.'s. If $\overline{X}_n = \frac{1}{n}(X_1 + \cdots + X_n)$ then of course the Law of Large Numbers does not hold, as X_n does not have a finite mathematical expectation. Kolmogorov's law however gives a more precise information about the behavior of the sequence $(\overline{X}_n)_n$: as the two sequences $(X_n)_n$ and $(-X_n)_n$ have the same joint laws, we have

$$\varliminf_{n\to\infty} \overline{X}_n \sim \varliminf_{n\to\infty} -\overline{X}_n = -\varlimsup_{n\to\infty} \overline{X}_n \qquad \text{a.s.}$$

As by the Kolmogorov strong law at least one among $\varliminf_{n\to\infty}\overline{X}_n$ and $\varlimsup_{n\to\infty}\overline{X}_n$ must be infinite, we derive that

$$\varliminf_{n\to\infty} \overline{X}_n = -\infty, \qquad \varlimsup_{n\to\infty} \overline{X}_n = +\infty \qquad \text{a.s.}$$

Hence the sequence of the empirical means takes infinitely many times very large and infinitely many times very small (i.e. negative and large in absolute value) values with larger and larger oscillations.

The law of Large Numbers is the theoretical justification for many algorithms of estimation and numerical approximation. The following example provides an instance of such an application. More insight about applications of the Law of Large Numbers is given in Sect. 6.1.

Example 3.14 (Histograms) Let $(X_n)_n$ be a sequence of real i.i.d. r.v.'s whose law has a density f with respect to the Lebesgue measure.

For a given bounded interval $[a, b]$, let us split it into subintervals $I_1, \ldots, I_k$, and let, for every $j = 1, \ldots, k$,

$$Z_j^{(n)} = \frac{1}{n}\sum_{i=1}^{n} 1_{I_j}(X_i) \,.$$

$\sum_{i=1}^{n} 1_{I_j}(X_i)$ is the number of r.v.'s (observations) X_i falling in the interval I_j, hence $Z_j^{(n)}$ is the proportion of the first n observations $X_1, \ldots, X_n$ whose values belong to the interval I_j.

It is usual to visualize the r.v.'s $Z_1^{(n)}, \ldots, Z_k^{(n)}$ by drawing above each interval I_j a rectangle of area proportional to $Z_j^{(n)}$; if the intervals I_j are equally spaced this means, of course, that the heights of the rectangles are proportional to $Z_j^{(n)}$. The resulting figure is called a *histogram*; this is a very popular method for visually presenting information concerning the common density of the observations $X_1, \ldots, X_n$.

The Law of Large Numbers states that

$$Z_j^{(n)} \xrightarrow[n \to \infty]{\text{a.s.}} E[1_{I_j}(X_i)] = P(X_i \in I_j) = \int_{I_j} f(x)\,dx .$$

If the intervals I_j are small enough, so that the variation of f on I_j is small, then the rectangles of the histogram will roughly have heights proportional to the corresponding values of f. Therefore for large n the histogram provides information about the density f. Figure 3.2 gives an example of a histogram for $n = 200$ independent observations of a $\Gamma(3, 1)$ law, compared with the true density.

This is a very rough and very initial instance of an important chapter of statistics: the estimation of a density.

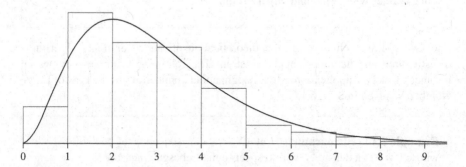

Fig. 3.2 Histogram of 200 independent $\Gamma(3, 1)$-distributed observations, compared with their density

3.4 Weak Convergence of Measures

We introduce now a notion of convergence of probability laws.

Let $(E, \mathscr{E})$ be a measurable space and $\mu, \mu_n, n \geq 1$, measures on $(E, \mathscr{E})$. A typical way (not the only one) of defining a convergence $\mu_n \to_{n\to\infty} \mu$ is the following: first fix a class $\mathscr{D}$ of measurable functions $f : E \to \mathbb{R}$ and then define that $\mu_n \to_{n\to\infty} \mu$ if and only if

$$\lim_{n\to\infty} \int_E f \, d\mu_n = \int_E f \, d\mu \qquad \text{for every } f \in \mathscr{D} .$$

Of course according to the choice of the class $\mathscr{D}$ we obtain different types of convergence (possibly mutually incomparable).

In the sequel, in order to simplify the notation we shall sometimes write $\mu(f)$ instead of $\int f \, d\mu$ (which reminds us that a measure can also be seen as a functional on functions, Proposition 1.24).

Definition 3.15 Let E be a topological space and $\mu, \mu_n, n \geq 1$, *finite* measures on $\mathscr{B}(E)$. We say that $(\mu_n)_n$ converges to μ *weakly* if and only if for every function $f \in C_b(E)$ (bounded continuous functions on E) we have

$$\lim_{n\to\infty} \int_E f \, d\mu_n = \int_E f \, d\mu . \tag{3.10}$$

A first important property of weak convergence is the following.

Remark 3.16 Let $\mu, \mu_n, n \geq 1$, be probabilities on the topological space E, let $\Phi : E \to G$ be a continuous map to another topological space G and let us denote by ν_n, ν the images of μ_n and μ under Φ respectively. If $\mu_n \to_{n\to\infty} \mu$ weakly, then also $\nu_n \to_{n\to\infty} \nu$ weakly.

Indeed if $f : G \to \mathbb{R}$ is bounded continuous, then $f \circ \Phi$ is also bounded continuous $E \to \mathbb{R}$. Hence, thanks to Proposition 1.27 (integration with respect to an image measure),

$$\nu_n(f) = \mu_n(f \circ \Phi) \underset{n\to\infty}{\to} \mu(f \circ \Phi) = \nu(f) .$$

Assume E to be a metric space. Then the weak limit is unique. Actually if simultaneously $\mu_n \to_{n\to\infty} \mu$ and $\mu_n \to_{n\to\infty} \nu$ weakly, then necessarily

$$\int_E f\,d\mu = \int_E f\,d\nu \tag{3.11}$$

for every $f \in C_b(E)$, and therefore μ and ν coincide (Proposition 1.25).

Proposition 3.17 *Let $\mathscr{D}$ be a vector space of bounded measurable functions on the measurable space $(E, \mathscr{E})$ and let $\mu, \mu_n, n \geq 1$, be probabilities on $(E, \mathscr{E})$. Then in order for the relation*

$$\mu_n(g) \xrightarrow[n\to\infty]{} \mu(g) \tag{3.12}$$

to hold for every $g \in \mathscr{D}$ it is sufficient for (3.12) to hold for every function g belonging to a set H that is total in $\mathscr{D}$.

Proof By definition H is total in $\mathscr{D}$ if and only if the vector space $\mathscr{H}$ of the linear combinations of functions of H is dense in $\mathscr{D}$ in the uniform norm.

If (3.12) holds for every $g \in H$, by linearity it also holds for every $g \in \mathscr{H}$. Let $f \in \mathscr{D}$ and let $g \in \mathscr{H}$ be such that $\|f - g\|_\infty \leq \varepsilon$; therefore for every n

$$\int_E |f - g|\,d\mu_n \leq \varepsilon, \qquad \int_E |f - g|\,d\mu \leq \varepsilon\,.$$

Let now n_0 be such that $|\mu_n(g) - \mu(g)| \leq \varepsilon$ for $n \geq n_0$; then for $n \geq n_0$

$$|\mu_n(f) - \mu(f)| \leq |\mu_n(f) - \mu_n(g)| + |\mu_n(g) - \mu(g)| + |\mu(g) - \mu(f)| \leq 3\varepsilon$$

and by the arbitrariness of ε the result follows.   ∎

If moreover E is also separable and locally compact then we have the following criterion.

Proposition 3.18 *Let $\mu, \mu_n, n \geq 1$, be finite measures on the locally compact separable metric space E, then $\mu_n \to_{n\to\infty} \mu$ weakly if and only if*

(a) $\mu_n(f) \to_{n\to\infty} \mu(f)$ *for every compactly supported continuous function,*
(b) $\mu_n(1) \to_{n\to\infty} \mu(1)$.

Proof Let us assume that (a) and (b) hold and let us prove that $(\mu_n)_n$ converges to μ weakly, the converse being obvious. Recall (Lemma 1.26) that there exists an increasing sequence $(h_n)_n$ of continuous compactly supported functions such that $h_n \uparrow 1$ as $n \to \infty$.

Let $f \in C_b(E)$, then $f h_k \to_{k \to \infty} f$ and the functions $f h_k$ are continuous and compactly supported. We have, for every k,

$$
\begin{aligned}
|\mu_n(f) - \mu(f)| &= \left|\mu_n\big((1 - h_k + h_k)f\big) - \mu\big((1 - h_k + h_k)f\big)\right| \\
&\leq \left|\mu_n\big((1 - h_k)f\big)\right| + \left|\mu\big((1 - h_k)f\big)\right| + \left|\mu_n(f h_k) - \mu(f h_k)\right| \qquad (3.13) \\
&\leq \|f\|_\infty \mu_n(1 - h_k) + \|f\|_\infty \mu(1 - h_k) + |\mu_n(f h_k) - \mu(f h_k)| \, .
\end{aligned}
$$

We have, adding and subtracting wisely,

$$
\begin{aligned}
\mu_n(1 - h_k) + \mu(1 - h_k) &= \mu_n(1) + \mu(1) - \mu_n(h_k) - \mu(h_k) \\
&= \mu_n(1) - \mu(1) + 2\mu(1) - 2\mu(h_k) + \mu(h_k) - \mu_n(h_k) \\
&= 2\mu(1 - h_k) + (\mu_n(1) - \mu(1)) + (\mu(h_k) - \mu_n(h_k))
\end{aligned}
$$

so that, going back to (3.13),

$$
|\mu_n(f) - \mu(f)|
$$
$$
\leq |\mu_n(f h_k) - \mu(f h_k)| + \|f\|_\infty \big(|\mu_n(h_k) - \mu(h_k)| + |\mu_n(1) - \mu(1)| + 2\mu(1 - h_k)\big) \, .
$$

Recalling that the functions h_k and $f h_k$ are compactly supported, if we choose k large enough so that $\mu(1 - h_k) \leq \varepsilon$, we have

$$
\varlimsup_{n \to \infty} |\mu_n(f) - \mu(f)| \leq 2\varepsilon \|f\|_\infty
$$

from which the result follows owing to the arbitrariness of ε. ∎

Remark 3.19 Putting together Propositions 3.17 and 3.18, if E is a locally compact separable metric space, in order to prove weak convergence we just need to check (3.10) for every $f \in C_K(E)$ or for every $f \in C_0(E)$ (functions vanishing at infinity) or indeed any family of functions that is total in $C_0(E)$.

If $E = \mathbb{R}^d$, a total family that we shall use in the sequel is that of the functions ψ_σ as in (2.51) for $\psi \in C_K(\mathbb{R}^d)$ and $\sigma > 0$, which is dense in $C_0(\mathbb{R}^d)$ thanks to Lemma 2.29.

Let $\mu, \mu_n, n \geq 1$, be probabilities on $\mathbb{R}^d$ and let us assume that $\mu_n \to_{n\to\infty} \mu$ weakly. Then clearly $\widehat{\mu}_n(\theta) \to_{n\to\infty} \widehat{\mu}(\theta)$: just note that for every $\theta \in \mathbb{R}^d$

$$\widehat{\mu}(\theta) = \int_{\mathbb{R}^d} e^{i\langle x, \theta \rangle} \, d\mu(x) \, ,$$

i.e. $\widehat{\mu}(\theta)$ is the integral with respect to μ of the bounded continuous function $x \mapsto e^{i\langle x,\theta \rangle}$. Therefore weak convergence, for probabilities on $\mathbb{R}^d$, implies pointwise convergence of the characteristic functions. The following result states that the converse also holds.

Theorem 3.20 (P. Lévy) *Let $\mu, \mu_n, n \geq 1$, be probabilities on $\mathbb{R}^d$. Then $(\mu_n)_n$ converges weakly to μ if and only if $\widehat{\mu}_n(\theta) \to_{n\to\infty} \widehat{\mu}(\theta)$ for every $\theta \in \mathbb{R}^d$.*

Proof Thanks to Remark 3.19 it suffices to prove that $\mu_n(\psi_\sigma) \to_{n\to\infty} \mu(\psi_\sigma)$ where ψ_σ is as in (2.51) with $\psi \in C_K(\mathbb{R}^d)$. Thanks to (2.53)

$$\int_{\mathbb{R}^d} \psi_\sigma(x) \, d\mu_n(x) = \frac{1}{(2\pi)^d} \int_{\mathbb{R}^d} \psi(y) \, dy \int_{\mathbb{R}^d} e^{-\frac{1}{2}\sigma^2 |\theta|^2} e^{-i\langle \theta, y \rangle} \widehat{\mu}_n(\theta) \, d\theta$$

$$= \int_{\mathbb{R}^d} \widehat{\mu}_n(\theta) H(\theta) \, d\theta \, , \tag{3.14}$$

where

$$H(\theta) = \frac{1}{(2\pi)^d} e^{-\frac{1}{2}\sigma^2 |\theta|^2} \int_{\mathbb{R}^d} \psi(y) e^{-i\langle \theta, y \rangle} \, dy \, .$$

The integrand of the integral on the right-hand side of (3.14) converges pointwise to $\widehat{\mu}H$ and is majorized in modulus by $\theta \mapsto (2\pi)^{-d} e^{-\frac{1}{2}\sigma^2 |\theta|^2} \int_{\mathbb{R}^d} |\psi(y)| \, dy$. We can therefore apply Lebesgue's Theorem, giving

$$\lim_{n\to\infty} \int_{\mathbb{R}^d} \psi_\sigma(x) \, d\mu_n(x) = \int_{\mathbb{R}^d} \widehat{\mu}(\theta) H(\theta) \, d\theta = \int_{\mathbb{R}^d} \psi_\sigma(x) \, d\mu(x) \, ,$$

which completes the proof. ∎

Actually P. Lévy proved a much deeper result: if $(\widehat{\mu}_n)_n$ converges pointwise to a function κ and if κ is continuous at 0, *then κ is the characteristic function of a probability μ and $(\mu_n)_n$ converges weakly to μ.* We will prove this sharper result in Theorem 6.21.

If $\mu_n \to_{n\to\infty} \mu$ weakly, what can be said of the behavior of $\mu_n(f)$ when f is not bounded continuous? And in particular when f is the indicator function of an event?

Theorem 3.21 (The "Portmanteau" Theorem) *Let $\mu, \mu_n, n \geq 1,$ be probabilities on the metric space E. Then $\mu_n \to_{n\to\infty} \mu$ weakly if and only if one of the following properties hold.*

(a) For every lower semi-continuous (l.s.c.) function $f : E \to \mathbb{R}$ bounded from below

$$\lim_{n\to\infty} \int_E f \, d\mu_n \geq \int_E f \, d\mu . \tag{3.15}$$

(b) For every upper semi-continuous (u.s.c.) function $f : E \to \mathbb{R}$ bounded from above

$$\overline{\lim_{n\to\infty}} \int_E f \, d\mu_n \leq \int_E f \, d\mu . \tag{3.16}$$

(c) For every bounded function f such that the set of its points of discontinuity is negligible with respect to μ

$$\lim_{n\to\infty} \int_E f \, d\mu_n = \int_E f \, d\mu . \tag{3.17}$$

Proof Clearly (a) and (b) are equivalent (if f is as in (a)), then $-f$ is as in (b) and together they imply weak convergence, as, if $f \in C_b(E)$, then to f we can apply simultaneously (3.15) and (3.16), obtaining (3.10).

Conversely, let us assume that $\mu_n \to_{n\to\infty} \mu$ weakly and that f is l.s.c. and bounded from below. Then (property of l.s.c. functions) there exists an increasing sequence of bounded continuous functions $(f_k)_k$ such that $\sup_k f_k = f$. As $f_k \leq f$, for every k we have

$$\int_E f_k \, d\mu = \lim_{n\to\infty} \int_E f_k \, d\mu_n \leq \underline{\lim_{n\to\infty}} \int_E f \, d\mu_n$$

and, taking the sup in k in this relation, by Beppo Levi's Theorem the term on the left-hand side increases to $\int_E f \, d\mu$ and we have (3.15).

Let us prove now that if $\mu_n \to_{n\to\infty} \mu$ weakly, then c) holds (the converse is obvious). Let f^* and f_* be the two functions defined as

$$f_*(x) = \lim_{y\to x} f(y) \qquad f^*(x) = \overline{\lim_{y\to x}} f(y) . \tag{3.18}$$

In the next Lemma 3.22 we prove that f_* is l.s.c. whereas f^* is u.s.c. Clearly $f_* \leq f \leq f^*$. Moreover these three functions coincide on the set C of continuity points of f; as we assume $\mu(C^c) = 0$ they are therefore bounded μ-a.s. and

$$\int_E f_* \, d\mu = \int_E f \, d\mu = \int_E f^* \, d\mu .$$

Now (3.15) and (3.16) give

$$\int_E f \, d\mu = \int_E f_* \, d\mu \leq \lim_{n\to\infty} \int_E f_* \, d\mu_n \leq \lim_{n\to\infty} \int_E f \, d\mu_n ,$$

$$\int_E f \, d\mu = \int_E f^* \, d\mu \geq \overline{\lim_{n\to\infty}} \int_E f^* \, d\mu_n \geq \overline{\lim_{n\to\infty}} \int_E f \, d\mu_n$$

which gives

$$\lim_{n\to\infty} \int_E f \, d\mu_n \geq \int_E f \, d\mu \geq \overline{\lim_{n\to\infty}} \int_E f \, d\mu_n ,$$

completing the proof.   ∎

Lemma 3.22 *The functions f_* and f^* in (3.18) are l.s.c. and u.s.c. respectively.*

Proof Let $x \in E$. We must prove that, for every $\delta > 0$, there exists a neighborhood U_δ of x such that $f_*(z) \geq f_*(x) - \delta$ for every $z \in U_\delta$. By the definition of $\underline{\lim}$, there exists a neighborhood V_δ of x such that $f(y) \geq f_*(x) - \delta$ for every $y \in V_\delta$.

If $z \in V_\delta$, there exists a neighborhood V of z such that $V \subset V_\delta$, so that $f(y) \geq f_*(x) - \delta$ for every $y \in V$. This implies that $f_*(z) = \underline{\lim}_{y\to z} f(y) \geq f_*(x) - \delta$. We can therefore choose $U_\delta = V_\delta$ and we have proved that f_* is l.s.c. Of course the argument for f^* is the same.   ∎

Assume that $\mu_n \to_{n\to\infty} \mu$ weakly and $A \in \mathcal{B}(E)$. Can we say that $\mu_n(A) \to_{n\to\infty} \mu(A)$? The portmanteau Theorem 3.21 gives some answers.

If $G \subset E$ is an open set, then its indicator function 1_G is l.s.c. and by (3.15)

$$\lim_{n \to \infty} \mu_n(G) = \lim_{n \to \infty} \int_E 1_G \, d\mu_n \geq \int_E 1_G \, d\mu = \mu(G) \,. \tag{3.19}$$

In order to give some intuition, think of a sequence of points $(x_n)_n \subset G$ and converging to some point $x \in \partial G$. It is easy to check that $\delta_{x_n} \to_{n \to \infty} \delta_x$ weakly (see also Example 3.24 below) and we would have $\delta_{x_n}(G) = 1$ for every n but $\delta_x(G) = 0$, as the limit point x does not belong to G.

Similarly if F is closed then 1_F is u.s.c. and

$$\overline{\lim_{n \to \infty}} \, \mu_n(F) = \overline{\lim_{n \to \infty}} \int_E 1_F \, d\mu_n \leq \int_E 1_F \, d\mu = \mu(F) \,. \tag{3.20}$$

Of course we have $\mu_n(A) \to_{n \to \infty} \mu(A)$, whether A is an open set or a closed one, if its boundary ∂A is μ-negligible: actually ∂A is the set of discontinuity points of 1_A.

Conversely if (3.19) holds for every open set G (resp. if (3.20) holds for every closed set F) it can be proved that $\mu_n \to_{n \to \infty} \mu$ (Exercise 3.17).

If $E = \mathbb{R}$ we have the following criterion.

Proposition 3.23 *Let μ, μ_n, $n \geq 1$, be probabilities on $\mathbb{R}$ and let us denote by F_n, F the respective distribution functions. Then $\mu_n \to_{n \to \infty} \mu$ weakly if and only if*

$$\lim_{n \to \infty} F_n(x) = F(x) \quad \text{for every continuity point } x \text{ of } F \,. \tag{3.21}$$

Proof Assume that $\mu_n \to_{n \to \infty} \mu$ weakly. We know that if x is a continuity point of F then $\mu(\{x\}) = 0$. As $\{x\}$ is the boundary of $] - \infty, x]$, by the portmanteau Theorem 3.21 c),

$$F_n(x) = \mu_n(] - \infty, x]) \underset{n \to \infty}{\to} \mu(] - \infty, x]) = F(x) \,.$$

Conversely let us assume that (3.21) holds. If a and b are continuity points of F then

$$\mu_n(]a, b]) = F_n(b) - F_n(a) \underset{n \to \infty}{\to} F(b) - F(a) = \mu(]a, b]) \,. \tag{3.22}$$

As the points of discontinuity of the increasing function F are at most countably many, (3.21) holds for x in a set D that is dense in $\mathbb{R}$. Thanks to Proposition 3.19 we just need to prove that $\mu_n(f) \to_{n \to \infty} \mu(f)$ for every $f \in C_K(\mathbb{R})$; this will follow from an adaptation of the argument of approximation of the integral with its Riemann sums.

As f is uniformly continuous, for fixed $\varepsilon > 0$ let $\delta > 0$ be such that $|f(x) - f(y)| < \varepsilon$ whenever $|x - y| < \delta$. Let $z_0 < z_1 < \cdots < z_N$ be a grid in an interval containing the support of f such that $z_k \in D$ and $|z_k - z_{k-1}| \leq \delta$. This is possible, D being dense in $\mathbb{R}$. If

$$S_n = \sum_{k=1}^{N} f(z_k)\big(F_n(z_k) - F_n(z_{k-1})\big),$$

$$S = \sum_{k=1}^{N} f(z_k)\big(F(z_k) - F(z_{k-1})\big)$$

then, as the z_k are continuity points of F, $\lim_{n\to\infty} S_n = S$. We have

$$\left| \int_{-\infty}^{+\infty} f\, d\mu_n - \int_{-\infty}^{+\infty} f\, d\mu \right|$$

$$\leq \left| \int_{-\infty}^{+\infty} f\, d\mu_n - S_n \right| + |S_n - S| + \left| \int_{-\infty}^{+\infty} f\, d\mu - S \right| \tag{3.23}$$

and

$$\left| \int_{-\infty}^{+\infty} f\, d\mu_n - S_n \right| = \left| \sum_{k=1}^{N} \int_{z_{k-1}}^{z_k} \big(f(x) - f(z_{k-1})\big)\, d\mu_n(x) \right|$$

$$\leq \sum_{k=1}^{N} \int_{z_{k-1}}^{z_k} |f(x) - f(z_{k-1})|\, d\mu_n(x)$$

$$\leq \varepsilon \sum_{k=1}^{N} \mu_n([z_{k-1}, z_k[) = \varepsilon\big(F(z_N) - F(z_0)\big) \leq \varepsilon.$$

Similarly

$$\left| \int_{-\infty}^{+\infty} f\, d\mu - S \right| \leq \varepsilon$$

and from (3.23) we obtain

$$\overline{\lim_{n\to\infty}} \left| \int_{-\infty}^{+\infty} f\, d\mu_n - \int_{-\infty}^{+\infty} f\, d\mu \right| \leq 2\varepsilon$$

and the result follows thanks to the arbitrariness of ε.   ∎

Example 3.24 (a) $\mu_n = \delta_{1/n}$ (Dirac mass at $\frac{1}{n}$). Then $\mu_n \to \delta_0$ weakly. Actually if $f \in C_b(\mathbb{R})$

$$\int_{\mathbb{R}} f \, d\mu_n = f(\tfrac{1}{n}) \underset{n\to\infty}{\to} f(0) = \int_{\mathbb{R}} f \, d\delta_0 \,.$$

Note that if $G =]0, 1[$, then $\mu_n(G) = 1$ for every n and therefore $\lim_{n\to\infty} \mu_n(G) = 1$ whereas $\delta_0(G) = 0$. Hence in this case $\mu_n(G) \not\to_{n\to\infty} \delta_0(G)$; note that $\partial G = \{0, 1\}$ and $\delta_0(\partial G) > 0$.

More generally, by the argument above, if $(x_n)_n$ is a sequence in the metric space E and $x_n \to_{n\to\infty} x$, then $\delta_{x_n} \to_{n\to\infty} \delta_x$ weakly.

(b) $\mu_n = \frac{1}{n} \sum_{k=0}^{n-1} \delta_{k/n}$. That is, μ_n is a sum of Dirac masses, each of weight $\frac{1}{n}$, placed at the locations $0, \frac{1}{n}, \ldots, \frac{n-1}{n}$.

Intuitively the total mass is crumbled into an increasing number of smaller and smaller, evenly spaced, Dirac masses. This suggests a limit that is uniform on the interval $[0, 1]$.

Formally, if $f \in C_b(\mathbb{R})$ then

$$\int_{\mathbb{R}} f \, d\mu_n = \sum_{k=0}^{n-1} \frac{1}{n} f(\tfrac{k}{n}) \,.$$

On the right-hand side we recognize, with some imagination, the Riemann sum of f on the interval $[0, 1]$ with respect to the partition $0, \frac{1}{n}, \ldots, \frac{n-1}{n}$. As f is continuous the Riemann sums converge to the integral and therefore

$$\lim_{n\to\infty} \int_{\mathbb{R}} f \, d\mu_n = \int_0^1 f(x) \, dx \,,$$

which proves that $(\mu_n)_n$ converges weakly to the uniform distribution on $[0, 1]$. The same result can also be obtained by computing the limit of the characteristic functions or of the d.f.'s.

(c) $\mu_n \sim B(n, \frac{\lambda}{n})$. Let us prove that $(\mu_n)_n$ converges to a Poisson law of parameter λ; i.e. the approximation of a binomial $B(n, p)$ law with a large parameter n and small p with a Poisson distribution is actually a weak convergence result. This can be seen in many ways. At this point we know of three methods to prove weak convergence:

- the definition;
- the convergence of the distribution functions, Proposition 3.23 (for probabilities on $\mathbb{R}$ only);
- the convergence of the characteristic functions (for probabilities on $\mathbb{R}^d$).

In this case, for instance, the d.f. F of the limit is continuous everywhere, the positive integers excepted. If $x > 0$, then

$$F_n(x) = \sum_{k=0}^{\lfloor x \rfloor} \binom{n}{k}\left(\frac{\lambda}{n}\right)^k \left(1 - \frac{\lambda}{n}\right)^{n-k} \xrightarrow[n\to\infty]{} \sum_{k=0}^{\lfloor x \rfloor} e^{-\lambda}\frac{\lambda^k}{k!} = F(x)$$

as in the sum only a finite number of terms appear ($\lfloor \ \rfloor$ denotes as usual the "integer part" function). If $x < 0$ there is nothing to prove as $F_n(x) = 0 = F(x)$. Note that in this case $F_n(x) \to_{n\to\infty} F(x)$ for every x, and not just for the x's that are continuity points. We might also compute the characteristic functions and their limit: recalling Example 2.25

$$\widehat{\mu}_n(\theta) = \left(1 - \frac{\lambda}{n} + \frac{\lambda}{n}e^{i\theta}\right)^n = \left(1 + \frac{\lambda}{n}(e^{i\theta} - 1)\right)^n \xrightarrow[n\to\infty]{} e^{\lambda(e^{i\theta}-1)} \, ,$$

which is the characteristic function of a Poisson law of parameter λ, and P. Lévy's Theorem 3.20 gives $\mu_n \to_{n\to\infty} \text{Poiss}(\lambda)$.

(d) $\mu_n \sim N(b, \frac{1}{n})$. Recall that the laws μ_n have a density given by bell shaped curves centered at b that become higher and narrower with n. This suggests that the μ_n tend to concentrate around b.

Also in this case in order to investigate the convergence we can compute either the limit of the d.f.'s or of the characteristic functions. The last method is the simplest one here:

$$\widehat{\mu}_n(\theta) = e^{ib\theta}e^{-\frac{1}{2n}\theta^2} \xrightarrow[n\to\infty]{} e^{ib\theta}$$

which is the characteristic function of a Dirac mass δ_b, in agreement with intuition.

(e) $\mu_n \sim N(0, n)$. The density of μ_n is

$$g_n(x) = \frac{1}{\sqrt{2\pi n}}\, e^{-\frac{1}{2n}x^2} \, .$$

As $g_n(x) \leq \frac{1}{\sqrt{2\pi n}}$ for every x, we have for every $f \in C_K(\mathbb{R})$

$$\lim_{n \to \infty} \int_{-\infty}^{+\infty} f(x) \, d\mu_n(x) = \lim_{n \to \infty} \int_{-\infty}^{+\infty} f(x) g_n(x) \, dx = 0 \,.$$

Hence $(\mu_n)_n$ cannot converge to a probability. This can also be proved via characteristic functions: indeed

$$\widehat{\mu}_n(\theta) = e^{-\frac{1}{2}n\theta^2} \underset{n \to \infty}{\to} \kappa(\theta) = \begin{cases} 1 & \text{if } \theta = 0 \\ 0 & \text{if } \theta \neq 0. \end{cases}$$

The limit κ is not continuous at 0 and cannot be a characteristic function.

Let μ_n, μ be probabilities on a σ-finite measure space $(E, \mathscr{E}, \rho)$ having densities f_n, f respectively with respect to ρ and assume that $f_n \to f$ pointwise as $n \to \infty$. What can be said about the weak convergence of $(\mu_n)_n$? Corollary 3.26 below gives an answer. It is a particular case of a more general statement that will also be useful in other situations.

Theorem 3.25 (Scheffé's Theorem) *Let $(E, \mathscr{E}, \rho)$ be a σ-finite measure space and $(f_n)_n$ a sequence of positive measurable functions such that*

(a) $f_n \to_{n \to \infty} f$ *ρ-a.e. for some measurable function f.*

(b) $\lim_{n \to \infty} \int_E f_n \, d\rho = \int_E f \, d\rho < +\infty.$

Then $f_n \to_{n \to \infty} f$ in $L^1(\rho)$.

Proof We have

$$\|f - f_n\|_1 = \int_E |f - f_n| \, d\rho = \int_E (f - f_n)^+ \, d\rho + \int_E (f - f_n)^- \, d\rho \,. \tag{3.24}$$

Let us prove that the two integrals on the right-hand side tend to 0 as $n \to \infty$.
   As f and f_n are positive we have

- If $f \geq f_n$ then $(f - f_n)^+ = f - f_n \leq f$.
- If $f \leq f_n$ then $(f - f_n)^+ = 0$.

In any case $(f - f_n)^+ \leq f$. As $(f - f_n)^+ \to_{n\to\infty} 0$ a.e. and f is integrable, by Lebesgue's Theorem,

$$\lim_{n\to\infty} \int_E (f - f_n)^+ d\rho = 0 .$$

As $f - f_n = (f - f_n)^+ - (f - f_n)^-$, we have also

$$\lim_{n\to\infty} \int_E (f - f_n)^- d\rho = \lim_{n\to\infty} \int_E (f - f_n)^+ d\rho - \lim_{n\to\infty} \int_E (f - f_n) d\rho = 0$$

and, going back to (3.24), the result follows.   ∎

Corollary 3.26 *Let $\mu, \mu_n, n \geq 1$ be probabilities on a topological space E and let us assume that there exists a σ-finite measure ρ on E such that μ and μ_n have densities f and f_n respectively with respect to ρ. Assume that*

$$\lim_{n\to\infty} f_n(x) = f(x) \qquad \rho\text{-a.e.}$$

Then $\mu_n \to_{n\to\infty} \mu$ weakly and also $\lim_{n\to\infty} \mu_n(A) = \mu(A)$ for every $A \in \mathcal{B}(E)$.

Proof As $\int f_n d\rho = \int f d\rho = 1$, conditions (a) and (b) of Theorem 3.25 are satisfied so that $f_n \to_{n\to\infty} f$ in L^1. If $\phi : E \to \mathbb{R}$ is bounded measurable then

$$\left| \int_E \phi \, d\mu_n - \int_E \phi \, d\mu \right| = \left| \int_E \phi(f - f_n) \, d\rho \right| \leq \|\phi\|_\infty \int_E |f - f_n| \, d\rho .$$

Hence

$$\lim_{n\to\infty} \int_E \phi \, d\mu_n = \int_E \phi \, d\mu$$

which proves weak convergence and, for $\phi = 1_A$, also the last statement.   ∎

3.5   Convergence in Law

Let $X, X_n, n \geq 1$, be r.v.'s with values in the same topological space E and let $\mu, \mu_n, n \geq 1$, denote their respective laws. The convergence of laws allows us to define a form of convergence of r.v.'s.

Definition 3.27 A sequence $(X_n)_n$ of r.v.'s with values in the topological space E is said to converge to X *in law* (and we write $X_n \to_{n\to\infty}^{\mathscr{L}} X$) if and only if $\mu_n \to_{n\to\infty} \mu$ weakly.

Remark 3.28 As

$$E[f(X_n)] = \int_E f(x)\, d\mu_n(x), \qquad E[f(X)] = \int_E f(x)\, d\mu(x), \qquad (3.25)$$

$X_n \to_{n\to\infty}^{\mathscr{L}} X$ if and only if

$$\lim_{n\to\infty} E[f(X_n)] = E[f(X)]$$

for every bounded continuous function $f : E \to \mathbb{R}$. If E is a locally compact separable metric space, it is sufficient to check (3.25) for every $f \in C_K(E)$ only (Proposition 3.19).

Let us compare convergence in law with the other forms of convergence.

Proposition 3.29 *Let $(X_n)_n$ be a sequence of r.v.'s with values in the metric space E. Then*

(a) *$X_n \to_{n\to\infty}^{P} X$ implies $X_n \to_{n\to\infty}^{\mathscr{L}} X$.*
(b) *If $X_n \to_{n\to\infty}^{\mathscr{L}} X$ and X is a constant r.v., i.e. such that $P(X = x_0) = 1$ for some $x_0 \in E$, then $X_n \to_{n\to\infty}^{P} X$.*

Proof (a) Keeping in mind Remark 3.28 let us prove that

$$\lim_{n\to\infty} E[f(X_n)] = E[f(X)] \qquad (3.26)$$

for every bounded continuous function $f : E \to \mathbb{R}$. Let us use Criterion 3.8 (the sub-sub-sequence criterion): (3.26) follows if it can be shown that from every subsequence of $(E[f(X_n)])_n$ we can extract a further subsequence along which (3.26) holds.

By Proposition 3.9(b), from every subsequence of $(X_n)_n$ a further subsequence $(X_{n_k})_k$ converging to X a.s. can be extracted. Therefore $\lim_{k\to\infty} f(X_{n_k}) = f(X)$ a.s. and, by Lebesgue's Theorem,

$$\lim_{k\to\infty} \mathrm{E}[f(X_{n_k})] = \mathrm{E}[f(X)].$$

(b) Let us denote by B_δ the open ball centered at x_0 with radius δ; then we can write $\mathrm{P}(d(X_n, x_0) \geq \delta) = \mathrm{P}(X_n \in B_\delta^c)$. B_δ^c is a closed set having probability 0 for the law of X, which is the Dirac mass δ_{x_0}. Hence by (3.20)

$$\overline{\lim_{n\to\infty}}\, \mathrm{P}(d(X_n, x_0) \geq \delta) \leq \mathrm{P}(d(X, x_0) \geq \delta) = 0.$$

∎

Convergence in law is therefore the weakest of all the convergences seen so far: a.s., in probability and in L^p. In addition note that, in order for it to take place, it is not even necessary for the r.v.'s to be defined on the same probability space.

Example 3.30 (Asymptotics of Student Laws) Let $(X_n)_n$ be a sequence of r.v.'s such that $X_n \sim t(n)$ (see p. 94). Let us prove that $X_n \xrightarrow[n\to\infty]{\mathscr{L}} X$ where $X \sim N(0, 1)$.

Let $Z, Y_n, n = 1, 2, \ldots$, be independent r.v.'s with $Z \sim N(0, 1)$ and $Y_n \sim \chi^2(1)$ for every n. Then $S_n = Y_1 + \cdots + Y_n \sim \chi^2(n)$ and S_n is independent of Z. Hence the r.v.

$$T_n := \frac{Z}{\sqrt{S_n}}\sqrt{n} = \frac{Z}{\sqrt{\dfrac{S_n}{n}}}$$

has a Student law $t(n)$. By the Law of Large Numbers $\frac{1}{n}S_n \xrightarrow[n\to\infty]{} \mathrm{E}(Y_1) = 1$ a.s. and therefore $T_n \xrightarrow[n\to\infty]{\text{a.s.}} Z$. As a.s. convergence implies convergence in law we have $T_n \xrightarrow[n\to\infty]{\mathscr{L}} Z$ and as $X_n \sim T_n$ for every n we have also $X_n \xrightarrow[n\to\infty]{\mathscr{L}} Z$.

This example introduces a sly method to determine the convergence in law of a sequence $(X_n)_n$: just construct another sequence $(W_n)_n$ such that

- $X_n \sim W_n$ for every n;
- $W_n \xrightarrow[n\to\infty]{} W$ a.s. (or in probability).

Then $(X_n)_n$ converges in law to W.

Example 3.31 The notion of convergence is closely related to that of approximation. As an application let us see a proof of the fact that the polynomials are dense in the space $C([0, 1])$ of real continuous functions on the interval $[0, 1]$ with respect to the uniform norm.

Let, for $x \in [0, 1]$, $(X_n^x)_n$ be a sequence of i.i.d. r.v.'s with a Bernoulli $B(1, x)$ law and let $S_n^x := X_1^x + \cdots + X_n^x$ so that $S_n^x \sim B(n, x)$. Let $f \in C([0, 1])$. Then

$$E\left[f\left(\tfrac{1}{n} S_n^x\right)\right] = \sum_{k=1}^{n} f\left(\tfrac{k}{n}\right)\binom{n}{k}x^k(1 - x)^{n-k} .$$

The right-hand side of the previous relation is a polynomial function of the variable x (the *Bernstein polynomial* of f of order n). Let us denote it by $P_n^f(x)$. By the Law of Large Numbers $\tfrac{1}{n} S_n^x \to_{n\to\infty} x$ a.s. hence also in law and

$$f(x) = \lim_{n\to\infty} E\left[f\left(\tfrac{1}{n} S_n^x\right)\right] = \lim_{n\to\infty} P_n^f(x) .$$

Therefore the sequence of polynomials $(P_n^f)_n$ converges pointwise to f. Let us demonstrate that the convergence is actually uniform. As f is uniformly continuous, for $\varepsilon > 0$ let $\delta > 0$ be such that $|f(x) - f(y)| \le \varepsilon$ whenever $|y - x| \le \delta$. Hence, for every $x \in [0, 1]$,

$$|P_n^f(x) - f(x)| \le E\left[|f\left(\tfrac{1}{n} S_n^x\right) - f(x)|\right]$$

$$= E\left[\underbrace{|f\left(\tfrac{1}{n} S_n^x\right) - f(x)| 1_{\{|\tfrac{1}{n} S_n^x - x| \le \delta\}}}_{\le \varepsilon}\right] + E\left[|f\left(\tfrac{1}{n} S_n^x\right) - f(x)| 1_{\{|\tfrac{1}{n} S_n^x - x| > \delta\}}\right]$$

$$\le \varepsilon + 2\|f\|_\infty P\left(|\tfrac{1}{n} S_n^x - x| > \delta\right) .$$

By Chebyshev's inequality and noting that $x(1 - x) \le \tfrac{1}{4}$ for $x \in [0, 1]$,

$$P\left(|\tfrac{1}{n} S_n^x - x| > \delta\right) \le \frac{1}{\delta^2}\mathrm{Var}\left(\tfrac{1}{n} S_n^x\right) \le \frac{1}{n\delta^2} x(1 - x) \le \frac{1}{4n\delta^2}$$

and therefore for n large

$$\|P_n^f - f\|_\infty \le 2\varepsilon .$$

See Fig. 3.3 for an example.

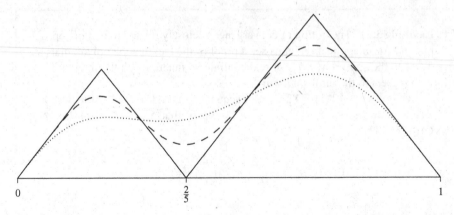

Fig. 3.3 Graph of some function f (solid) and of the approximating Bernstein polynomials of order $n = 10$ (dots) and $n = 40$ (dashes)

3.6 Uniform Integrability

Definition 3.32 A family $\mathscr{H}$ of m-dimensional r.v.'s is *uniformly integrable* if

$$\lim_{R \to +\infty} \sup_{Y \in \mathscr{H}} \int_{\{|Y| > R\}} |Y| \, d\mathrm{P} = 0 \, .$$

The set formed of a single integrable r.v. Y is the simplest example of a uniformly integrable family: actually $\lim_{R \to +\infty} |Y| 1_{\{|Y| > R\}} = 0$ a.s. and, as $|Y| 1_{\{|Y| > R\}} \leq |Y|$, by Lebesgue's Theorem,

$$\lim_{R \to +\infty} \int_{\{|Y| > R\}} |Y| \, d\mathrm{P} = 0 \, .$$

By a similar argument, if there exists a real integrable r.v. Z such that $Z \geq |Y|$ a.s. for every $Y \in \mathscr{H}$ then $\mathscr{H}$ is uniformly integrable, as in this case $\{|Y| > R\} \subset \{Z > R\}$ a.s. and

$$\int_{\{|Y| > R\}} |Y| \, d\mathrm{P} \leq \int_{\{Z > R\}} Z \, d\mathrm{P} \qquad \text{for every } Y \in \mathscr{H} \, .$$

Note however that in order for a family of r.v.'s $\mathcal{H}$ to be uniformly integrable it is not necessary for them to be defined on the same probability space: actually

$$\int_{\{|Y|>R\}} |Y|\, d\mathrm{P} = \int_{\{|y|>R\}} |y|\, d\mu_Y(y)$$

so that uniform integrability is a condition concerning *only the laws* of the r.v.'s of $\mathcal{H}$.

Note that a uniformly integrable family $\mathcal{H}$ is necessarily bounded in L^1: if $R > 0$ is such that

$$\sup_{Y\in\mathcal{H}} \int_{\{|Y|>R\}} |Y|\, d\mathrm{P} \le 1\,,$$

then, for every $Y \in \mathcal{H}$,

$$\mathrm{E}\big(|Y|\big) = \int_{\{|Y|\le R\}} |Y|\, d\mathrm{P} + \int_{\{|Y|>R\}} |Y|\, d\mathrm{P} \le R + 1\,.$$

The next proposition gives a useful characterization of uniform integrability.

Proposition 3.33 *A family $\mathcal{H}$ of r.v.'s. is uniformly integrable if and only if*

(i) *$\mathcal{H}$ is bounded in L^1 and*
(ii) *for every $\varepsilon > 0$ there exists a $\delta > 0$ such that for every $Y \in \mathcal{H}$*

$$\int_A |Y|\, d\mathrm{P} \le \varepsilon \qquad \text{whenever } \mathrm{P}(A) \le \delta\,. \tag{3.27}$$

Proof If $\mathcal{H}$ is uniformly integrable we already know that it is bounded in L^1. Also, for every event A we have

$$\int_A |Y|\, d\mathrm{P} = \int_{A\cap\{|Y|<R\}} |Y|\, d\mathrm{P} + \int_{A\cap\{|Y|<R\}} |Y|\, d\mathrm{P}$$

$$\le \int_{\{|Y|\ge R\}} |Y|\, d\mathrm{P} + R\,\mathrm{P}(Y \in A) = I_1 + I_2$$

and now just choose R so that $I_1 \le \frac{\varepsilon}{2}$ and then $\delta \le \frac{\varepsilon}{2R}$.

Conversely, if (i) and (ii) hold, let M be an upper bound of the L^1 norms of the r.v.'s of $\mathcal{H}$. Let $\varepsilon > 0$. Then by Markov's inequality, for every $Y \in \mathcal{H}$,

$$P(|Y| \geq R) \leq \frac{E(|Y|)}{R} \leq \frac{M}{R}$$

so that if $R \geq \frac{M}{\delta}$ then $P(|Y| \geq R) \leq \delta$ and, by (3.27),

$$\int_{\{|Y| \geq R\}} |Y| \, dP \leq \varepsilon$$

for every $Y \in \mathcal{H}$, thus proving uniform integrability. ∎

Let us note that, as a consequence of Proposition 3.33, a sequence $(Y_n)_n$ converging in L^1 to an r.v. Y is uniformly integrable. Indeed it is also bounded in L^1, so that, denoting by M an upper bound of the L^1 norms of the Y_n, by Markov's inequality,

$$P(|Y_n| \geq R) \leq \frac{M}{R} \cdot \tag{3.28}$$

As $|Y_n| \leq |Y_n - Y| + |Y|$,

$$E\big(|Y_n| 1_{\{|Y_n| \geq R\}}\big) \leq E\big(|Y_n - Y| 1_{\{|Y_n| \geq R\}}\big) + E\big(|Y| 1_{\{|Y_n| \geq R\}}\big) . \tag{3.29}$$

Let δ be such that $E(|Y| 1_A) \leq \varepsilon$ if $P(A) \leq \delta$ (the family $\{Y\}$ is uniformly integrable). Then for $R \geq \frac{M}{\delta}$ by (3.28) we have $P(|Y_n| \geq R) \leq \delta$ and therefore $E(|Y| 1_{\{|Y_n| \geq R\}}) \leq \varepsilon$.

Let now n_0 be such that $\|Y_n - Y\|_1 \leq \varepsilon$ for $n > n_0$, then (3.29) gives

$$E\big(|Y_n| 1_{\{|Y_n| \geq R\}}\big) \leq \|Y_n - Y\|_1 + E\big(|Y| 1_{\{|Y_n| \geq R\}}\big) \leq 2\varepsilon \tag{3.30}$$

for $n > n_0$. As each of the r.v.'s Y_k is, individually, uniformly integrable, there exist $R_1, \ldots, R_{n_0}$ such that $E(|Y_i| 1_{\{|Y_i| \geq R_i\}}) \leq \varepsilon$ for $i = 1, \ldots, n_0$ and, possibly replacing R with the largest among $R_1, \ldots, R_{n_0}, R$, we have $E(|Y_n| 1_{\{|Y_n| \geq R\}}) \leq 2\varepsilon$ for every n.

The following theorem is an extension of Lebesgue's Theorem. Note that it gives a necessary and sufficient condition.

Theorem 3.34 *Let $(Y_n)_n$ be a sequence of r.v.'s on a probability space $(\Omega, \mathcal{F}, P)$ converging a.s. to Y. Then the convergence takes place in L^1 if and only if $(Y_n)_n$ is uniformly integrable.*

Proof The only if part is already proved. Conversely, let us assume $(Y_n)_n$ is uniformly integrable. Then by Fatou's Lemma

$$E(|Y|) \leq \lim_{n \to \infty} E(|Y_n|) \leq M ,$$

where M is an upper bound of the L^1 norms of the Y_n. Moreover, for every $\varepsilon > 0$,

$$E(|Y - Y_n|) = E(|Y - Y_n|1_{\{|Y-Y_n|\leq\varepsilon\}}) + E(|Y - Y_n|1_{\{|Y-Y_n|>\varepsilon\}})$$
$$\leq \varepsilon + E(|Y_n|1_{\{|Y_n-Y|>\varepsilon\}}) + E(|Y|1_{\{|Y_n-Y|>\varepsilon\}}) .$$

As a.s. convergence implies convergence in probability, we have, for large n, $P(|Y_n - Y| > \varepsilon) \leq \delta$ (δ as in the statement of Proposition 3.33) so that

$$E(|Y_n|1_{\{|Y_n-Y|>\varepsilon\}}) \leq \varepsilon, \qquad E(|Y|1_{\{|Y_n-Y|>\varepsilon\}}) \leq \varepsilon$$

and for large n we have $E(|Y - Y_n|) \leq 3\varepsilon$. ∎

The following is a useful criterion for uniform integrability.

Proposition 3.35 *Let $\mathcal{H}$ be a family of r.v.'s and assume that there exists a measurable map $\Phi : \mathbb{R}^+ \to \mathbb{R}$, bounded below, such that $\lim_{t\to+\infty} \frac{1}{t}\Phi(t) = +\infty$ and*

$$\sup_{Y \in \mathcal{H}} E[\Phi(|Y|)] < +\infty .$$

Then $\mathcal{H}$ is uniformly integrable.

Proof Let Φ be as in the statement of the theorem. We can assume that Φ is positive, otherwise if $\Phi \geq -r$ just replace Φ with $\Phi + r$.

Let $K > 0$ be such that $E[\Phi(|Y|)] \leq K$ for every $Y \in \mathcal{H}$ and let $\varepsilon > 0$ be fixed. Let R_0 be such that $\frac{1}{R}\Phi(R) \geq \frac{K}{\varepsilon}$ for $R > R_0$, i.e. $|Y| \leq \frac{\varepsilon}{K}\Phi(|Y|)$ for $|Y| \geq R_0$ for every $Y \in \mathcal{H}$. Then, for every $Y \in \mathcal{H}$,

$$\int_{\{|Y|>R_0\}} |Y| \, dP \leq \frac{\varepsilon}{K} \int_{\{|Y|>R_0\}} \Phi(|Y|) \, dP \leq \frac{\varepsilon}{K} \int \Phi(|Y|) \, dP \leq \varepsilon .$$

∎

In particular, taking $\Phi(t) = t^p$, bounded subsets of L^p, $p > 1$, are uniformly integrable.

Actually there is a converse to Proposition 3.35: if $\mathcal{H}$ is uniformly integrable then there exists a function Φ as in Proposition 3.35 (and convex in addition to that). See [9], Theorem 22, p. 24, for a proof of this converse.

Therefore the criterion of Proposition 3.35 is actually a characterization of uniform integrability.

3.7 Convergence in a Gaussian World

In this section we see that, concerning convergence, Gaussian r.v.'s enjoy some special properties. The first result is stability of Gaussianity under convergence in law.

Proposition 3.36 *Let $(X_n)_n$ be a sequence of d-dimensional Gaussian r.v.'s converging in law to an r.v. X. Then X is Gaussian and the means and covariance matrices of the X_n converge to the mean and covariance matrix of X. In particular, $(X_n)_n$ is bounded in L^2.*

Proof Let us first assume the X_n's are real-valued. Their characteristic functions are of the form

$$\phi_n(\theta) = e^{ib_n\theta} e^{-\frac{1}{2}\sigma_n^2\theta^2} \tag{3.31}$$

and, by assumption, $\phi_n(\theta) \to_{n\to\infty} \phi(\theta)$ for every θ, where by ϕ we denote the characteristic function of the limit X.

Let us prove that ϕ is the characteristic function of a Gaussian r.v. The heart of the proof is that pointwise convergence of $(\phi_n)_n$ implies convergence of the sequences $(b_n)_n$ and $(\sigma_n^2)_n$. Taking the complex modulus in (3.31) we obtain

$$|\phi_n(\theta)| = e^{-\frac{1}{2}\sigma_n^2\theta^2} \quad \underset{n\to\infty}{\to} \quad |\phi(\theta)| .$$

This implies that the sequence $(\sigma_n^2)_n$ is bounded: otherwise there would exist a subsequence $(\sigma_{n_k}^2)_k$ converging to $+\infty$ and we would have $|\phi(\theta)| = 0$ for $\theta \neq 0$ and $|\phi(\theta)| = 1$ for $\theta = 0$, impossible because ϕ is necessarily continuous.

Let us show that the sequence $(b_n)_n$ of the means is also bounded. As the X_n's are Gaussian, if $\sigma_n^2 > 0$ then $P(X_n \geq b_n) = \frac{1}{2}$. If instead $\sigma_n^2 = 0$, then the law of X_n is the Dirac mass at b_n. In any case $P(X_n \geq b_n) \geq \frac{1}{2}$. If the means b_n were not bounded there would exist a subsequence $(b_{n_k})_k$ converging, say, to $+\infty$ (if $b_{n_k} \to -\infty$ the argument would be the same). Then, for every $M \in \mathbb{R}$ we would have $b_{n_k} \geq M$ for k large and therefore (the first inequality follows from

Theorem 3.21, the portmanteau theorem, as $[M, +\infty[$ is a closed set)

$$P(X \geq M) \geq \overline{\lim_{k \to \infty}} \, P(X_{n_k} \geq M) \geq \overline{\lim_{k \to \infty}} \, P(X_{n_k} \geq b_{n_k}) \geq \frac{1}{2} ,$$

which is not possible as $\lim_{M \to \infty} P(X \geq M) = 0$.

Hence both $(b_n)_n$ and $(\sigma_n^2)_n$ are bounded and for a subsequence we have $b_{n_k} \to b$ and $\sigma_{n_k}^2 \to \sigma^2$ as $k \to \infty$ for some numbers b and σ^2. Therefore

$$\phi(\theta) = \lim_{k \to \infty} e^{i b_{n_k} \theta} e^{-\frac{1}{2} \sigma_{n_k}^2 \theta^2} = e^{i b \theta} e^{-\frac{1}{2} \sigma^2 \theta^2} ,$$

which is the characteristic function of a Gaussian law.

A closer look at the argument above indicates that we have proved that from every subsequence of $(b_n)_n$ and of $(\sigma_n^2)_n$ a further subsequence can be extracted converging to b and σ^2 respectively. Hence by the sub-sub-sequence criterion, (Criterion 3.8), the means and the variances of the X_n converge to the mean and the variance of the limit and $(X_n)_n$ is bounded in L^2.

If the X_n's are d-dimensional, note that, for every $\xi \in \mathbb{R}^d$, the r.v.'s $Z_n = \langle \xi, X_n \rangle$ are Gaussian, being linear functions of Gaussian r.v.'s, and real-valued. Obviously $Z_n \xrightarrow[n \to \infty]{\mathcal{L}} \langle \xi, X \rangle$, which turns out to be Gaussian by the first part of the proof. As this holds for every $\xi \in \mathbb{R}^m$, this implies that X is Gaussian itself (see Exercise 2.49).

Let us prove convergence of means and covariance matrices in the multidimensional case. Let us denote by C_n, C the covariance matrices of X_n and X respectively. Thanks again to the first part of the proof the means and the variances of the r.v.'s $Z_n = \langle \xi, X_n \rangle$ converge to the mean and the variance of $\langle \xi, X \rangle$. Note that the mean of $\langle \xi, X_n \rangle$ is $\langle \xi, b_n \rangle$, whereas the variance is $\langle C_n \xi, \xi \rangle$. As this occurs for every vector $\xi \in \mathbb{R}^m$, we deduce that $b_n \xrightarrow[n \to \infty]{} b$ and $C_n \xrightarrow[n \to \infty]{} C$. ∎

As L^2 convergence implies convergence in law, the Gaussian r.v.'s on a probability space form a closed subset of L^2. But not a vector subspace ... (see Exercise 2.56).

An important feature of Gaussian r.v.'s is that the moment of order 2 controls all the moments of higher order. If $X \sim N(0, \sigma^2)$, then $X = \sigma Z$ for some $N(0, 1)$-distributed r.v. Z. Hence, as $\sigma^2 = E(|X|^2)$,

$$E(|X|^p) = \sigma^p \underbrace{E(|Z|^p)}_{:=c_p} = c_p E(|X|^2)^{p/2} .$$

If X is not centered the L^p norm of X can still be controlled by the L^2 norm, but this requires more care. Of course we can assume $p \geq 2$ as for $p \leq 2$ the L^2 norm is always larger than the L^p norm, thanks to Jensen's inequality. The key tools are, for positive numbers $x_1, \dots, x_n$, the inequalities

$$x_1^p + \cdots + x_n^p \le (x_1 + \cdots + x_n)^p \le n^{p-1}(x_1^p + \cdots + x_n^p) \tag{3.32}$$

that hold for every $n \ge 2$ and $p \ge 1$. If $X \sim N(b, \sigma^2)$ then $X \sim b + \sigma Z$ with $Z \sim N(0, 1)$ and

$$|X|^p = |b + \sigma Z|^p \le (|b| + \sigma|Z|)^p \le 2^{p-1}(|b|^p + \sigma^p|Z|^p) \tag{3.33}$$

hence, if now $c_p = 2^{p-1}(1 + E(|Z|^p))$,

$$E(|X|^p) \le 2^{p-1}(|b|^p + \sigma^p E(|Z|^p)) \le c_p(|b|^p + \sigma^p) .$$

Again by (3.32) (the inequality on the left-hand side for $\frac{p}{2}$) we have

$$|b|^p + \sigma^p = (|b|^2)^{p/2} + (\sigma^2)^{p/2} \le (|b|^2 + \sigma^2)^{p/2}$$

and, in conclusion,

$$E(|X|^p) \le c_p(|b|^2 + \sigma^2)^{p/2} = c_p E(|X|^2)^{p/2} . \tag{3.34}$$

A similar inequality also holds if X is d-dimensional Gaussian: as

$$|X|^p = (X_1^2 + \cdots + X_d^2)^{p/2} \le d^{p/2-1}(|X_1|^p + \cdots + |X_d|^p)$$

we have, using repeatedly (3.32) and (3.34),

$$E(|X|^p) \le d^{p/2-1} \sum_{k=1}^{d} E(|X_k|^p) \le c_p d^{p/2-1} \sum_{k=1}^{d} E(|X_k|^2)^{p/2}$$

$$\le c_p d^{p/2-1} \Big(\sum_{k=1}^{d} E(|X_k|^2) \Big)^{p/2} = c_p d^{p/2-1} E(|X|^2)^{p/2} .$$

This inequality together with Proposition 3.36 gives

Corollary 3.37 *A sequence of Gaussian r.v.'s converging in law is bounded in L^p for every $p \ge 1$.*

This is the key point of another important feature of the Gaussian world: a.s. convergence implies convergence in L^p for every $p > 0$.

Theorem 3.38 *Let $(X_n)_n$ be a sequence of Gaussian d-dimensional r.v.'s on a probability space $(\Omega, \mathscr{F}, P)$ converging a.s. to an r.v. X. Then the convergence takes place in L^p for every $p > 0$.*

Proof Let us first assume that $d = 1$. Thanks to Corollary 3.37, as a.s. convergence implies convergence in law, the sequence is bounded in L^p for every p. This implies also that $X \in L^p$ for every p: if by M_p we denote an upper bound of the L^p norms of the X_n then, by Fatou's Lemma,

$$E(|X|^p) \leq \lim_{n \to \infty} E(|X_n|^p) \leq M_p^p$$

(this is the same as in Exercise 1.15 a1)). We have for every $q > p$

$$\left(|X_n - X|^p\right)^{q/p} = |X_n - X|^q \leq 2^{q-1}\left(|X_n|^q + |X|^q\right).$$

The sequence $(|X_n - X|^p)_n$ converges to 0 a.s. and is bounded in $L^{q/p}$. As $\frac{q}{p} > 1$, it is uniformly integrable by Theorem 3.35 and Theorem 3.34 gives

$$\lim_{n \to \infty} \|X_n - X\|_p = \lim_{n \to \infty} E\left(|X_n - X|^p\right)^{1/p} = 0.$$

In general, if $d \geq 1$, we have obviously L^p convergence of the components of X_n to the components of X. The result then follows thanks to the inequalities (3.32). ∎

As a consequence we have the following result stating that for Gaussian r.v.'s all L^p convergences are equivalent.

Corollary 3.39 *Let $(X_n)_n$ be a sequence of Gaussian d-dimensional r.v.'s converging to an r.v. X in L^p for some $p > 0$. Then the convergence takes place in L^p for every $p > 0$.*

Proof As $(X_n)_n$ converges also in probability, by Theorem 3.9 there exists a subsequence $(X_{n_k})_k$ converging a.s. to X, hence also in L^p for every p by the previous Theorem 3.38. The result then follows by the precious sub-sub-sequence Criterion 3.8. ∎

3.8   The Central Limit Theorem

We now present the most classical result of convergence in law.

> **Theorem 3.40 (The Central Limit Theorem)**   *Let $(X_n)_n$ be a sequence of d-dimensional i.i.d. r.v.'s, with mean b and covariance matrix C and let*
>
> $$S_n^* := \frac{X_1 + \cdots + X_n - nb}{\sqrt{n}} \, .$$
>
> *Then S_n^* converges in law to a Gaussian multivariate $N(0, C)$ distribution.*

Proof The proof boils down to the computation of the limit of the characteristic functions of the r.v.'s S_n^*, and then applying P. Lévy's Theorem 3.20.

If $Y_k = X_k - b$, then the Y_k's are centered, have the same covariance matrix C and $S_n^* = \frac{1}{\sqrt{n}}(Y_1 + \cdots + Y_n)$. Let us denote by ϕ the common characteristic function of the Y_k's. Then, recalling the formulas of the characteristic function of a sum of independent r.v.'s, (2.38), and of their transformation under linear maps, (2.39),

$$\phi_{S_n^*}(\theta) = \phi\left(\frac{\theta}{\sqrt{n}}\right)^n = \left(1 + \left(\phi\left(\frac{\theta}{\sqrt{n}}\right) - 1\right)\right)^n \, .$$

This is a classical 1^∞ form. Let us compute the Taylor expansion to the second order of ϕ at $\theta = 0$: recalling that

$$\phi'(0) = i\mathrm{E}(Y_1) = 0, \qquad \mathrm{Hess}\,\phi(0) = -C \, ,$$

we have

$$\phi(\theta) = 1 - \frac{1}{2}\langle C\theta, \theta\rangle + o(|\theta|^2) \, .$$

Therefore, as $n \to +\infty$,

$$\phi\left(\frac{\theta}{\sqrt{n}}\right) - 1 = -\frac{1}{2n}\langle C\theta, \theta\rangle + o\left(\frac{1}{n}\right)$$

and, as $\log(1 + z) \sim z$ for $z \to 0$,

$$\lim_{n \to \infty} \phi_{S_n^*}(\theta) = \lim_{n \to \infty} \exp\left[n \log\left(1 + \left(\phi\left(\tfrac{\theta}{\sqrt{n}}\right) - 1\right)\right)\right]$$

$$= \lim_{n \to \infty} \exp\left[n\left(\phi\left(\tfrac{\theta}{\sqrt{n}}\right) - 1\right)\right] = \lim_{n \to \infty} \exp\left[n\left(-\frac{1}{2n}\langle C\theta, \theta\rangle + o(\tfrac{1}{n})\right)\right] = \mathrm{e}^{-\frac{1}{2}\langle C\theta, \theta\rangle},$$

which is the characteristic function of an $N(0, C)$ distribution. ∎

Corollary 3.41 *Let $(X_n)_n$ be a sequence of real i.i.d. r.v.'s with mean b and variance σ^2. Then if*

$$S_n^* = \frac{X_1 + \cdots + X_n - nb}{\sigma\sqrt{n}}$$

we have $S_n^* \xrightarrow[n \to \infty]{\mathscr{L}} N(0, 1)$.

The Central Limit Theorem has a long history, made of a streak of increasingly sharper and sophisticated results. The first of these is the De Moivre-Laplace Theorem (1738), which concerns the case where the X_n are Bernoulli r.v.'s, so that the sums S_n are binomial-distributed, and it is elementary (but not especially fun) to directly estimate their d.f. by using Stirling's formula for the factorials,

$$n! = \sqrt{2\pi n}\left(\frac{n}{e}\right)^n + o(n!).$$

The Central Limit Theorem states that, for large n, the law of S_n^* can be approximated by an $N(0, 1)$ law. How large must n be for this to be a reasonable approximation?

In spite of the fact that $n = 30$ (or sometimes $n = 50$) is often claimed to be acceptable, in fact there is no all-purpose rule for n.

Actually, whatever the value of n, if $X_k \sim \text{Gamma}(\tfrac{1}{n}, 1)$, then S_n would be exponential and S_n^* would be far from being Gaussian.

An accepted empirical rule is that we have a good approximation, also for small values of n, if the law of the X_i's is symmetric with respect to its mean: see Exercise 3.27 for an instance of a very good approximation for $n = 12$. In the case of asymmetric distributions it is better to be cautious and require larger values of n. Figures 3.4 and 3.5 give some visual evidence (see Exercise 2.25 for a possible way of "measuring" the symmetry of an r.v.).

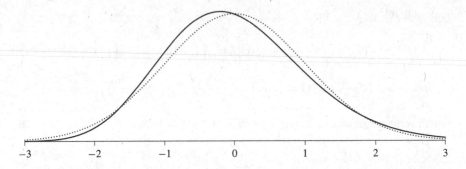

Fig. 3.4 Graph of the density of S_n^* for sums of Gamma($\frac{1}{2}$, 1)-distributed r.v.'s (solid) to be compared with the $N(0, 1)$ density (dots). Here $n = 50$: despite this relatively large value, the two graphs are rather distant

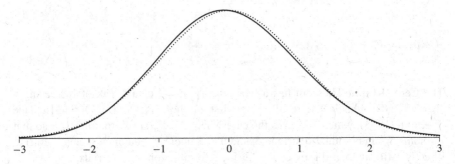

Fig. 3.5 This is the graph of the S_n^* density for sums of Gamma(7, 1)-distributed r.v.'s (solid) compared with the $N(0, 1)$ density (dots). Here $n = 30$. Despite a smaller value of n, we have a much better approximation. The Gamma(7, 1) law is much more symmetric than the Gamma($\frac{1}{2}$, 1): in Exercise 2.25 b) we found that the skewness of these distributions are respectively $2^{3/2}7^{-1/2} = 1.07$ and $2^{3/2} = 2.83$. Note however that the Central Limit Theorem, Theorem 3.40, guarantees weak convergence of the laws, not pointwise convergence of the densities. In this sense there are more refined results (see e.g. [13], Theorem XV.5.2)

3.9 Application: Pearson's Theorem, the χ^2 Test

We now present a classical application of the Central Limit Theorem.

Let $(X_n)_n$ be a sequence of i.i.d. r.v.'s with values in a finite set with cardinality m, which we shall assume to be $\{1, \ldots, m\}$, and let $p_i = P(X_1 = i), i = 1, \ldots, m$. Assume that $p_i > 0$ for every $i = 1, \ldots, m$, and let, for $n > 0, i = 1, \ldots, m$,

$$N_i^{(n)} = 1_{\{X_1 = i\}} + \cdots + 1_{\{X_n = i\}}, \qquad \overline{p}_i^{(n)} = \frac{1}{n} N_i^{(n)}.$$

$\overline{p}_i$ is therefore the proportion, up to time n, of observations $X_1, \ldots, X_n$ that have taken the value i. Of course $\sum_{i=1}^{m} N_i^{(n)} = n$ and $\sum_{i=1}^{m} \overline{p}_i^{(n)} = 1$. Note that by the strong Law of Large Numbers

$$\overline{p}_i^{(n)} = \frac{1}{n} \sum_{k=1}^{n} 1_{\{X_k = i\}} \overset{\text{a.s.}}{\underset{n \to \infty}{\to}} E\left(1_{\{X_k = i\}}\right) = p_i . \tag{3.35}$$

Let, for every n,

$$T_n = \sum_{i=1}^{m} \frac{1}{np_i} (N_i^{(n)} - np_i)^2 = n \sum_{i=1}^{m} \frac{(\overline{p}_i^{(n)} - p_i)^2}{p_i}$$

Pearson's Statistics: The quantity T_n is a measure of the disagreement between the probabilities p and $\overline{p}_i^{(n)}$. Let us keep in mind that, whereas $p \in \mathbb{R}^m$ is a deterministic quantity, the $\overline{p}_i^{(n)}$ form a random vector (they are functions of the observations $X_1, \ldots, X_n$). In the sequel, for simplicity, we shall omit the index $^{(n)}$ and write $N_i, \overline{p}_i$.

Theorem 3.42 (Pearson)

$$T_n \overset{\mathscr{L}}{\underset{n \to \infty}{\to}} \chi^2(m - 1) . \tag{3.36}$$

Proof Let Y_n be the m-dimensional random vector with components

$$Y_{n,i} = \frac{1}{\sqrt{p_i}} 1_{\{X_n = i\}} , \tag{3.37}$$

so that

$$Y_{1,i} + \cdots + Y_{n,i} = \frac{1}{\sqrt{p_i}} N_i = n \frac{\overline{p}_i}{\sqrt{p_i}} . \tag{3.38}$$

Let us denote by N, p and $\sqrt{p}$ the vectors of $\mathbb{R}^m$ having components N_i, p_i and $\sqrt{p_i}, i = 1, \ldots, m$, respectively; therefore the vector $\sqrt{p}$ has modulus $= 1$. Clearly the random vectors Y_n are independent, being functions of independent r.v.'s, and $E(Y_n) = \sqrt{p}$ (recall that $\sqrt{p}$ is a vector).

The covariance matrix $C = (c_{ij})_{ij}$ of Y_n is computed easily from (3.37): keeping in mind that $P(X_n = i, X_n = j) = 0$ if $i \neq j$ and $P(X_n = i, X_n = j) = p_i$ if $i = j$,

$$c_{ij} = \frac{1}{\sqrt{p_i p_j}} \Big(E\big(1_{\{X_n=i\}} 1_{\{X_n=j\}}\big) - E(1_{\{X_n=i\}}) E(1_{\{X_n=j\}}) \Big)$$

$$= \frac{1}{\sqrt{p_i p_j}} \Big(P(X_n = i, X_n = j) - P(X_n = i) P(X_n = j) \Big)$$

$$= \delta_{ij} - \sqrt{p_i p_j} \,,$$

so that, for $x \in \mathbb{R}^d$, $(Cx)_i = x_i - \sqrt{p_i} \sum_{j=1}^m \sqrt{p_j} \, x_j$, i.e.

$$Cx = x - \langle \sqrt{p}, x \rangle \sqrt{p} \,. \tag{3.39}$$

By the Central Limit Theorem the sequence

$$W_n := \frac{1}{\sqrt{n}} \sum_{k=1}^n \big(Y_k - E(Y_k) \big) = \frac{1}{\sqrt{n}} \sum_{k=1}^n \big(Y_k - \sqrt{p} \big)$$

converges in law, as $n \to \infty$, to an $N(0, C)$-distributed r.v., V say. Note however that (recall (3.38)) W_n is a random vector whose i-th component is

$$\frac{1}{\sqrt{n}} \Big(n \frac{\overline{p}_i}{\sqrt{p_i}} - n\sqrt{p_i} \Big) = \sqrt{n} \, \frac{\overline{p}_i - p_i}{\sqrt{p_i}} \,,$$

so that $|W_n|^2 = T_n$ and

$$T_n \xrightarrow[n \to \infty]{\mathscr{L}} |V|^2 \,.$$

Therefore we must just compute the law of $|V|^2$, i.e. the law of the square of the modulus of an $N(0, C)$-distributed r.v. As $|V|^2 = |OV|^2$ for every rotation O and the covariance matrix of OV is O^*CO, we can assume the covariance matrix C to be diagonal, so that

$$|V|^2 = \sum_{k=1}^m \lambda_k V_k^2 \,, \tag{3.40}$$

where $\lambda_1, \ldots, \lambda_m$ are the eigenvalues of the covariance matrix C and the r.v.'s V_k^2 are $\chi^2(1)$-distributed (see Exercise 2.53 for a complete argument).

Let us determine the eigenvalues λ_k. Going back to (3.39) we note that $C\sqrt{p} = 0$, whereas $Cx = x$ for every x that is orthogonal to $\sqrt{p}$. Therefore one of the λ_i is equal to 0 and the $m - 1$ other eigenvalues are equal to 1 (C is the projector on the subspace orthogonal to $\sqrt{p}$, which has dimension $m - 1$).

Hence the law of the r.v. in (3.40) is the sum of $m - 1$ independent $\chi^2(1)$-distributed r.v.'s and has a $\chi^2(m - 1)$ distribution. ∎

Let us look at some applications of Pearson's Theorem. Imagine we have n independent observations $X_1, \ldots, X_n$ of some random quantity taking the possible values $\{1, \ldots, m\}$. Is it possible to check whether their law is given by some vector p, i.e. $P(X_n = i) = p_i$?

For instance imagine that a die has been thrown 2000 times with the following outcomes

$$
\begin{array}{cccccc}
1 & 2 & 3 & 4 & 5 & 6 \\
388 & 322 & 314 & 316 & 344 & 316
\end{array}
\tag{3.41}
$$

can we decide whether the die is a fair one, meaning that the outcome of a throw is uniform on $\{1, 2, 3, 4, 5, 6\}$?

Pearson's Theorem provides a way of checking this hypothesis.

Actually under the hypothesis that $P(X_n = i) = p_i$ the r.v. T_n is approximately $\chi^2(m - 1)$-distributed, whereas if the law of the X_n was given by another vector $q = (q_1, \ldots, q_m)$, $q \neq p$, we would have, as $n \to \infty$, $\overline{p}_i \to_{n \to \infty} q_i$ by the Law of Large Numbers so that, as $\sum_{i=1}^{m} \frac{(q_i - p_i)^2}{p_i} > 0$,

$$
\lim_{n \to \infty} T_n = \lim_{n \to \infty} n \sum_{i=1}^{m} \frac{(q_i - p_i)^2}{p_i} = +\infty .
$$

In other words under the assumption that the observations follow the law given by the vector p, the statistic T_n is asymptotically $\chi^2(m - 1)$-distributed, otherwise T_n will tend to take large values.

Example 3.43 Let us go back to the data (3.41). There are some elements of suspicion: indeed the outcome 1 has appeared more often than the others: the frequencies are

$$
\begin{array}{cccccc}
\overline{p}_1 & \overline{p}_2 & \overline{p}_3 & \overline{p}_4 & \overline{p}_5 & \overline{p}_6 \\
0.196 & 0.161 & 0.157 & 0.158 & 0.172 & 0.158
\end{array}
\tag{3.42}
$$

How do we establish whether this discrepancy is significant (and the die is loaded)? Or are these normal fluctuations and the die is fair?

Under the hypothesis that the die is a fair one, thanks to Pearson's Theorem the (random) quantity

$$T_n = 2000 \times \sum_{i=1}^{6} (\overline{p}_i - \tfrac{1}{6})^2 \times 6 = 12.6$$

is approximately $\chi^2(5)$-distributed, whereas if the hypothesis was not true T_n would have a tendency to take large values. The question hence boils down to the following: can the observed value of T_n be considered a typical value for a $\chi^2(5)$-distributed r.v.? Or is it too large?

We can argue in the following way: let us fix a threshold α ($\alpha = 0.05$, for instance). If we denote by $\chi^2_{1-\alpha}(5)$ the quantile of order $1 - \alpha$ of the $\chi^2(5)$ law, then, for a $\chi^2(5)$-distributed r.v. X, $P(X > \chi^2_{1-\alpha}(5)) = \alpha$. We shall decide to reject the hypothesis that the die is a fair one if the observed value of T_n is larger than $\chi^2_{1-\alpha}(5)$ as, if the die was a fair one, the probability of observing a value exceeding $\chi^2_{1-\alpha}(5)$ would be too small.

Any suitable software can provide the quantiles of the χ^2 distribution and it turns out that $\chi^2_{0.95}(5) = 11.07$. We conclude that the die cannot be considered a fair one. In the language of Mathematical Statistics, Pearson's Theorem allows us to reject the hypothesis that the die is a fair one at the level 5%. The value 12.6 corresponds to the quantile of order 97.26% of the $\chi^2(5)$ law. Hence if the die was a fair one, a value of T_n larger than 12.6 would appear with probability 2.7%.

The data of this example were simulated with probabilities $q_1 = 0.2$, $q_2 = \ldots = q_6 = 0.16$.

Pearson's Theorem is therefore the theoretical foundation of important applications in hypothesis testing in Statistics, when it is required to check whether some data are in agreement with a given theoretical distribution.

However we need to inquire how large n should be in order to assume that T_n has a law close to a $\chi^2(m - 1)$. A practical rule, that we shall not discuss here, requires that $np_i \geq 5$ for every $i = 1, \ldots, m$. In the case of Example 3.43 this requirement is clearly satisfied, as in this case $np_i = \tfrac{1}{6} \times 2000 = 333.33$.

Table 3.1 The Geissler data

k	N_k	p_k	$\overline{p}_k$
0	3	0.000244	0.000491
1	24	0.002930	0.003925
2	104	0.016113	0.017007
3	286	0.053711	0.046770
4	670	0.120850	0.109567
5	1033	0.193359	0.168929
6	1343	0.225586	0.219624
7	1112	0.193359	0.181848
8	829	0.120850	0.135568
9	478	0.053711	0.078168
10	181	0.016113	0.029599
11	45	0.002930	0.007359
12	7	0.000244	0.001145

Example 3.44 At the end of the nineteenth century the German doctor and statistician A. Geissler investigated the problem of modeling the outcome (male or female) of the subsequent births in a family. Geissler collected data on the composition of large families.

The data of Table 3.1 concern 6115 families of 12 children. For every k, $k = 0, 1, \ldots, 12$, it displays the number, N_k, of families having k sons and the corresponding empirical probabilities $\overline{p}_k$. A natural hypothesis is to assume that every birth gives rise to a son or a daughter with probability $\frac{1}{2}$, and moreover that the outcomes of different births are independent. Can we say that this hypothesis is not rejected by the data?

Under this hypothesis, the r.v. X ="number of sons" is distributed according to a binomial $B(12, \frac{1}{2})$ law, i.e. the probability of observing a family with k sons would be

$$p_k = \binom{12}{k}\left(\frac{1}{2}\right)^k\left(1 - \frac{1}{2}\right)^{12-k} = \binom{12}{k}\left(\frac{1}{2}\right)^{12}.$$

Do the observed values $\overline{p}_k$ agree with the p_k? Or are the discrepancies appearing in Table 3.1 significant? This is a typical application of Pearson's Theorem. However, the condition of applicability of Pearson's Theorem is not satisfied, as for $i = 0$ or $i = 12$ we have $p_i = 2^{-12}$ and

$$np_0 = np_{12} = 6115 \cdot 2^{-12} = 1.49 \,,$$

which is smaller than 5 and therefore not large enough to apply Pearson's approximation. This difficulty can be overcome with the trick of merging classes: let us consider a new r.v. Y defined as

$$Y = \begin{cases} 1 & \text{if } X = 0 \text{ or } 1 \\ k & \text{if } X = k \text{ for } k = 2, \ldots, 10 \\ 11 & \text{if } X = 11 \text{ or } 12 . \end{cases}$$

In other words Y coincides with X if $X = 1, \ldots, 11$ and takes the value 1 also on $\{X = 0\}$ and 11 also on $\{X = 12\}$. Clearly the law of Y is

$$P(Y = k) = q_k := \begin{cases} p_0 + p_1 & \text{if } k = 1 \\ p_k & \text{if } k = 2, \ldots, 10 \\ p_{11} + p_{12} & \text{if } k = 11 . \end{cases}$$

It is clear now that if we *group* together the observations of the classes 0 and 1 and of the classes 11 and 12, under the hypothesis (i.e. that the number of sons in a family follows a binomial law) the new empirical distributions thus obtained should follow the same distribution as Y. In other words, we shall compare, using Pearson's Theorem, the distributions

k	q_k	$\overline{q}_k$
1	0.003174	0.004415
2	0.016113	0.017007
3	0.053711	0.046770
4	0.120850	0.109567
5	0.193359	0.168929
6	0.225586	0.219624
7	0.193359	0.181848
8	0.120850	0.135568
9	0.053711	0.078168
10	0.016113	0.029599
11	0.003174	0.008504

where the $\overline{q}_k$ are obtained by grouping the empirical distributions: $\overline{q}_1 = \overline{p}_0 + \overline{p}_1$, $\overline{q}_k = \overline{p}_k$ for $k = 2, \ldots, 10$, $\overline{q}_{11} = \overline{p}_{11} + \overline{p}_{12}$. Now the products nq_1 and nq_{11} are equal to $6115 \cdot 0.003174 = 19.41 > 5$ and Pearson's approximation is applicable. The numerical computation now gives

$$T = 6115 \cdot \sum_{i=1}^{11} \frac{(\overline{q}_i - q_i)^2}{q_i} = 242.05 ,$$

which is much larger than the usual quantiles of the $\chi^2(10)$ distribution, as $\chi_{0.95}(10) = 18.3$. The hypothesis that the data follow a $B(12, \frac{1}{2})$ distribution is therefore rejected with strong evidence.

By the way, some suspicion in this direction should already have been raised by the histogram comparing expected and empirical values, provided in Fig. 3.6.

Indeed, rather than large discrepancies between expected and empirical values, the suspicious feature is that the empirical values exceed the expected ones for extreme values (0, 1, 2 and 8, 9, 10, 11, 12) but are smaller for central values. If the differences were ascribable to random fluctuations (as opposed to inadequacy of the model) a greater irregularity in the differences would be expected.

The model suggested so far, with the assumption of

- independence of the outcomes of different births and
- equiprobability of daughter/son,

must therefore be rejected.

This confronts us with the problem of finding a more adequate model. What can we do?

A first, simple, idea is to change the assumption of equiprobability of daughter/son at birth. But this is not likely to improve the adequacy of the model. Actually, for values of p larger than $\frac{1}{2}$ we can expect an increase of the values q_k for k close to 11, but also, at the other extreme, a decrease for those that are close to 1. And the other way round if we choose $p < \frac{1}{2}$.

By the way, there is some literature concerning the construction of a reasonable model for Geissler's data. We shall come back to these data later in Example 4.18 where we shall try to put together a more successful model.

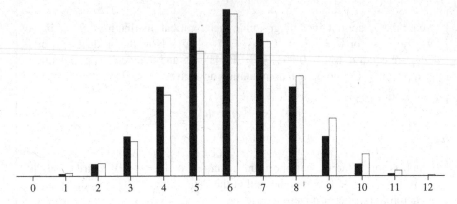

Fig. 3.6 The white bars are for the empirical values $\overline{p}_k$, the black ones for the expected values p_k

3.10 Some Useful Complements

Let us consider some transformations that preserve the convergence in law. A first result of this type has already appeared in Remark 3.16.

Lemma 3.45 (Slutsky's Lemma) *Let $Z_n, U_n, n \geq 1$, be respectively $\mathbb{R}^d$- and $\mathbb{R}^m$-valued r.v.'s on some probability space $(\Omega, \mathcal{F}, P)$ and let us assume that $Z_n \xrightarrow[n \to \infty]{\mathscr{L}} Z$, $U_n \xrightarrow[n \to \infty]{\mathscr{L}} U$ where U is a constant r.v. taking the value $u_0 \in \mathbb{R}^m$ with probability 1. Then*

(a) $(Z_n, U_n) \xrightarrow[n \to \infty]{\mathscr{L}} (Z, u_0)$.

(b) *If $\Phi : \mathbb{R}^m \times \mathbb{R}^d \to \mathbb{R}^l$ is a continuous map then $\Phi(Z_n, U_n) \xrightarrow[n \to \infty]{\mathscr{L}} \Phi(Z, u_0)$. In particular*

(b1) *if $d = m$ then $Z_n + U_n \xrightarrow[n \to \infty]{\mathscr{L}} Z + u_0$;*

(b2) *if $m = 1$ (i.e. the sequence $(U_n)_n$ is real-valued) then $Z_n U_n \xrightarrow[n \to \infty]{\mathscr{L}} Zu_0$.*

Proof (a) If $\xi \in \mathbb{R}^d$, $\theta \in \mathbb{R}^m$, then the characteristic function of (Z_n, U_n) computed at $(\xi, \theta) \in \mathbb{R}^{d+m}$ is

$$\mathrm{E}\big(e^{i\langle \xi, Z_n \rangle} e^{i\langle \theta, U_n \rangle}\big) = \mathrm{E}\big(e^{i\langle \xi, Z_n \rangle} e^{\langle \theta, u_0 \rangle}\big) + \mathrm{E}\big[e^{i\langle \xi, Z_n \rangle} \big(e^{i\langle \theta, U_n \rangle} - e^{i\langle \theta, u_0 \rangle}\big)\big].$$

The first term on the right-hand side converges to $E(e^{i\langle \xi, Z \rangle} e^{i\langle \theta, u_0 \rangle})$; it will therefore be sufficient to prove that the other term tends to 0. Indeed

$$\left| E[e^{i\langle \xi, Z_n \rangle}(e^{i\langle \theta, U_n \rangle} - e^{i\langle \theta, u_0 \rangle})] \right| \le E\left[|e^{i\langle \xi, Z_n \rangle}(e^{i\langle \theta, U_n \rangle} - e^{i\langle \theta, u_0 \rangle})| \right]$$

$$= E\left(|e^{i\langle \theta, U_n \rangle} - e^{i\langle \theta, u_0 \rangle}| \right) = E[f(U_n)],$$

where $f(x) = |e^{i\langle \theta, x \rangle} - e^{i\langle \theta, u_0 \rangle}|$; we have $E[f(U_n)] \to_{n \to \infty} E[f(U)] = f(u_0) = 0$, as f is a bounded continuous function.

(b) Follows from (a) and Remark 3.28. ∎

Note that in Slutsky's Lemma no assumption of independence between the Z_n's and the U_n's is made. This makes it a very useful tool, as highlighted in the next example.

Example 3.46 Let us go back to the situation of Pearson's Theorem 3.42 and recall the definition of relative entropy (or Kullback-Leibler divergence) between the common distribution of the r.v.'s and their empirical distribution $\overline{p}_n$ (Exercise 2.24)

$$H(\overline{p}_n; p) = \sum_{i=1}^{m} \left(\frac{\overline{p}_n(i)}{p_i} \log \frac{\overline{p}_n(i)}{p_i} \right) p_i.$$

Recall that relative entropy is also a measure of the discrepancy between probabilities and note first that, as by the Law of Large Numbers (see relation (3.35)) $\overline{p}_n \to p$, we have $\overline{p}_n(i)/p_i \to_{n \to \infty} 1$ for every i and therefore $H(\overline{p}_n; p) \to_{n \to \infty} 0$, since the function $x \mapsto x \log x$ vanishes at 1.

What can be said of the limit $n\, H(\overline{p}_n; p) \xrightarrow{\mathscr{L}}_{n \to \infty}$? It turns out that Pearson's statistics T_n is closely related to relative entropy.

The Taylor expansion of $x \mapsto x \log x$ at $x_0 = 1$ gives

$$x \log x = (x - 1) + \frac{1}{2}(x - 1)^2 - \frac{1}{6\xi^2}(x - 1)^3,$$

where ξ is a number between x and 1. Therefore

$$n\, H(\overline{p}_n; p)$$

$$= n \sum_{i=1}^{m} \left(\frac{\overline{p}_n(i)}{p_i} - 1 \right) p_i + \frac{1}{2} n \sum_{i=1}^{m} \left(\frac{\overline{p}_n(i)}{p_i} - 1 \right)^2 p_i - n \sum_{i=1}^{m} \frac{1}{6\xi_{i,n}^2} \left(\frac{\overline{p}_n(i)}{p_i} - 1 \right)^3 p_i$$

$$= I_1 + I_2 + I_3.$$

Of course $I_1 = 0$ for every n as

$$\sum_{i=1}^{m} \left(\frac{\overline{p}_n(i)}{p_i} - 1 \right) p_i = \sum_{i=1}^{m} \overline{p}_n(i) - \sum_{i=1}^{m} p_i = 1 - 1 = 0 \,.$$

By Pearson's Theorem,

$$2I_2 = n \sum_{i=1}^{m} \frac{(\overline{p}_n(i) - p_i)^2}{p_i} = T_n \xrightarrow[n \to \infty]{\mathscr{L}} \chi^2(m-1) \,.$$

Finally

$$|I_3| \le n \sum_{i=1}^{m} \left(\frac{\overline{p}_n(i)}{p_i} - 1 \right)^2 p_i \times \max_{i=1,\dots,m} \left(\frac{1}{6\xi_{i,n}^2} \left| \frac{\overline{p}_n(i)}{p_i} - 1 \right| \right)$$

$$= T_n \times \max_{i=1,\dots,m} \left(\frac{1}{6\xi_{i,n}^2} \left| \frac{\overline{p}_n(i)}{p_i} - 1 \right| \right) \,.$$

As mentioned above, by the Law of Large Numbers $\frac{\overline{p}_n(i)}{p_i} \to_{n \to \infty} 1$ a.s. for every $i = 1, \dots, m$ hence also $\xi_{i,n}^2 \to_{n \to \infty} 1$ a.s. ($\xi_{i,n}$ is a number between $\frac{\overline{p}_n(i)}{p_i}$ and 1), so that $|I_3|$ turns out to be the product of a term converging in law to a $\chi^2(m-1)$ distribution and a term converging to 0. By Slutsky's Lemma therefore $I_3 \to_{n \to \infty}^{\mathscr{L}} 0$ and, by Slutsky again,

$$n \times 2H(\overline{p}_n; p) \xrightarrow[n \to \infty]{\mathscr{L}} \chi^2(m-1) \,.$$

In some sense Pearson's statistics T_n is the first order term in the expansion of the relative entropy H around p multiplied by 2 (see Fig. 3.7).

Another useful application of Slutsky's Lemma is the following.

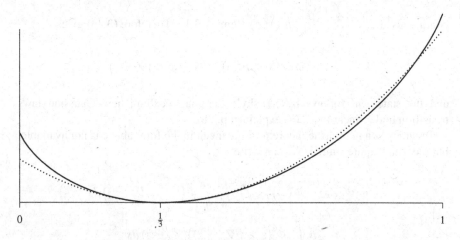

Fig. 3.7 Comparison between the graphs, as a function of q, of the relative entropy of a Bernoulli $B(q, 1)$ distribution with respect to a $B(p, 1)$ with $p = \frac{1}{3}$ multiplied by 2 and of the corresponding Pearson's statistics (dots)

Theorem 3.47 (The Delta Method) *Let $(Z_n)_n$ be a sequence of $\mathbb{R}^d$-valued r.v.'s, such that*

$$\sqrt{n}\,(Z_n - z) \xrightarrow[n \to \infty]{\mathcal{L}} Z \sim N(0, C)\,.$$

Let $\Phi : \mathbb{R}^d \to \mathbb{R}^m$ be a differentiable map with a continuous derivative at z. Then

$$\sqrt{n}\,\big(\Phi(Z_n) - \Phi(z)\big) \xrightarrow[n \to \infty]{\mathcal{L}} N\big(0, \Phi'(z)\, C\, \Phi'(z)^*\big)\,.$$

Proof Thanks to Slutski's Lemma 3.45(b), we have

$$Z_n - z = \frac{1}{\sqrt{n}} \times \sqrt{n}\,(Z_n - z) \xrightarrow[n \to \infty]{\mathcal{L}} 0 \cdot Z = 0\,.$$

Hence, by Proposition 3.29(b), $Z_n \xrightarrow[n \to \infty]{P} z$. Let us first prove the statement for $m = 1$, so that Φ is real-valued. By the theorem of the mean, we can write

$$\sqrt{n}\,\big(\Phi(Z_n) - \Phi(z)\big) = \sqrt{n}\,\Phi'(\widetilde{Z}_n)(Z_n - z)\,, \qquad (3.43)$$

where $\widetilde{Z}_n$ is a (random) vector in the segment between z and Z_n so that $|\widetilde{Z}_n - z| \le |Z_n - z|$. It follows that $|\widetilde{Z}_n - z| \to_{n \to \infty} 0$ in probability and in law. Since Φ' is continuous at z, $\Phi'(\widetilde{Z}_n) \overset{\mathscr{L}}{\underset{n\to\infty}{\to}} \Phi'(z)$ by Remark 3.16. Therefore (3.43) gives

$$\sqrt{n}\left(\Phi(Z_n) - \Phi(z)\right) \overset{\mathscr{L}}{\underset{n\to\infty}{\to}} \Phi'(z)\, Z$$

and the statement follows by Slutsky's Lemma, recalling how Gaussian laws transform under linear maps (as explained p. 88).

In dimension $m > 1$ the theorem of the mean in the form above is not available, but the idea is quite similar. We can write

$$\sqrt{n}\left(\Phi(Z_n) - \Phi(z)\right) = \sqrt{n} \int_0^1 \frac{d}{ds} \Phi\left(z + s(Z_n - z)\right) ds$$

$$= \sqrt{n} \int_0^1 \Phi'\left(z + s(Z_n - z)\right)(Z_n - z)\, ds$$

$$= \sqrt{n}\, \Phi'(z)(Z_n - z) + \underbrace{\sqrt{n} \int_0^1 \left(\Phi'(z + s(Z_n - z)) - \Phi'(z)\right)(Z_n - z)\, ds}_{:=I_n}.$$

We have

$$\sqrt{n}\, \Phi'(z)(Z_n - z) \overset{\mathscr{L}}{\underset{n\to\infty}{\to}} N\left(0, \Phi'(z)\, C\, \Phi'(z)^*\right),$$

so that, by Slutsky's lemma, the proof is complete if we prove that $I_n \to_{n\to\infty} 0$ in probability. We have

$$|I_n| \leq |\sqrt{n}(Z_n - z)| \times \sup_{0 \leq s \leq 1} \left|\Phi'(z + s(Z_n - z)) - \Phi'(z)\right|.$$

Now $|\sqrt{n}(Z_n - z)| \to |Z|$ in law and the result will follow from Slutsky's lemma again if we can show that

$$\sup_{0 \leq s \leq 1} \left|\Phi'(z + s(Z_n - z)) - \Phi'(z)\right| \overset{\mathscr{L}}{\underset{n\to\infty}{\to}} 0.$$

Let $\varepsilon > 0$. As Φ' is assumed to be continuous at z, let $\delta > 0$ be such that $\left|\Phi'(z + x) - \Phi'(z)\right| \leq \varepsilon$ whenever $|x| \leq \delta$. Then we have

$$P\left(\sup_{0 \leq s \leq 1} \left|\Phi'(z + s(Z_n - z)) - \Phi'(z)\right| > \varepsilon \right)$$

$$= P\Big(\underbrace{\sup_{0 \leq s \leq 1} \left|\Phi'(z + s(Z_n - z)) - \Phi'(z)\right| > \varepsilon, |Z_n - z| \leq \delta}_{=\emptyset} \Big)$$

$$+P\Big(\sup_{0\le s\le 1}\big|\Phi'(z+s(Z_n-z))-\Phi'(z)\big|>\varepsilon,\,|Z_n-z|>\delta\Big)$$

$$\le P\big(|Z_n-z|>\delta\big)$$

so that

$$\varlimsup_{n\to\infty} P\Big(\sup_{0\le s\le 1}\big|\Phi'(z+s(Z_n-z))-\Phi'(z)\big|>\varepsilon\Big)\le \lim_{n\to\infty} P\big(|Z_n-z|>\delta\big)=0.$$

∎

Exercises

3.1 (p. 317) Let $(X_n)_n$ be a sequence of real r.v.'s converging to X in L^p, $p\ge 1$.

(a) Prove that

$$\lim_{n\to\infty} E(X_n)=E(X).$$

(b) Prove that if two sequences $(X_n)_n$, $(Y_n)_n$, defined on the same probability space, converge in L^2 to X and Y respectively, then the product sequence $(X_nY_n)_n$ converges to XY in L^1.

(c1) Prove that if $X_n\to_{n\to\infty} X$ in L^2 then also

$$\lim_{n\to\infty} \mathrm{Var}(X_n)=\mathrm{Var}(X).$$

(c2) Prove that if $(X_n)_n$ and X are $\mathbb{R}^d$-valued and $X_n\to_{n\to\infty} X$ in L^2, then the covariance matrices converge.

3.2 (p. 317) Let $(X_n)_n$ be a sequence of real r.v.'s on $(\Omega,\mathscr{F},P)$ and δ a real number. Which of the following is true?

(a)

$$\varlimsup_{n\to\infty}\{X_n\ge\delta\}=\Big\{\varlimsup_{n\to\infty} X_n\ge\delta\Big\}.$$

(b)

$$\varlimsup_{n\to\infty}\{X_n<\delta\}\subset\Big\{\varlimsup_{n\to\infty} X_n\le\delta\Big\}.$$

3.3 (p. 317)

(a) Let X be an r.v. uniform on $[0, 1]$ and let

$$A_n = \{X \le \tfrac{1}{n}\} .$$

(a1) Compute $\sum_{n=1}^{\infty} P(A_n)$.
(a2) Compute $P(\lim_{n \to \infty} A_n)$.
(b) Let $(X_n)_n$ be a sequence of *independent* r.v.'s uniform on $[0, 1]$.
(b1) Let

$$B_n = \{X_n \le \tfrac{1}{n}\} .$$

Compute $P(\overline{\lim}_{n \to \infty} B_n)$.
(b2) And if

$$B_n = \{X_n \le \tfrac{1}{n^2}\} ?$$

3.4 (p. 318) Let $(X_n)_n$ be a sequence of independent r.v.'s having exponential law respectively of parameter $a_n = (\log(n + 1))^{\alpha}$, $\alpha > 0$. Note that the sequence $(a_n)_n$ is increasing so that the r.v.'s X_n "become smaller" as n increases.

(a) Determine $P(\overline{\lim}_{n \to \infty} \{X_n \ge 1\})$ according to the value of α.
(b1) Compute $\overline{\lim}_{n \to \infty} X_n$ according to the value of α.
(b2) Compute $\underline{\lim}_{n \to \infty} X_n$ according to the value of α.
(c) For which values of α (recall that $\alpha > 0$) does the sequence $(X_n)_n$ converge a.s.?

3.5 (p. 319) (Recall Remark 2.1) Let $(Z_n)_n$ be a sequence of i.i.d. positive r.v.'s.

(a) Prove the inequalities

$$\sum_{n=1}^{\infty} P(Z_1 \ge n) \le E(Z_1) \le \sum_{n=0}^{\infty} P(Z_1 \ge n) .$$

(b) Prove that
(b1) if $E(Z_1) < +\infty$ then $P(Z_n \ge n$ infinitely many times$) = 0$;
(b2) if $E(Z_1) = +\infty$ then $Z_n \ge n$ infinitely many times with probability 1.
(c) Let $(X_n)_n$ be a sequence of i.i.d. real r.v.'s and let

$$x_2 = \sup\{\theta; E(e^{\theta X_n}) < +\infty\}$$

be the right convergence abscissa of the Laplace transform of the X_n.

(c1) Prove that if $x_2 < +\infty$ then

$$\overline{\lim_{n\to\infty}} \frac{X_n}{\log n} = \frac{1}{x_2}$$

with the understanding $\frac{1}{x_2} = +\infty$ if $x_2 = 0$.

(c2) Assume that $X_n \sim N(0, 1)$. Compute

$$\overline{\lim_{n\to\infty}} \frac{|X_n|}{\sqrt{\log n}}.$$

3.6 (p. 320) Let $(X_n)_n$ be a sequence of i.i.d. r.v.'s such that $0 < \mathrm{E}(|X_1|) < +\infty$. For every $\omega \in \Omega$ let us consider the power series

$$\sum_{n=1}^{\infty} X_n(\omega) x^n$$

and let $R(\omega) = \left(\overline{\lim}_{n\to\infty} |X_n(\omega)|^{1/n}\right)^{-1}$ be its radius of convergence.

(a) Prove that R is an a.s. constant r.v.

(b) Prove that there exists an $a > 0$ such that

$$P(|X_n| \geq a \text{ for infinitely many indices } n) = 1$$

and deduce that $R \leq 1$ a.s.

(c) Let $b > 1$. Prove that $\sum_{n=1}^{\infty} P(|X_n| \geq b^n) < +\infty$ and deduce the value of R a.s.

3.7 (p. 321) Let $(X_n)_n$ be a sequence of r.v.'s with values in the metric space E. Prove that $\lim_{n\to\infty} X_n = X$ in probability if and only if

$$\lim_{n\to\infty} \mathrm{E}\left[\frac{d(X_n, X)}{1 + d(X_n, X)}\right] = 0 . \tag{3.44}$$

Beware, sub-sub-sequences…

3.8 (p. 322) Let $(X_n)_n$ be a sequence of r.v.'s on the probability space $(\Omega, \mathscr{F}, P)$ such that

$$\sum_{k=1}^{\infty} \mathrm{E}(|X_k|) < +\infty . \tag{3.45}$$

(a) Prove that the series

$$\sum_{k=1}^{\infty} X_k \tag{3.46}$$

converges in L^1.

(b1) Prove that the series $\sum_{k=1}^{\infty} X_k^+$ converges a.s.

(b2) Prove that in (3.46) convergence also takes place a.s.

3.9 (p. 322) (Lebesgue's Theorem for convergence in probability) If $(X_n)_n$ is a sequence of r.v.'s that is bounded in absolute value by an integrable r.v. Z and such that $X_n \to_{n\to\infty}^P X$, then

$$\lim_{n\to\infty} E(X_n) = E(X) .$$

Sub-sub-sequences...

3.10 (p. 323) Let $(X_n)_n$ be a sequence of i.i.d. Gamma(1, 1)-distributed (i.e. exponential of parameter 1) r.v.'s and

$$U_n = \min(X_1, \ldots, X_n) .$$

(a1) What is the law of U_n?

(a2) Prove that $(U_n)_n$ converges in law and determine the limit law.

(b) Does the convergence also take place a.s.?

(c) Let, for $\alpha > 1$, $V_n = U_n^\alpha$. Let $1 < \beta < \alpha$. Compute $P(V_n \geq \frac{1}{n^\beta})$ and prove that the series

$$\sum_{n=1}^{\infty} V_n$$

converges a.s.

3.11 (p. 323) Let $(X_n)_n$ be a sequence of i.i.d. square integrable centered r.v.'s with common variance σ^2.

(a1) Does the r.v. $X_1 X_2$ have finite mathematical expectation? Finite variance? In the affirmative, what are their values?

(a2) If $Y_n := X_n X_{n+1}$, what is the value of $Cov(Y_k, Y_m)$ for $k \neq m$?

(b) Does the sequence

$$\frac{1}{n} \left(X_1 X_2 + X_2 X_3 + \cdots + X_n X_{n+1} \right)$$

converge a.s.? If yes, to which limit?

3.12 (p. 324) Let $(X_n)_n$ be a sequence of i.i.d. r.v.'s having a Laplace law of parameter λ. Discuss the a.s. convergence of the sequences

$$\frac{1}{n}\left(X_1^4 + X_2^4 + \cdots + X_n^4\right), \qquad \frac{X_1^2 + X_2^2 + \cdots + X_n^2}{X_1^4 + X_2^4 + \cdots + X_n^4}.$$

3.13 (p. 324) (Estimation of the variance) Let $(X_n)_n$ be a sequence of square integrable real i.i.d. r.v.'s with variance σ^2 and let

$$S_n^2 = \frac{1}{n}\sum_{k=1}^{n}(X_k - \overline{X}_n)^2,$$

where $\overline{X}_n = \frac{1}{n}\sum_{k=1}^{n} X_k$ are the empirical means.

(a) Prove that $(S_n^2)_n$ converges a.s. to a limit to be determined.
(b) Compute $E(S_n^2)$.

3.14 (p. 325) Let $(\mu_n)_n$, $(\nu_n)_n$ be sequences of probabilities on $\mathbb{R}^d$ and $\mathbb{R}^m$ respectively converging weakly to the probabilities μ and ν respectively.

(a) Prove that, weakly,

$$\lim_{n\to\infty} \mu_n \otimes \nu_n = \mu \otimes \nu. \qquad (3.47)$$

(b1) Prove that if $d = m$ then, weakly,

$$\lim_{n\to\infty} \mu_n * \nu_n = \mu * \nu.$$

(b2) If ν_n denotes an $N(0, \frac{1}{n}I)$ probability, prove that $\mu * \nu_n \to_{n\to\infty} \mu$ weakly.

3.15 (p. 326) (First have a look at Exercise 2.5)

(a) Let $f : \mathbb{R}^d \to \mathbb{R}$ be a differentiable function with bounded derivatives and μ a probability on $\mathbb{R}^d$. Prove that the function

$$\mu * f(x) := \int_{\mathbb{R}^d} f(x - y)\, d\mu(y)$$

is differentiable.
(b1) Let g_n be the density of a d-dimensional $N(0, \frac{1}{n}I)$ law. Prove that its derivatives of order α are of the form $P_\alpha(x)\, e^{-n|x|^2/2}$, where P_α is a polynomial, and that they are therefore bounded.
(b2) Prove that there exists a sequence $(f_n)_n$ of C^∞ probability densities on $\mathbb{R}^d$ such that, if $d\mu_n := f_n\, dx$ then $\mu_n \to_{n\to\infty} \mu$ weakly.

3.16 (p. 327) Let $(E, \mathscr{B}(E))$ be a topological space and ρ a σ-finite measure on $\mathscr{B}(E)$. Let f_n, $n \geq 1$, be densities with respect to ρ and let $d\mu_n = f_n \, d\rho$ be the probability on $(E, \mathscr{B}(E))$ having density f_n with respect to ρ.

(a) Assume that $f_n \to_{n\to\infty} f$ in $L^1(\rho)$.

(a1) Prove that f is itself a density.

(a2) Prove that, if $d\mu = f \, d\rho$, then $\mu_n \to_{n\to\infty} \mu$ weakly and moreover that, for every $A \in \mathscr{B}(E)$, $\mu_n(A) \to_{n\to\infty} \mu(A)$.

(b) On $(\mathbb{R}, \mathscr{B}(\mathbb{R}))$ let

$$f_n(x) = \begin{cases} 1 + \cos(2n\pi x) & \text{if } 0 \leq x \leq 1 \\ 0 & \text{otherwise.} \end{cases}$$

(b1) Prove that the f_n's are probability densities with respect to the Lebesgue measure of $\mathbb{R}$.

(b2) Prove that the probabilities $d\mu_n(x) = f_n(x) \, dx$ converge weakly to a probability μ having a density f to be determined.

(b3) Prove that the sequence $(f_n)_n$ *does not* converge to f in L^1 (with respect to the Lebesgue measure).

3.17 (p. 328) Let $(E, \mathscr{B}(E))$ be a topological space and μ_n, μ probabilities on it. We know (this is (3.19)) that if $\mu_n \to_{n\to\infty} \mu$ weakly then

$$\lim_{n\to\infty} \mu_n(G) \geq \mu(G) \quad \text{for every open set } G \subset E . \tag{3.48}$$

Prove the converse, i.e. that, if (3.48) holds, then $\mu_n \to_{n\to\infty} \mu$ weakly.
Recall Remark 2.1. Of course a similar criterion holds with closed sets.

3.18 (p. 329) Let $(X_n)_n$ be a sequence of r.v.'s (no assumption of independence) with $X_n \sim \chi^2(n)$, $n \geq 1$. What is the behavior of the sequence $(\frac{1}{n} X_n)_n$? Does it converge in law? In probability?

3.19 (p. 330) Let $(X_n)_n$ be a sequence of r.v.'s having respectively a geometric law of parameter $p_n = \frac{\lambda}{n}$. Show that the sequence $(\frac{1}{n} X_n)_n$ converges in law and determine its limit.

3.20 (p. 331) Let $(X_n)_n$ be a sequence of real independent r.v.'s having respectively density, with respect to the Lebesgue measure, $f_n(x) = 0$ for $x < 0$ and

$$f_n(x) = \frac{n}{(1 + nx)^2} \quad \text{for } x > 0 .$$

(a) Investigate the convergence in law and in probability of $(X_n)_n$.

(b) Prove that $(X_n)_n$ does not converge a.s. and compute $\overline{\lim}$ and $\underline{\lim}$ of $(X_n)_n$.

3.21 (p. 331) Let $(X_n)_n$ be a sequence of i.i.d. r.v.'s uniform on $[0, 1]$ and let

$$Z_n = \min(X_1, \ldots, X_n) .$$

(a) Does the sequence $(Z_n)_n$ converge in law as $n \to \infty$? In probability? A.s.?
(b) Prove that the sequence $(n Z_n)_n$ converges in law as $n \to \infty$ and determine the limit law. Give an approximation of the probability

$$P\left(\min(X_1, \ldots, X_n) \le \tfrac{2}{n} \right)$$

for n large.

3.22 (p. 332) Let, for every $n \ge 1$, $U_1^{(n)}, \ldots, U_n^{(n)}$ be i.i.d. r.v.'s uniform on $\{0, 1, \ldots, n\}$ respectively and

$$M_n = \min_{k \le n} U_k^{(n)} .$$

Prove that $(M_n)_n$ converges in law and determine the limit law.

3.23 (p. 332)

(a) Let μ_n be the probability on $\mathbb{R}$

$$\mu_n = (1 - a_n)\delta_0 + a_n\delta_n$$

where $0 \le a_n \le 1$. Prove that if $\lim_{n\to\infty} a_n = 0$ then $(\mu_n)_n$ converges weakly and compute its limit.
(b) Construct an example of a sequence $(\mu_n)_n$ converging weakly but such that the means or the variances of the μ_n do not converge to the mean and the variance of the limit (see however Exercise 3.30 below).
(c) Prove that, in general, if $X_n \to_{n\to\infty} X$ in law then $\underline{\lim}_{n\to\infty} E(|X_n|) \ge E(|X|)$ and $\underline{\lim}_{n\to\infty} E(X_n^2) \ge E(X^2)$.

3.24 (p. 333) Let $(X_n)_n$ be a sequence of r.v.'s with $X_n \sim \text{Gamma}(1, \lambda_n)$ with $\lambda_n \to_{n\to\infty} 0$.

(a) Prove that $(X_n)_n$ *does not* converge in law.
(b) Let $Y_n = X_n - \lfloor X_n \rfloor$. Prove that $(Y_n)_n$ converges in law and determine its limit ($\lfloor \ \rfloor$ = the integer part function).

3.25 (p. 334) Let $(X_n)_n$ be a sequence of $\mathbb{R}^d$-valued r.v.'s. Prove that $X_n \to^{\mathscr{L}}_{n\to\infty} X$ if and only if, for every $\theta \in \mathbb{R}^d$, $\langle \theta, X_n \rangle \to^{\mathscr{L}}_{n\to\infty} \langle \theta, X \rangle$.

3.26 (p. 334) Let $(X_n)_n$ be a sequence of i.i.d. r.v.'s with mean 0 and variance σ^2.
Prove that the sequence

$$Z_n = \frac{(X_1 + \cdots + X_n)^2}{n}$$

converges in law and determine the limit law.

3.27 (p. 334) In the FORTRAN libraries in use in the 1970s (but also nowadays...),
in order to generate an $N(0, 1)$-distributed random number the following procedure
was implemented. If $X_1, \ldots, X_{12}$ are independent r.v.'s uniform on $[0, 1]$, then the
number

$$W = X_1 + \cdots + X_{12} - 6 \tag{3.49}$$

is (approximately) $N(0, 1)$-distributed.

(a) Can you give a justification of this procedure?
(b) Let $Z \sim N(0, 1)$. What is the value of $E(Z^4)$? And of $E(W^4)$? What do you
    think of this procedure?

3.28 (p. 336) Let $(\Omega, \mathscr{F}, P)$ be a probability space.

(a) Let $(A_n)_n \subset \mathscr{F}$ be a sequence of events and assume that, for some $\alpha > 0$,
    $P(A_n) \geq \alpha$ for infinitely many indices n. Prove that

$$P\left(\varlimsup_{n \to \infty} A_n \right) \geq \alpha .$$

(b) Let Q be another probability on $(\Omega, \mathscr{F})$ such that $Q \ll P$. Prove that, for every
    $\varepsilon > 0$ there exists a $\delta > 0$ such that, for every $A \in \mathscr{F}$, if $P(A) \leq \delta$ then
    $Q(A) \leq \varepsilon$.

3.29 (p. 337) Let $(X_n)_n$ be a sequence of m-dimensional r.v.'s converging a.s. to an
r.v. X. Assume that $(X_n)_n$ is bounded in L^r for some $r > 1$ and let M be an upper
bound for the L^r norms of the X_n.

(a) Prove that $X \in L^r$.
(b) Prove that, for every $p < r$, $X_n \to_{n \to \infty} X$ in L^p. What if we assumed
    $X_n \to_{n \to \infty} X$ in probability instead of a.s.?

3.30 (p. 337) Let $(X_n)_n$ be a sequence of real r.v.'s converging in law to an r.v. X.
In general convergence in law does not imply convergence of the means, as the
function $x \mapsto x$ is not bounded and Exercise 3.23 provides some examples. But if
we add the assumption of uniform integrability...

(a) Let $\psi_R(x) := x \left(d(x, [-(R+1), R+1]^c) \wedge 1 \right)$; ψ_R is a continuous function
    that coincides with $x \mapsto x$ on $[-R, R]$ and vanishes outside the interval

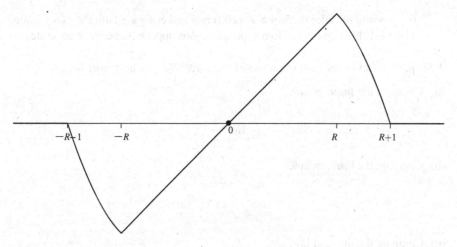

Fig. 3.8 The graph of ψ_R

$[-(R+1), R+1]$ (see Fig. 3.8). Prove that, $\mathrm{E}[\psi_R(X_n)] \to_{n\to\infty} \mathrm{E}[\psi_R(X_n)]$, for every $R > 0$.

(b) Prove that if, in addition, the sequence $(X_n)_n$ is uniformly integrable then X is integrable and $\mathrm{E}(X_n) \to_{n\to\infty} \mathrm{E}(X)$.

• In particular, if $(X_n)_n$ is bounded in L^p $p > 1$, then $\mathrm{E}(X_n) \to_{n\to\infty} \mathrm{E}(X)$.

3.31 (p. 338) In this exercise we see two approximations of the d.f. of a $\chi^2(n)$ distribution for large n using the Central Limit Theorem, the first one naive, the other more sophisticated.

(a) Prove that if $X_n \sim \chi^2(n)$ then

$$\frac{X_n - n}{\sqrt{2n}} \xrightarrow[n\to\infty]{\mathcal{L}} N(0, 1) .$$

(b1) Prove that

$$\lim_{n\to\infty} \frac{\sqrt{2n}}{\sqrt{2X_n} + \sqrt{2n}} = \frac{1}{2} \quad \text{a.s.}$$

(b2) (Fisher's approximation) Prove that

$$\sqrt{2X_n} - \sqrt{2n-1} \xrightarrow[n\to\infty]{\mathcal{L}} N(0, 1) . \tag{3.50}$$

(c) Derive first from (a) and then from (b) an approximation of the d.f. of the $\chi^2(n)$ laws for n large. Use them in order to obtain approximate values of the quantile of order 0.95 of a $\chi^2(100)$ law and compare with the exact value 124.34. Which one of the two approximations appears to be more accurate?

3.32 (p. 340) Let $(X_n)_n$ be a sequence of r.v.'s with $X_n \sim \text{Gamma}(n, 1)$.

(a) Compute the limit, in law,

$$\lim_{n \to \infty} \frac{1}{n} X_n \ .$$

(b) Compute the limit, in law,

$$\lim_{n \to \infty} \frac{1}{\sqrt{n}} (X_n - n) \ .$$

(c) Compute the limit, in law,

$$\lim_{n \to \infty} \frac{1}{\sqrt{X_n}} (X_n - n) \ .$$

3.33 (p. 341) Let $(X_n)_n$ be a sequence of i.i.d. r.v.'s with $P(X_n = \pm 1) = \frac{1}{2}$ and let $\overline{X}_n = \frac{1}{n}(X_1 + \cdots + X_n)$. Compute the limits in law of the sequences

(a) $(\sqrt{n} \sin \overline{X}_n)_n$.
(b) $(\sqrt{n} (1 - \cos \overline{X}_n))_n$.
(c) $(n (1 - \cos \overline{X}_n))_n$.

Chapter 4
Conditioning

4.1 Introduction

Let $(\Omega, \mathcal{F}, P)$ be a probability space. The following definition is well-known.

Let $B \in \mathcal{F}$ be a non-negligible event. The *conditional probability of* P *given* B is the probability P_B on $(\Omega, \mathcal{F})$ defined as

$$P_B(A) = \frac{P(A \cap B)}{P(B)} \quad \text{for every } A \in \mathcal{F} . \tag{4.1}$$

The fact that P_B is a probability on $(\Omega, \mathcal{F})$ is immediate.

From a modeling point of view: at the beginning we know that every event $A \in \mathcal{F}$ can occur with probability $P(A)$. If, afterwards, we acquire the information that the event B has taken place, we shall replace the probability P with P_B, in order to take into account the new information.

Similarly, let X be a real r.v. and Z an r.v. taking values in a countable set E such that $P(Z = z) > 0$ for every $z \in E$. For every Borel set $A \subset \mathbb{R}$ and every $z \in E$ let

$$n(z, A) = P(X \in A | Z = z) = \frac{P(X \in A, Z = z)}{P(Z = z)} .$$

The set function $A \mapsto n(z, A)$ is, for every $z \in E$, a probability on $\mathbb{R}$: it is the *conditional law of* X *given* $Z = z$. This probability has an intuitive meaning not dissimilar to the one above: $A \mapsto n(z, A)$ is the law that is reasonable to appoint to X if we acquire the information that the event $\{Z = z\}$ has occurred.

© The Author(s), under exclusive license to Springer Nature Switzerland AG 2023
P. Baldi, *Probability*, Universitext, https://doi.org/10.1007/978-3-031-38492-9_4

The *conditional expectation of X given Z = z* is defined as the mean, if it exists, of this law:

$$E(X|Z = z) = \int_{\mathbb{R}} x \, n(z, dx) = \frac{1}{P(Z = z)} \int_{\{Z=z\}} X \, dP = \frac{E(X 1_{\{Z=z\}})}{P(Z = z)} \, .$$

These are very important notions, as we shall see throughout. It is therefore important to extend them to the case of a general r.v. Z (i.e. without the assumption that Z takes at most countably many values). This is the goal of this chapter, where we shall also see some applications.

The idea is to characterize the quantity $h(z) = E(X|Z = z)$ in a way that also makes sense if Z is not discretely valued. For every $B \subset E$ we have

$$\int_{\{Z \in B\}} h(Z) \, dP = \sum_{z \in B} E(X|Z = z) P(Z = z) = \sum_{z \in B} E(X 1_{\{Z=z\}})$$

$$= E(X 1_{\{Z \in B\}}) = \int_{\{Z \in B\}} X.dP \, ,$$

i.e. the integrals of $h(Z)$, which is $\sigma(Z)$-measurable, and of X on the events of $\sigma(Z)$ coincide. We shall see that this property characterizes the conditional expectation.

In the sequel we shall proceed contrariwise with respect to this section: we will first define conditional expectations and then return to the conditional laws at the end.

4.2   Conditional Expectation

Recall (see p. 14) that for a real r.v. X, if $X = X^+ - X^-$ is its decomposition into positive and negative parts, X is lower semi-integrable (l.s.i.) if X^- is integrable and that in this case we can define the mathematical expectation $E(X) = E(X^+) - E(X^-)$ (possibly $E(X) = +\infty$). In the sequel we shall need the following result.

Lemma 4.1 *Let $(\Omega, \mathscr{F}, P)$ be a probability space, $\mathscr{D} \subset \mathscr{F}$ a sub-σ-algebra, X and Y real l.s.i. $\mathscr{D}$-measurable r.v.'s*

(a) *If*

$$E(X 1_D) \ge E(Y 1_D) \qquad \text{for every } D \in \mathscr{D} \tag{4.2}$$

    then $X \ge Y$ a.s.

(continued)

Lemma 4.1 (continued)
(b) *If*

$$E(X1_D) = E(Y1_D) \qquad for\ every\ D \in \mathcal{D}$$

then $X \overset{\cdot}{=} Y$ a.s.

Proof (a) Let $D_{r,q} = \{X \le r < q \le Y\} \in \mathcal{D}$. Note that $\{X < Y\} = \bigcup_{r,q \in \mathbb{Q}} D_{r,q}$, which is a countable union, so that it is enough to show that if (4.2) holds then $P(D_{r,q}) = 0$ for every $r < q$. But if $P(D_{r,q}) > 0$ for some $r, q, r < q$, then we would have

$$\int_{D_{r,q}} X\,dP \le rP(D_{r,q}) < qP(D_{r,q}) \le \int_{D_{r,q}} Y\,dP,$$

contradicting the assumption.

(b) Follows from (a) exchanging the roles of X and Y. ∎

Note that for integrable r.v.'s the lemma is a consequence of Exercise 1.9 (c).

Definition and Theorem 4.2 Let X be an l.s.i. r.v. and $\mathcal{D} \subset \mathcal{F}$ a sub-σ algebra. Then there exists an l.s.i. r.v. Y which is

(a) $\mathcal{D}$-measurable
(b) such that

$$\int_D Y\,dP = \int_D X\,dP \qquad for\ every\ D \in \mathcal{D}. \tag{4.3}$$

We shall denote $E(X\,|\,\mathcal{D})$ such an Y. $E(X\,|\,\mathcal{D})$ is the *conditional expectation* of X given $\mathcal{D}$.

Proof Let us assume first that X is square integrable. Let $K = L^2(\Omega, \mathcal{D}, P)$ denote the subspace of L^2 of the square integrable r.v.'s that are $\mathcal{D}$-measurable. Or, to be precise, recalling that the elements of L^2 are equivalent classes of functions, K is the space of these classes that contain a function that is $\mathcal{D}$-measurable. As L^2 convergence implies a.s. convergence for a subsequence and a.s. convergence preserves measurability (recall Remark 1.15), K is a closed subspace of L^2.

Going back to Proposition 2.40, let $Y = PX$ the orthogonal projection of X on $L^2(\Omega, \mathcal{D}, \mathrm{P})$. We can write $X = Y + QX$ where $QX = X - PX$. As QX is orthogonal to $L^2(\Omega, \mathcal{D}, \mathrm{P})$ (Proposition 2.40 again), we have, for every $D \in \mathcal{D}$,

$$\int_D X \, d\mathrm{P} = \int_D Y \, d\mathrm{P} + \int_D QX \, d\mathrm{P} = \int_D Y \, d\mathrm{P} + \int 1_D QX \, d\mathrm{P} = \int_D Y \, d\mathrm{P},$$

and PX satisfies (a) and (b) in the statement. We now drop the assumption that X is square integrable. Let us assume X to be positive and let $X_n = X \wedge n$. Then, for every n, X_n is square integrable and $X_n \uparrow X$ a.s. If $Y_n := PX_n$ then, for every $D \in \mathcal{D}$,

$$\int_D Y_n \, d\mathrm{P} = \int_D X_n \, d\mathrm{P} \le \int_D X_{n+1} \, d\mathrm{P} = \int_D Y_{n+1} \, d\mathrm{P}$$

and therefore, thanks to Lemma 4.1, also $(Y_n)_n$ is an a.s. increasing sequence. By Beppo Levi's Theorem, twice, we obtain

$$\int_D X \, d\mathrm{P} = \lim_{n \to \infty} \int_D X_n \, d\mathrm{P} = \lim_{n \to \infty} \mathrm{E}(X_n 1_D) = \lim_{n \to \infty} \mathrm{E}(Y_n 1_D) = \int_D Y \, d\mathrm{P}.$$

Taking $D = \Omega$ in the previous relation, if X is integrable then also $\mathrm{E}(X | \mathcal{D}) := Y$ is integrable, hence a.s. finite. If $X = X^+ - X^-$ is l.s.i., then we can just define

$$\mathrm{E}(X | \mathcal{D}) = \mathrm{E}(X^+ | \mathcal{D}) - \mathrm{E}(X^- | \mathcal{D})$$

with no danger of encountering a $+\infty - \infty$ situation as $\mathrm{E}(X^- | \mathcal{D})$ is integrable, hence a.s. finite.

Uniqueness follows immediately from Lemma 4.1.   ∎

We shall deal mostly with the conditional expectation of *integrable* r.v.'s. It is however useful to have this notion defined in the more general l.s.i. setting. In particular, $\mathrm{E}(X | \mathcal{D})$ is always defined if $X \ge 0$. See also Proposition 4.6 (d).

By linearity (4.3) is equivalent to

$$\mathrm{E}(ZW) = \mathrm{E}(XW) \tag{4.4}$$

for every r.v. W that is the linear combination with positive coefficients of indicator functions of events of $\mathcal{D}$, hence (Proposition 1.6) for every $\mathcal{D}$-measurable bounded positive r.v. W.

We shall often have to prove statements of the type "a certain r.v. Z is equal to $\mathrm{E}(X | \mathcal{D})$". On the basis of Theorem 4.2 this requires us to prove two things, namely that

(a) Z is $\mathcal{D}$-measurable
(b) $\mathrm{E}(Z 1_D) = \mathrm{E}(X 1_D)$ for every $D \in \mathcal{D}$.

Actually requirement (b) can be weakened considerably (but not surprisingly), as explained in the following remark, which we only state in the case when X is integrable.

Remark 4.3 If X is integrable then $Z = E(X|\mathcal{D})$ if and only if

(a) Z is integrable and $\mathcal{D}$-measurable
(b) $E(Z1_D) = E(X1_D)$ as D ranges over a class $\mathscr{C} \subset \mathcal{D}$ generating $\mathcal{D}$, stable with respect to finite intersections and containing Ω.

Actually let us prove that the family $\mathcal{M} \subset \mathcal{D}$ of the events D such that $E(Z1_D) = E(X1_D)$ is a monotone class.
• If $A, B \in \mathcal{M}$ with $A \subset B$ then

$$E(Z1_{B\setminus A}) = E(Z1_B) - E(Z1_A) = E(X1_B) - E(X1_A) = E(X1_{B\setminus A})$$

and therefore also $B \setminus A \in \mathcal{M}$. Note that the previous relation requires X to be integrable, so that both $E(X1_B)$ and $E(X1_A)$ are finite.
• Let $(D_n)_n \subset \mathcal{M}$ be an increasing sequence of events and $D = \bigcup_n D_n$. Then $X1_{D_n} \to_{n\to\infty} X1_D$, $Z1_{D_n} \to_{n\to\infty} Z1_D$ and, as $|X|1_{D_n} \le |X|$ and $|Z|1_{D_n} \le |Z|$, we can apply Lebesgue's Theorem (twice) and obtain that

$$E(Z1_D) = \lim_{n\to\infty} E(Z1_{D_n}) = \lim_{n\to\infty} E(X1_{D_n}) = E(X_D) \,.$$

$\mathcal{M}$ is therefore a monotone class containing $\mathscr{C}$ that is stable with respect to finite intersections. By the Monotone Class Theorem 1.2, $\mathcal{M}$ contains also $\sigma(\mathscr{C}) = \mathcal{D}$.

Remark 4.4 The conditional expectation operator is monotone, i.e. if X and Y are l.s.i. and $X \ge Y$ a.s., then $E(X|\mathcal{D}) \ge E(Y|\mathcal{D})$ a.s. Indeed for every $D \in \mathcal{D}$

$$E\big[E(X|\mathcal{D})1_D\big] = E(X1_D) \ge E(Y1_D) = E\big[E(Y|\mathcal{D})1_D\big]$$

and the property follows by Lemma 4.1.

The following two statements provide further elementary, but important, properties of the conditional expectation operator.

Proposition 4.5 *Let X, Y be integrable r.v.'s and α, $\beta \in \mathbb{R}$. Then*

(a) $\mathrm{E}(\alpha X + \beta Y \,|\, \mathscr{D}) = \alpha \, \mathrm{E}(X \,|\, \mathscr{D}) + \beta \, \mathrm{E}(Y \,|\, \mathscr{D})$ *a.s.*

(b) $\mathrm{E}\big[\mathrm{E}(X \,|\, \mathscr{D})\big] = \mathrm{E}(X)$.

(c) *If $\mathscr{D}' \subset \mathscr{D}$, $\mathrm{E}\big[\mathrm{E}(X \,|\, \mathscr{D}) \,|\, \mathscr{D}'\big] = \mathrm{E}(X \,|\, \mathscr{D}')$ a.s. (i.e. to condition first with respect to $\mathscr{D}$ and then to the smaller σ-algebra $\mathscr{D}'$ is the same as conditioning directly with respect to $\mathscr{D}'$).*

(d) *If Z is bounded and $\mathscr{D}$-measurable then $\mathrm{E}(ZX \,|\, \mathscr{D}) = Z \, \mathrm{E}(X \,|\, \mathscr{D})$ a.s. (i.e. bounded $\mathscr{D}$-measurable r.v.'s can go in and out of the conditional expectation, as if they were constants).*

(e) *If X is independent of $\mathscr{D}$ then $\mathrm{E}(X \,|\, \mathscr{D}) = \mathrm{E}(X)$ a.s.*

Proof These are immediate applications of the definition and boil down to the validation of the two conditions (a) and (b) p. 180; let us give the proofs of the last three points.

(c) The r.v. $\mathrm{E}\big[\mathrm{E}(X \,|\, \mathscr{D}) \,|\, \mathscr{D}'\big]$ is $\mathscr{D}'$-measurable; moreover if W is bounded $\mathscr{D}'$-measurable then

$$\mathrm{E}\big[W\mathrm{E}(\mathrm{E}(X \,|\, \mathscr{D}) \,|\, \mathscr{D}')\big] = \mathrm{E}\big[W\mathrm{E}(X \,|\, \mathscr{D})\big] = \mathrm{E}(WX) \,,$$

where the first equality comes from the definition of conditional expectation with respect to $\mathscr{D}'$ and the last one from the fact that W is also $\mathscr{D}$-measurable.

(d) We must prove that the r.v. $Z \, \mathrm{E}(X \,|\, \mathscr{D})$ is $\mathscr{D}$-measurable (which is immediate) and that, for every bounded $\mathscr{D}$-measurable r.v. W,

$$\mathrm{E}(WZX) = \mathrm{E}\big[WZ\mathrm{E}(X \,|\, \mathscr{D})\big] \,. \tag{4.5}$$

But this is immediate as W is bounded $\mathscr{D}$-measurable and therefore so is ZW.

(e) The r.v. $\omega \mapsto \mathrm{E}(X)$ is constant and therefore $\mathscr{D}$-measurable. If W is $\mathscr{D}$-measurable then it is independent of X and

$$\mathrm{E}(WX) = \mathrm{E}(W)\mathrm{E}(X) = \mathrm{E}\big[W\mathrm{E}(X)\big]$$

and therefore $\mathrm{E}(X \,|\, \mathscr{D}) = \mathrm{E}(X)$ a.s. ∎

It is easy to extend Proposition 4.5 to the case of r.v.'s that are only l.s.i. Note however that (a) holds only if α, $\beta \geq 0$ (otherwise $\alpha X + \beta Y$ might not be l.s.i. anymore) and that (d) holds only if Z is bounded *positive* (again ZX might turn out not to be l.s.i.).

The next statement concerns the behavior of the conditional expectation with respect to convergence.

Proposition 4.6 *Let X, X_n, $n = 1, 2, \ldots$, be real l.s.i. r.v.'s. Then*

(a) *(Beppo Levi) if $X_n \uparrow X$ as $n \to \infty$ a.s. then $E(X_n | \mathcal{D}) \uparrow E(X | \mathcal{D})$ as $n \to \infty$ a.s.*

(b) *(Fatou) If $\underline{\lim}_{n \to \infty} X_n = X$ a.s. and the r.v.'s X_n are bounded from below by the same integrable r.v. then*

$$\varliminf_{n \to \infty} E(X_n | \mathcal{D}) \geq E(X | \mathcal{D}) \qquad a.s.$$

(c) *(Lebesgue) If $|X_n| \leq Z$ for some integrable r.v. Z for every n and $X_n \to_{n \to \infty} X$ a.s. then*

$$\lim_{n \to \infty} E(X_n | \mathcal{D}) = E(X | \mathcal{D}) \qquad a.s.$$

(d) *(Jensen's inequality) If $\Phi : \mathbb{R}^d \to \overline{\mathbb{R}}^+$ is a lower semi-continuous convex function and $X = (X_1, \ldots, X_d)$ is a d-dimensional integrable r.v. then $\Phi(X)$ is l.s.i. and*

$$E\big[\Phi(X) | \mathcal{D}\big] \geq \Phi \circ E(X | \mathcal{D}) \qquad a.s.$$

denoting by $E(X | \mathcal{D})$ the d-dimensional r.v. with components $E(X_k | \mathcal{D})$, $k = 1, \ldots, d$.

Proof (a) As the sequence $(X_n)_n$ is a.s. increasing, $(E(X_n | \mathcal{D}))_n$ is also a.s. increasing thanks to Remark 4.4; the r.v. $Z := \overline{\lim}_{n \to \infty} E(X_n | \mathcal{D})$ is $\mathcal{D}$-measurable and $E(X_n | \mathcal{D}) \uparrow Z$ as $n \to \infty$ a.s. If $D \in \mathcal{D}$, by Beppo Levi's Theorem applied twice,

$$\int_D Z \, dP = \lim_{n \to \infty} \int_D E(X_n | \mathcal{D}) \, dP = \lim_{n \to \infty} \int_D X_n \, dP = \int_D X \, dP$$

and therefore $Z = E(X | \mathcal{D})$ a.s.

(b) If $Y_n = \inf_{k \geq n} X_k$ then

$$\lim_{n \to \infty} \uparrow Y_n = \varliminf_{n \to \infty} X_n = X \, .$$

As $(Y_n)_n$ is increasing and $Y_n \leq X_n$, (a) gives

$$E(X | \mathcal{D}) = \lim_{n \to \infty} E(Y_n | \mathcal{D}) \leq \varliminf_{n \to \infty} E(X_n | \mathcal{D}) \, .$$

(c) Immediate consequence of (b), applied both to the r.v.'s X_n and $-X_n$.

(d) Same as the proof of Jensen's inequality: recall, see (2.17), that a convex l.s.c. function Φ is equal to the supremum of all affine-linear functions minorizing Φ. If $f(x) = \langle a, x \rangle + b$ is an affine function minorizing Φ, then $\langle a, X \rangle + b$ is an integrable r.v. minorizing $\Phi(X)$ so that the latter is l.s.i. and

$$E\big[\Phi(X) \,|\, \mathcal{D}\big] \geq E\big[f(X) \,|\, \mathcal{D}\big] = E(\langle a, X \rangle + b \,|\, \mathcal{D})$$
$$= \langle a, E(X \,|\, \mathcal{D}) \rangle + b = f(E(X \,|\, \mathcal{D})) \,.$$

Now just take the supremum in f among all affine-linear functions minorizing Φ. ∎

Example 4.7 If $\mathcal{D} = \{\Omega, \emptyset\}$ is the trivial σ-algebra, then

$$E(X \,|\, \mathcal{D}) = E(X) \,.$$

Actually the only $\mathcal{D}$-measurable r.v.'s are constant and, if $c = E(X \,|\, \mathcal{D})$, then the constant c is determined by the relation $c = E[E(X \,|\, \mathcal{D})] = E(X)$. Mathematical expectation appears therefore to be a particular case of conditional expectation.

Example 4.8 Let $B \in \mathcal{F}$ be an event having strictly positive probability and let $\mathcal{D} = \{B, B^c, \Omega, \emptyset\}$ be the σ-algebra generated by B. Then $E(X \,|\, \mathcal{D})$, which is $\mathcal{D}$-measurable, is a real r.v. that is constant on B and on B^c. If we denote by c_B the value of $E(X \,|\, \mathcal{D})$ on B, from the relation

$$c_B P(B) = E\big[1_B E(X \,|\, \mathcal{D})\big] = \int_B X \, dP$$

and by the similar one for B^c we obtain

$$E(X \,|\, \mathcal{D}) = \begin{cases} \dfrac{1}{P(B)} \displaystyle\int_B X \, dP & \text{on } B \\[2mm] \dfrac{1}{P(B^c)} \displaystyle\int_{B^c} X \, dP & \text{on } B^c \,. \end{cases}$$

In particular, $E(X \,|\, \mathcal{D})$ is equal to $\int X \, dP_B$ on B, where P_B is as in (4.1), and equal to $\int X \, dP_{B^c}$ on B^c.

Remark 4.9 As is apparent in the proof of Theorem 4.2, if X is square integrable then $E(X|\mathcal{D})$ is the best approximation in L^2 of X with a $\mathcal{D}$-measurable r.v. and moreover the r.v.'s

$$E(X|\mathcal{D}) \quad \text{and} \quad X - E(X|\mathcal{D})$$

are orthogonal. As a consequence, as $X = \big(X - E(X|\mathcal{D})\big) + E(X|\mathcal{D})$, we have (Pythagoras's theorem)

$$E(X^2) = E\big[(X - E(X|\mathcal{D}))^2\big] + E\big[E(X|\mathcal{D})^2\big]$$

and the useful relation

$$E\big(|X - E(X|\mathcal{D})|^2\big) = E(X^2) - E[E(X|\mathcal{D})^2] \,. \tag{4.6}$$

Remark 4.10 (Conditional Expectations and L^p Spaces) It is immediate that, if $X = X'$ a.s., then $E(X|\mathcal{D}) = E(X'|\mathcal{D})$ a.s.: for every $D \in \mathcal{D}$ we have

$$E\big[E(X|\mathcal{D})1_D\big] = E(X1_D) = E(X'1_D) = E\big[E(X'|\mathcal{D})1_D\big]$$

and the property follows by Proposition 4.1 (b).

Conditional expectation is therefore defined on equivalence classes of r.v.'s. In particular, it is defined on L^p spaces, whose elements are equivalence classes.

Proposition 4.6 (d) (Jensen), applied to the convex function $x \mapsto |x|^p$ with $p \geq 1$, gives

$$E\big(|E(X|\mathcal{D})|^p\big) \leq E\big[E(|X|^p|\mathcal{D})\big] = E(|X|^p) \,. \tag{4.7}$$

Hence conditional expectation is a *continuous* linear map $L^p \to L^p$, $p \geq 1$; its norm is actually ≤ 1, i.e. it is a *contraction*. The image of L^p under the operator $X \mapsto E(X|\mathcal{D})$ is the subspace of L^p, that we shall denote $L^p(\mathcal{D})$, that is formed by the equivalence classes of r.v.'s that contain at least a $\mathcal{D}$-measurable representative.

In particular, if $p \geq 1$, $X_n \xrightarrow[n \to \infty]{L^p} X$ implies $E(X_n|\mathcal{D}) \xrightarrow[n \to \infty]{L^p} E(X|\mathcal{D})$.

If Y is an r.v. taking values in some measurable space $(E, \mathcal{E})$, sometimes we shall write $E(X|Y)$ instead of $E[X|\sigma(Y)]$. We know that all real $\sigma(Y)$-measurable r.v.'s are of the form $g(Y)$, where $g : E \to \mathbb{R}$ is a measurable function (this is Doob's criterion, Lemma 1.7). Hence there exists a measurable function $g : E \to \mathbb{R}$ such that $E(X|Y) = g(Y)$ a.s. Sometimes we shall denote, in a suggestive way, such a function g by

$$g(y) = E(X|Y = y) .$$

As every real $\sigma(Y)$-measurable r.v. is of the form $\psi(Y)$ for some measurable function $\psi : E \to \mathbb{R}$, g must satisfy the relation

$$E\big[X\psi(Y)\big] = E\big[g(Y)\psi(Y)\big] \tag{4.8}$$

for every bounded measurable function $\psi : E \to \mathbb{R}$.

If X is square integrable, by Remark 4.9 $g(Y)$ is "the best approximation of X by a function of Y" (in the sense of L^2). Compare with Example 2.24, the regression line.

The computation of a conditional expectation is an operation that we are led to perform quite often and that, sometimes, is even our goal. The next lemma can be very helpful.

Let $\mathcal{G} \subset \mathcal{F}$ be a σ-algebra and X an $\mathcal{G}$-measurable r.v. If Z is an r.v. independent of $\mathcal{G}$, we know that, if X and Z are integrable, then also their product XZ is integrable and

$$E(XZ|\mathcal{G}) = X \, E(Z|\mathcal{G}) = X \, E(Z) . \tag{4.9}$$

This formula is a particular case of the following lemma.

Lemma 4.11 (The "Freezing Lemma") *Given a probability space* $(\Omega, \mathcal{F}, P)$ *let*

- $(E, \mathcal{E})$ *be a measurable space,*
- $\mathcal{G}, \mathcal{H}$ *independent sub-σ-algebras of* $\mathcal{F}$,
- X *a $\mathcal{G}$-measurable $(E, \mathcal{E})$-valued r.v.,*
- $\Psi : E \times \Omega \to \mathbb{R}$ *an $\mathcal{E} \otimes \mathcal{H}$-measurable function such that* $\omega \mapsto \Psi(X(\omega), \omega)$ *is integrable.*

Then

$$E\big[\Psi(X, \cdot)|\mathcal{G}\big] = \Phi(X) , \tag{4.10}$$

where $\Phi(x) = E[\Psi(x, \cdot)]$.

Proof The proof uses the usual arguments of measure theory. Let us denote by $\mathcal{V}^+$ the family of $\mathcal{E} \otimes \mathcal{H}$-measurable positive functions $\Psi : E \times \Omega \to \mathbb{R}$ satisfying (4.10). It is immediate that $\mathcal{V}^+$ is stable with respect to increasing limits: if $(\Psi_n)_n \subset \mathcal{V}^+$ and $\Psi_n \uparrow \Psi$ as $n \to \infty$ then

$$E[\Psi_n(X, \cdot)|\mathcal{G}] \uparrow E[\Psi(X, \cdot)|\mathcal{G}],$$
$$E[\Psi_n(x, \cdot)] \uparrow E[\Psi(x, \cdot)],$$

(4.11)

so that $\Psi \in \mathcal{V}^+$.

Next let us denote by $\mathcal{M}$ the class of sets $\Lambda \in \mathcal{E} \otimes \mathcal{H}$ such that $\Psi(x, \omega) = 1_\Lambda(x, \omega)$ belongs to $\mathcal{V}^+$. It is immediate that it is stable with respect to increasing limits, thanks to (4.11), and to relative complementation, hence it is a monotone class (Definition 1.1). $\mathcal{M}$ contains the rectangle sets $\Lambda = A \times \Lambda_1$ with $A \in \mathcal{E}$, $\Lambda_1 \in \mathcal{H}$ as

$$E[1_\Lambda(X, \cdot)|\mathcal{G}] = E[1_A(X)1_{\Lambda_1}|\mathcal{G}] = 1_A(X)P(\Lambda_1)$$

and $\Phi(x) := E[1_A(x)1_{\Lambda_1}] = 1_A(x)P(\Lambda_1)$. By the Monotone class theorem, Theorem 1.2, $\mathcal{M}$ contains the whole σ-algebra generated by the rectangles, i.e. all $\Lambda \in \mathcal{E} \otimes \mathcal{H}$.

By linearity, $\mathcal{V}^+$ contains all elementary $\mathcal{E} \otimes \mathcal{H}$-measurable functions and, by Proposition 1.6, every positive $\mathcal{E} \otimes \mathcal{H}$-measurable function. Then we just have to decompose Ψ as in the statement of the lemma into positive and negative parts. ∎

Let us now present some applications of the freezing Lemma 4.11.

Example 4.12 Let $(X_n)_n$ be a sequence of i.i.d. r.v.'s with $P(X_n = \pm 1) = \frac{1}{2}$ and let $S_n = X_1 + \cdots + X_n$ for $n \geq 1$, $S_0 = 0$. Let T be a geometric r.v. of parameter p, independent of $(X_n)_n$. How can we compute the mean, variance and characteristic function of $Z = S_T$?

Intuitively S_n models the evolution of a random motion (a *stochastic process*, as we shall see more precisely in the next chapter) where at every iteration a step to the left or to the right is made with probability $\frac{1}{2}$; we want to find information concerning its position when it is stopped at a random time independent of the motion and geometrically distributed.

Let us first compute the mean. Let $\Psi : \mathbb{N} \times \Omega \to \mathbb{Z}$ be defined as $\Psi(n, \omega) = S_n(\omega)$. We have then $S_T(\omega) = \Psi(T, \omega)$ and we are in the situation of Lemma 4.11 with $\mathcal{H} = \sigma(T)$ and $\mathcal{G} = \sigma(X_1, X_2, \dots)$. By the freezing Lemma 4.11

$$E(S_T) = E[\Psi(T, \cdot)] = E[E(\Psi(T, \cdot)|\sigma(T))] = E[\Phi(T)],$$

where $\Phi(n) = E[\Psi(n, \cdot)] = E(S_n) = 0$, so that $E(S_T) = 0$. For the second order moment the argument is the same: let $\Psi(n, \omega) = S_n^2(\omega)$ so that

$$E(S_T^2) = E\big[E\big(\Psi(T, \omega)|\sigma(T)\big)\big] = E[\Phi(T)],$$

where now $\Phi(n) = E[\Psi(n, \cdot)] = E(S_n^2) = n\,\mathrm{Var}(X_1) = n$. Hence

$$E(S_T^2) = E(T) = \frac{1}{p}.$$

In the same way, with $\Psi(n, \omega) = e^{i\theta S_n(\omega)}$,

$$E(e^{i\theta S_T}) = E\big[E\big(e^{i\theta S_T}|\sigma(T)\big)\big] = E[\Phi(T)],$$

where now

$$\Phi(n) = E(e^{i\theta S_n}) = E(e^{i\theta X_1})^n = \left[\frac{1}{2}\,(e^{i\theta} + e^{-i\theta})\right]^n = \cos^n\theta$$

and therefore

$$E(e^{i\theta S_T}) = E[(\cos T)^n] = p\sum_{n=0}^{\infty}(1 - p)^n \cos^n\theta = \frac{p}{1 - (1 - p)\cos\theta}.$$

This example clarifies how to use the freezing lemma, but also the method of computing a mathematical expectation by "inserting" in the computation a conditional expectation and taking advantage of the fact that the expectation of a conditional expectation is the same as taking the expectation directly (Proposition 4.5 (b)).

4.3 Conditional Laws

In this section we investigate conditional distributions, extending to a general space the definition that we have seen in Sect. 4.1.

> **Definition 4.13** Let X, Y be r.v.'s taking values in the measurable spaces $(E, \mathscr{E})$ and $(G, \mathscr{G})$ respectively and let us denote by μ the law of X. A family of probabilities $(n(x, dy))_{x \in E}$ on $(G, \mathscr{G})$ is a *conditional law of* Y *given* $X = x$ if:
>
> (a) For every $A \in \mathscr{G}$, the map $x \mapsto n(x, A)$ is $\mathscr{E}$-measurable.
> (b) For every $A \in \mathscr{G}$ and $B \in \mathscr{E}$,
>
> $$P(Y \in A, X \in B) = \int_B n(x, A) \mu(dx) . \qquad (4.12)$$

Intuitively $n(x, \cdot)$ is "the distribution of Y taking into account the information that $X = x$".

Relation (4.12) can be written

$$E\big[1_A(Y)1_B(X)\big] = \int_E 1_B(x)n(x, A) \mu(dx) = \int_E 1_B(x) \mu(dx) \int_G 1_A(y) n(x, dy) .$$

The usual application of Proposition 1.6, approximating f and g with linear combinations of indicator functions, implies that, if $f : E \to \mathbb{R}^+$ and $g : G \to \mathbb{R}^+$ are positive measurable functions, then

$$E\big[f(X)g(Y)\big] = \int_E f(x)\mu(dx) \int_G g(y) n(x, dy) . \qquad (4.13)$$

With the usual decomposition into the difference of positive and negative parts we obtain that (4.13) holds if $f : E \to \mathbb{R}$ and $g : G \to \mathbb{R}$ are measurable and bounded or at least such that $f(X)g(Y)$ is integrable.

Note that (4.13) can also be written as

$$E\big[f(X)g(Y)\big] = E\big[f(X)h(X)\big] ,$$

where

$$h(x) := \int_G g(y) n(x, dy) .$$

Comparing with (4.8), this means precisely that

$$E\big[g(Y)|X = x\big] = h(x) . \qquad (4.14)$$

Hence if Y is a real integrable r.v.

$$E(Y \mid X = x) = \int_G y\, n(x, dy) \qquad (4.15)$$

and we recover the relation from which we started in Sect. 4.1: the conditional expectation is the mean of the conditional law.

Remark 4.14 Let X, Y be as in Definition 4.13. Assume that the conditional law $(n(x, dy))_{x \in E}$ of Y given $X = x$ does not depend on x, i.e. there exists a probability v on $(E, \mathscr{E})$ such that $n(x, \cdot) = v$ for every $x \in E$. Then from (4.13) first taking $g \equiv 1$ we find that v is the law of Y and then that the joint law of X and Y is the product $\mu \otimes v$, so that X and Y are independent. Note that this is consistent with intuition.

Let us now present some results which will allow us to actually compute conditional distributions. The following statement is very useful in this direction: its intuitive content is almost immediate, but a formal proof is required.

Lemma 4.15 (The Second Freezing Lemma) *Let $(E, \mathscr{E})$, $(H, \mathscr{H})$ and $(G, \mathscr{G})$ be measurable spaces, X, Z independent r.v.'s with values in E and H respectively and $\Psi : E \times H \to G$. Let $Y = \Psi(X, Z)$.*

Then the conditional law of Y given $X = x$ is the law, $\overline{v}_x$, of the r.v. $\Psi(x, Z)$.

Proof This is just a rewriting of the freezing Lemma 4.11. Let us denote by μ the law of X. We must prove that, for every pair of bounded measurable functions $f : E \to \mathbb{R}$ and $g : G \to \mathbb{R}$,

$$E\big[f(X)g(Y)\big] = \int_E f(x)\, d\mu(x) \int_G g(y)\, d\overline{v}_x(y) . \qquad (4.16)$$

We have

$$E\big[f(X)g(Y)\big] = E\big[f(X)g(\Psi(X, Z))\big] = E\big[E[f(X)g(\Psi(X, Z)) \mid X]\big] . \qquad (4.17)$$

As Z is independent of X, by the freezing lemma

$$E\big[f(X)g(\Psi(X, Z)) \mid X\big] = \Phi(X)$$

where

$$\Phi(x) = \mathrm{E}\big[f(x)g(\Psi(x, Z))\big] = f(x) \int_G g(y)\, d\overline{v}_x(y)$$

and, going back to (4.17), we have

$$\mathrm{E}\big[f(X)g(Y)\big] = \mathrm{E}[\Phi(X)] = \int_E f(x)\, d\mu(x) \int_G g(y)\, d\overline{v}_x(y)$$

i.e. (4.16).   ∎

As mentioned above, this lemma is rather intuitive: the information $X = x$ tells us that we can replace X by x in the relation $Y = \Psi(X, Z)$, whereas it does not give any information on the value of Z, which is independent of X.

The next example recalls a general situation where the computation of the conditional law is easy.

Example 4.16 Let X, Y be r.v.'s with values in the measurable spaces $(E, \mathscr{E})$ and $(G, \mathscr{G})$ respectively. Let ρ, γ be σ-finite measures on $(E, \mathscr{E})$ and $(G, \mathscr{G})$ respectively and assume that the pair (X, Y) has a density h with respect to the product measure $\rho \otimes \gamma$ on $(E \times G, \mathscr{E} \otimes \mathscr{G})$. Let

$$h_X(x) = \int_E h(x, y)\, \gamma(dy)$$

be the density of the law of X with respect to ρ and let $Q = \{x;\, h_X(x) = 0\} \in \mathscr{E}$. Clearly the event $\{X \in Q\}$ is negligible as $P(X \in Q) = \int_Q h_X(x)\rho(dx) = 0$. Let

$$\overline{h}(y; x) := \begin{cases} \dfrac{h(x, y)}{h_X(x)} & \text{if } x \notin Q \\[2mm] \text{any density} & \text{if } x \in Q, \end{cases} \tag{4.18}$$

and $n(x, dy) = \overline{h}(y; x)\, d\gamma(y)$. Let us prove that n is a conditional law of Y given $X = x$.

Indeed, for any pair f, g of real bounded measurable functions on $(E, \mathscr{E})$ and $(G, \mathscr{G})$ respectively,

$$\mathrm{E}\big[f(X)g(Y)\big] = \int_E f(x)\, d\rho(x) \int_G g(y)h(x, y)\, d\gamma(y)$$

$$= \int_E f(x)h_X(x)\, d\rho(x) \int_G g(y)\overline{h}(y; x)\, d\gamma(y)$$

which means precisely that the conditional law of Y given $X = x$ is $n(x, dy) = \overline{h}(y; x) d\gamma(y)$. In particular, for every bounded measurable function g,

$$E(g(Y)|X = x) = \int_G g(y)\overline{h}(y; x) d\gamma(y) .$$

Conversely, note that, if the conditional density $\overline{h}(\cdot; x)$ of Y with respect to X and the density of X are known, then the joint law of (X, Y) has density $(x, y) \mapsto h_X(x)\overline{h}(y; x)$ with respect to $\rho \otimes \gamma$ and the density, h_Y, of Y with respect to γ is

$$h_Y(y) = \int_E h(x, y) d\rho(x) = \int_E \overline{h}(y; x)h_X(x) d\rho(x) . \qquad (4.19)$$

Example 4.17 Let us take advantage of the second freezing lemma, Lemma 4.15, in order to compute the density of a Student law $t(n)$. Recall that this is the law of an r.v. of the form $T := \frac{X}{\sqrt{Y}} \sqrt{n}$, where X and Y are independent and $N(0, 1)$- and $\chi^2(n)$-distributed respectively.

Thanks to the second freezing lemma, the conditional law of T given $Y = y$ is the law of $\frac{X}{\sqrt{y}} \sqrt{n}$, i.e. an $N(0, \frac{n}{y})$, so that

$$\overline{h}(t; y) = \frac{\sqrt{y}}{\sqrt{2\pi n}} e^{-\frac{1}{2n} y t^2} .$$

By (4.19), the density of T is

$$h_T(t) = \int \overline{h}(t; y)h_Y(y) dy$$

$$= \frac{1}{2^{n/2}\Gamma(\frac{n}{2})\sqrt{2\pi n}} \int_0^{+\infty} \sqrt{y}\, y^{n/2-1} e^{-y/2} e^{-\frac{1}{2n} y t^2} dy$$

$$= \frac{1}{2^{n/2}\Gamma(\frac{n}{2})\sqrt{2\pi n}} \int_0^{+\infty} y^{\frac{1}{2}(n+1)-1} e^{-\frac{y}{2}(1+\frac{1}{n} t^2)} dy .$$

We recognize in the last integral, but for the constant, a Gamma(α, λ) density with $\alpha = \frac{1}{2}(n + 1)$ and $\lambda = \frac{1}{2}(1 + \frac{1}{n} t^2)$, so that

$$h_T(t) = \frac{1}{2^{n/2}\Gamma(\frac{n}{2})\sqrt{2\pi n}} \frac{\Gamma(\frac{n}{2} + \frac{1}{2})}{(\frac{1}{2} + \frac{1}{2n}t^2)^{\frac{n+1}{2}}} = \frac{\Gamma(\frac{n}{2} + \frac{1}{2})}{\Gamma(\frac{n}{2})\sqrt{\pi n}} \frac{1}{(1 + \frac{t^2}{n})^{\frac{n+1}{2}}} .$$

The $t(n)$ densities have a shape similar to the Gaussian (see Figs. 4.1 and 4.2 below) but they go to 0 at infinity only polynomially fast. Also $t(1)$ is the Cauchy law.

Let us now tackle the question of the existence of a conditional expectation. So far we know the existence in the following situations.

- When X and Y are independent: just choose $n(x, dy) = \nu(dy)$ for every x, ν denoting the law of Y.
- When $Y = \Psi(X, Z)$ with Z independent of X as in the second freezing lemma, Lemma 4.15,
- When the joint law of X and Y has a density with respect to a σ-finite product measure, as in Remark 4.16.

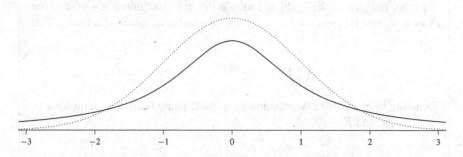

Fig. 4.1 Comparison between an $N(0, 1)$ density (dots) and a $t(1)$ (i.e. Cauchy) density

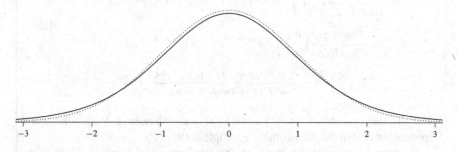

Fig. 4.2 Comparison between an $N(0, 1)$ density (dots) and a $t(9)$ density. Recall that (Example 3.30), as $n \to \infty$, $t(n)$ converges weakly to $N(0, 1)$

In general it can be proved that if the spaces E and G of Definition 4.13 are Polish, i.e. metric, complete and separable, conditional laws do exist and a uniqueness result holds. A proof can be found in most of the references. In particular, see Theorem 1.1.6, p. 13, in [23].

Conditional laws appear in a natural way also in the modeling of random phenomena, as often the data of the problem provide the law of some r.v. X and the conditional law of some other r.v. Y given $X = x$.

Example 4.18 A coin is chosen at random from a heap of possible coins and tossed n times. Let us denote by Y the number of tails obtained.

Assume that it is not known whether the chosen coin is a fair one. Let us actually make the assumption that the coin gives tail with a probability p that is itself random and Beta(α, β)-distributed. What is the value of $P(Y = k)$? What is the law of Y? How many tails appear in n throws on average?

If we denote by X the Beta(α, β)-distributed r.v. that models the choice of the coin, the data of the problem indicate that the conditional law of Y given $X = x, \bar{v}_x$ say, is binomial $B(n, x)$ (the total number of throws n is fixed). That is

$$\bar{v}_x(k) = \binom{n}{k} x^k (1 - x)^{n-k}, \qquad k = 0, \dots, n .$$

Denoting by μ the Beta distribution of X, (4.12) here becomes, again for $k = 0, 1, \dots, n$, and $B = [0, 1]$,

$$P(Y = k) = \int \bar{v}_x(k) \, \mu(dx)$$

$$= \frac{\Gamma(\alpha + \beta)}{\Gamma(\alpha)\Gamma(\beta)} \binom{n}{k} \int_0^1 x^{\alpha-1}(1 - x)^{\beta-1} x^k (1 - x)^{n-k} \, dx$$

$$= \frac{\Gamma(\alpha + \beta)}{\Gamma(\alpha)\Gamma(\beta)} \binom{n}{k} \int_0^1 x^{\alpha+k-1}(1 - x)^{\beta+n-k-1} \, dx \qquad (4.20)$$

$$= \binom{n}{k} \frac{\Gamma(\alpha + \beta)\Gamma(\alpha + k)\Gamma(n + \beta - k)}{\Gamma(\alpha)\Gamma(\beta)\Gamma(\alpha + \beta + n)} .$$

This discrete probability law is known as *Skellam's binomial*. We shall see an application of it in the forthcoming Example 4.19.

In order to compute the mean of this law, instead of using the definition $E(Y) = \sum_{k=0}^n kP(Y = k)$ that leads to unpredictable computations, recall that the mean is also the expectation of the conditional expectation (Proposition 4.5 (b)), i.e.

$$E(Y) = E\big[E(Y \,|\, X)\big] .$$

Now $E(Y|X = x) = nx$, as the conditional law of Y given $X = x$ is $B(n, x)$ and, recalling that the mean of a Beta(α, β)-distributed r.v. is $\frac{\alpha}{\alpha+\beta}$,

$$E(Y) = E(nX) = \frac{n\alpha}{\alpha + \beta} \cdot$$

Example 4.19 Let us go back to Geissler's data. In Example 3.44 we have seen that a binomial model is not able to explain them. Might Skellam's model above be a more fitting alternative? This would mean, intuitively, that every family has its own "propensity" to a male offspring which follows a Beta distribution.

Let us try to fit the data with a Skellam binomial. Now we play with two parameters, i.e. α and β. For instance, with the choice of $\alpha = 34.13$ and $\beta = 31.61$ let us compare the observed values $\bar{q}_k$ with those, r_k, of the Skellam binomial with the parameters above (q_k are the values produced by the "old" binomial model):

k	q_k	r_k	$\bar{q}_k$
1	0.003174	0.004074	0.004415
2	0.016113	0.017137	0.017007
3	0.053711	0.050832	0.046770
4	0.120850	0.107230	0.109567
5	0.193359	0.169463	0.168929
6	0.225586	0.205732	0.219624
7	0.193359	0.193329	0.181848
8	0.120850	0.139584	0.135568
9	0.053711	0.075529	0.078168
10	0.016113	0.029081	0.029599
11	0.003174	0.008008	0.008504

The value of Pearson's T statistics now is $T = 13.9$ so that the Skellam model gives a much better approximation. However Pearson's Theorem cannot be applied here, at least in the form of Theorem 3.42, as the parameters α and β above were estimated from the data.

How the values α and β above were estimated from the data and how the statement of Pearson's theorem should be modified in this situation is left to a more advanced course in statistics.

4.4   The Conditional Laws of Gaussian Vectors

In this section we investigate conditional laws (and therefore also conditional expectations) when the r.v. Y (whose conditional law we want to compute) and X (the conditioning r.v.) are jointly Gaussian. It is possible to take advantage of the method of Example 4.16, taking the quotient between the joint density and the other marginal, but now we shall see a much quicker and efficient method. Moreover, let us not forget that for a Gaussian vector existence of the joint density is not guaranteed.

Let Y, X be Gaussian vectors $\mathbb{R}^m$- and $\mathbb{R}^d$-valued respectively. Assume that their joint law on the product space $(\mathbb{R}^{m+d}, \mathscr{B}(\mathbb{R}^{m+d}))$ is Gaussian of mean and covariance matrix respectively

$$\begin{pmatrix} b_Y \\ b_X \end{pmatrix} \qquad \begin{pmatrix} C_Y & C_{YX} \\ C_{XY} & C_X \end{pmatrix}$$

where C_Y and C_X are the covariance matrices of Y and X respectively and $C_{YX} = \mathrm{E}[(Y - \mathrm{E}(Y))(X - \mathrm{E}(X))^*] = C_{XY}^*$ is the $m \times d$ matrix of the covariances of the components of Y and those of X; let us assume moreover that C_X is strictly positive definite (and therefore invertible).

Let us first look for a $m \times d$ matrix A such that the r.v.'s $Y - AX$ and X are independent.

Let $Z = Y - AX$. The pair (Y, X) is Gaussian as well as (Z, X), which is a linear function of the former. Hence, as seen in Sect. 2.8, p. 90, independence of Z and X follows as soon as $\mathrm{Cov}(Z_i, X_j) = 0$ for every $i = 1, \ldots, m$, $j = 1, \ldots, d$. First, to simplify the notation, let us assume that the means b_Y and b_X vanish. The condition of absence of correlation between the components of Z and those of X can then be written

$$0 = \mathrm{E}(ZX^*) = \mathrm{E}[(Y - AX)X^*] = \mathrm{E}(YX^*) - A\mathrm{E}(XX^*) = C_{YX} - AC_X .$$

Hence $A = C_{YX}C_X^{-1}$. Without the assumptions that the means vanish, just make the same computation with Y and X replaced by $Y - b_Y$ and $X - b_X$. We can write now

$$Y = AX + (Y - AX),$$

where the r.v.'s $Y - AX$ and X are independent. Hence by Lemma 4.15 (the second freezing lemma) the conditional law of Y given $X = x$ is the law of $Ax + Y - AX$. As $Y - AX$ is Gaussian, the law of $Ax + Y - AX$ is determined by its mean

$$Ax + b_Y - Ab_X = b_Y - C_{YX}C_X^{-1}(b_X - x) \qquad (4.21)$$

and its covariance matrix

$$
\begin{aligned}
C_{Y-AX} &= \mathrm{E}\big[(Y - b_Y - A(X - b_X))(Y - b_Y - A(X - b_X))^*\big] \\
&\quad \mathrm{E}\big[(Y - b_Y)(Y - b_Y)^*\big] - \mathrm{E}\big[(Y - b_Y)(X - b_X)^* A^*\big] \\
&\quad\quad -\mathrm{E}\big[A(X - b_X)(Y - b_Y)^*\big] + \mathrm{E}\big[A(X - b_X)(X - b_X)^* A^*\big] \\
&= C_Y - C_{YX} A^* - A C_{XY} + A C_X A^* \\
&= C_Y - C_{YX} C_X^{-1} C_{YX}^* - C_{YX} C_X^{-1} C_{YX}^* + C_{YX} C_X^{-1} C_X C_X^{-1} C_{YX}^* \\
&= C_Y - C_{YX} C_X^{-1} C_{YX}^* ,
\end{aligned}
$$

$$(4.22)$$

where we have taken advantage of the fact that C_X is symmetric and of the relation $C_{XY} = C_{YX}^*$. In particular, from (4.21) we obtain the conditional expectation

$$
\mathrm{E}(Y \mid X = x) = b_Y - C_{YX} C_X^{-1} (b_X - x) . \tag{4.23}
$$

When both Y and X are real r.v.'s, (4.23) and (4.22) give for the values of the mean and the variance of the conditional distribution, respectively

$$
b_Y - \frac{\mathrm{Cov}(Y, X)}{\mathrm{Var}(X)} (b_X - x) , \tag{4.24}
$$

which is equal to $\mathrm{E}(Y \mid X = x)$ and

$$
\mathrm{Var}(Y) - \frac{\mathrm{Cov}(Y, X)^2}{\mathrm{Var}(X)} . \tag{4.25}
$$

Note that the variance of the conditional law is always smaller than the variance of Y, which is a general fact already noted in Remark 4.9.

Let us point out some important features.

Remark 4.20 (a) The conditional laws of a Gaussian vector are also Gaussian.

(b) If Y and X are jointly Gaussian, the conditional expectation of Y given X is an affine-linear function of X and (therefore) coincides with the regression line. Recall (Remark 2.24) that the conditional expectation is the best approximation in L^2 of Y by a function of X whereas the regression line provides the best approximation of Y by an *affine-linear* function of X.

(c) Only the mean of the conditional law depends on the value of the conditioning variable X. The covariance matrix of the conditional law *does not* depend on the value of X.

Exercises

4.1 (p. 342) Let X, Y be i.i.d. r.v.'s with a $B(1, p)$ law, i.e. Bernoulli with parameter p and let $Z = 1_{\{X+Y=0\}}$, $\mathscr{G} = \sigma(Z)$.

(a) What are the events of the σ-algebra $\mathscr{G}$?

b) Compute $E(X|\mathscr{G})$ and $E(Y|\mathscr{G})$ and determine their law. Are these r.v.'s also independent?

4.2 (p. 342) Let $(\Omega, \mathscr{F}, P)$ be a probability space and $\mathscr{G} \subset \mathscr{F}$ a sub-σ-algebra.

(a) Let $A \in \mathscr{F}$ and $B = \{E(1_A|\mathscr{G}) = 0\}$. Show that $B \subset A^c$ a.s.

(b) Let X be a *positive* r.v. Prove that

$$\{E(X|\mathscr{G}) = 0\} \subset \{X = 0\} \qquad \text{a.s.}$$

i.e. the zeros of a positive r.v. shrink under conditioning.

4.3 (p. 343) Let X be a real integrable r.v. on a probability space $(\Omega, \mathscr{F}, P)$ and $\mathscr{G} \subset \mathscr{F}$ a sub-σ-algebra. Let $\mathscr{D} \subset \mathscr{F}$ be another σ-algebra independent of X and independent of $\mathscr{G}$.

(a) Is it true that

$$E(X|\mathscr{G} \vee \mathscr{D}) = E(X|\mathscr{G}) \text{ ?} \qquad (4.26)$$

(b) Prove that if $\mathscr{D}$ is independent of $\sigma(X) \vee \mathscr{G}$, then (4.26) holds.

• Recall Remarks 2.12 and 4.3.

4.4 (p. 343) Let $(\Omega, \mathscr{F}, P)$ be a probability space and $\mathscr{G} \subset \mathscr{F}$ a sub-σ-algebra. A non-empty event $A \in \mathscr{G}$ is an *atom* of $\mathscr{G}$ if there is no event of $\mathscr{G}$ which is strictly contained in A save $\emptyset$. Let E be a Hausdorff topological space and X an E-valued r.v.

(a) Prove that $\{X = x\}$ is an atom of $\sigma(X)$.

(b) Prove that if X is $\mathscr{G}$-measurable then it is constant on the atoms of $\mathscr{G}$.

(c) Let Z be a real r.v. Prove that if $P(X = x) > 0$ then $E(Z|X)$ is constant on $\{X = x\}$ and on this event takes the value

$$\frac{1}{P(X = x)} \int_{\{X=x\}} Z \, dP \, . \qquad (4.27)$$

• Recall that in a Hausdorff topological space the sets formed by a single point are closed, hence Borel sets.

4.5 (p. 344)

(a) Let X, Y be r.v.'s with values in a measurable space $(E, \mathscr{E})$ and Z another r.v. taking values in some other measurable space. Assume that the pairs (X, Z) and (Y, Z) have the same law (in particular X and Y have the same law). Prove that, if $h : E \to \mathbb{R}$ is a measurable function such that $h(X)$ is integrable, then

$$E[h(X)|Z] = E[h(Y)|Z] \qquad \text{a.s.}$$

(b) Let $T_1, \ldots, T_n$ be real i.i.d. integrable r.v.'s and $T = T_1 + \cdots + T_n$.

(b1) Prove that the pairs $(T_1, T), (T_2, T), \ldots, (T_n, T)$ have the same law.

(b2) Prove that

$$E(T_1|T) = \frac{T}{n} .$$

4.6 (p. 344) Let X, Y be independent r.v.'s both with a Laplace distribution of parameter 1.

(a) Prove that X and XY have the same joint distribution as $-X$ and XY.

(b1) Compute $E(X|XY = z)$.

(b2) What if X and Y were both $N(0, 1)$-distributed instead?

(b3) And with a Cauchy distribution?

4.7 (p. 345) Let X be an m-dimensional r.v. having density f with respect to the Lebesgue measure of $\mathbb{R}^m$ of the form $f(x) = g(|x|)$, where $g : \mathbb{R}^+ \to \mathbb{R}^+$.

(a) Prove that the real r.v. $|X|$ has a density with respect to the Lebesgue measure and compute it.

(b) Let $\psi : \mathbb{R}^m \to \mathbb{R}$ be a bounded measurable function. Compute $E\big[\psi(X)\,\big|\,|X|\big]$.

4.8 (p. 346) (Conditional expectations under a change of probability) Let Z be a positive r.v. defined on the probability space $(\Omega, \mathscr{F}, P)$ and $\mathscr{G} \subset \mathscr{F}$ a sub-σ-algebra. Recall (Exercise 4.2) that $\{Z = 0\} \supset \{E(Z|\mathscr{G}) = 0\}$ a.s.

(a) Note that $Z1_{\{E(Z|\mathscr{G})>0\}} = Z$ a.s. and deduce that, for every r.v. Y such that YZ is integrable, we have

$$E(ZY|\mathscr{G}) = E(ZY|\mathscr{G})1_{\{E(Z|\mathscr{G})>0\}} \qquad \text{a.s.} \qquad (4.28)$$

(b1) Assume moreover that $E(Z) = 1$. Let Q be the probability on $(\Omega, \mathscr{F})$ having density Z with respect to P and let us denote by E^Q the mathematical expectation with respect to Q. Prove that $E(Z|\mathscr{G}) > 0$ Q-a.s. (E still denotes the expectation with respect to P).

(b2) Prove that if Y is integrable with respect to Q, then

$$E^Q(Y|\mathscr{G}) = \frac{E(YZ|\mathscr{G})}{E(Z|\mathscr{G})} \qquad \text{Q-a.s.} \qquad (4.29)$$

• Note that if the density Z is itself $\mathcal{G}$-measurable, then $\mathrm{E}^Q(Y|\mathcal{G}) = \mathrm{E}(Y|\mathcal{G})$ Q-a.s.

4.9 (p. 347) Let T be an r.v. having density, with respect to the Lebesgue measure, given by

$$f(t) = 2t, \qquad 0 \le t \le 1$$

and $f(t) = 0$ for $t \notin [0, 1]$. Let Z be an $N(0, 1)$-distributed r.v. independent of T.

(a) Compute the Laplace transform and characteristic function of $X = ZT$. What are the convergence abscissas?
(b) Compute the mean and variance of X.
(c) Prove that for every $R > 0$ there exists a constant c_R such that

$$P(|X| \ge x) \le c_R \, e^{-Rx} .$$

4.10 (p. 348) (A useful independence criterion) Let X be an m-dimensional r.v. on the probability space $(\Omega, \mathcal{F}, P)$ and $\mathcal{G} \subset \mathcal{F}$ a sub-σ-algebra.

(a) Prove that if X is independent of $\mathcal{G}$, then

$$\mathrm{E}(e^{i \langle \theta, X \rangle}|\mathcal{G}) = \mathrm{E}(e^{i \langle \theta, X \rangle}) \qquad \text{for every } \theta \in \mathbb{R}^m . \tag{4.30}$$

(b) Assume that (4.30) holds.
(b1) Let Y be a real $\mathcal{G}$-measurable r.v. and compute the characteristic function of (X, Y).
(b2) Prove that if (4.30) holds, then X is independent of $\mathcal{G}$.

4.11 (p. 348) Let X, Y be independent r.v.'s Gamma$(1, \lambda)$- and $N(0, 1)$-distributed respectively.

(a) Compute the characteristic function of $Z = \sqrt{X} \, Y$.
(b) Compute the characteristic function of an r.v. W having a Laplace law of parameter α, i.e. having density with respect to the Lebesgue measure

$$f(x) = \frac{\alpha}{2} e^{-\alpha|x|} .$$

(c) Prove that Z has a density with respect to the Lebesgue measure and compute it.

4.12 (p. 349) Let X, Y be independent $N(0, 1)$-distributed r.v.'s and let, for $\lambda \in \mathbb{R}$,

$$Z = e^{-\frac{1}{2}\lambda^2 Y^2 + \lambda XY} .$$

(a) Prove that $\mathrm{E}(Z) = 1$.

(b) Let Q be the probability on $(\Omega, \mathcal{F})$ having density Z with respect to P. What is the law of X with respect to Q?

4.13 (p. 349) Let X and Y be independent $N(0, 1)$-distributed r.v.'s.

(a) Compute, for $t \in \mathbb{R}$, the Laplace transform

$$L(t) := E(e^{tXY}).$$

(b) Let $|t| < 1$ and let Q be the new probability $dQ = \sqrt{1 - t^2}\, e^{tXY}\, dP$. Determine the joint law of X and Y under Q. Compute $\mathrm{Var}_Q(X)$ and $\mathrm{Cov}_Q(X, Y)$.

4.14 (p. 350) Let $(X_n)_n$ be a sequence of independent $\mathbb{R}^d$-valued r.v.'s, defined on the same probability space. Let $S_0 = 0$, $S_n = X_1 + \cdots + X_n$ and $\mathcal{F}_n = \sigma(S_k, k \leq n)$. Show that, for every bounded Borel function $f : \mathbb{R}^d \to \mathbb{R}$,

$$E\big[f(S_{n+1})|\mathcal{F}_n\big] = E\big[f(S_{n+1})|S_n\big] \tag{4.31}$$

and express this quantity in terms of the law μ_n of X_n.

• This exercise proves rigorously a rather intuitive feature: as $S_{n+1} = X_{n+1} + S_n$ and X_{n+1} is independent of $X_1, \ldots, X_n$ hence also of $S_1, \ldots, S_n$, in order to foresee the value of S_{n+1}, once the value of S_n is known, the additional knowledge of $S_1, \ldots, S_n$ does not provide any additional information. In the world of stochastic processes (4.31) means that $(S_n)_n$ enjoys the *Markov property*.

4.15 (p. 350) Compute the mean and variance of a Student $t(n)$ law.

4.16 (p. 351) Let X, Y, Z be independent r.v.'s with $X, Y \sim N(0, 1)$, $Z \sim$ Beta(α, β).

(a) Let $W = ZX + \sqrt{1 - Z^2}\, Y$. What is the conditional law of W given $Z = z$? What is the law of W?

(b) Are W and Z independent?

4.17 (p. 351) (Multivariate Student t's) A multivariate (centered) $t(n, d, C)$ distribution is the law of the r.v.

$$\frac{X}{\sqrt{Y}}\sqrt{n},$$

where X and Y are independent, $Y \sim \chi^2(n)$ and X is d-dimensional $N(0, C)$-distributed with a covariance matrix C that is assumed to be invertible.

Prove that a $t(n, d, C)$ law has a density with respect to the Lebesgue measure and compute it.

Try to reproduce the argument of Example 4.17.

4.18 (p. 352) Let X, Y be $N(0, 1)$-distributed r.v.'s. and W another real r.v. Let us assume that X, Y, Z are independent and let

$$Z = \frac{X + YW}{\sqrt{1 + W^2}} \, .$$

(a) What is the conditional law of Z given $W = w$?
(b) What is the law of Z?

4.19 (p. 352) A family $\{X_1, \dots, X_n\}$ of r.v.'s, defined on the same probability space $(\Omega, \mathcal{F}, \mathrm{P})$ and taking values in the measurable space $(E, \mathcal{E})$, is said to be *exchangeable* if and only if the law of $X = (X_1, \dots, X_n)$ is the same as the law of $X_\sigma = (X_{\sigma_1}, \dots, X_{\sigma_n})$, where $\sigma = (\sigma_1, \dots, \sigma_n)$ is any permutation of $(1, \dots, n)$.

(a) Prove that if $\{X_1, \dots, X_n\}$ is exchangeable then the r.v.'s $X_1, \dots, X_n$ have the same law; and also that the law of (X_i, X_j) does not depend on $i, j, i \neq j$.
(b) Prove that if $X_1, \dots, X_n$ are i.i.d. then they are exchangeable.
(c) Assume that $X_1, \dots, X_n$ are real-valued and that their joint distribution has a density with respect to the Lebesgue measure of $\mathbb{R}^n$ of the form

$$f(x) = g(|x|) \tag{4.32}$$

for some measurable function $g : \mathbb{R}^+ \to \mathbb{R}^+$. Then $\{X_1, \dots, X_n\}$ is exchangeable.
(d1) Assume that there exists an r.v. Y defined on $(\Omega, \mathcal{F}, \mathrm{P})$ and taking values in some measurable space $(G, \mathcal{G})$ such that the r.v.'s $X_1, \dots, X_n$ are conditionally independent and identically distributed given $Y = y$, i.e. such that the conditional law of $(X_1, \dots, X_n)$ given $Y = y$ is a product $\overline{\mu}_y \otimes \cdots \otimes \overline{\mu}_y$. Prove that the family $\{X_1, \dots, X_n\}$ is exchangeable.
(d2) Let $X = (X_1, \dots, X_d)$ be a multidimensional Student $t(n, d, I)$-distributed r.v. (see Exercise 4.17), with $I =$ the identity matrix. Prove that $\{X_1, \dots, X_d\}$ is exchangeable.

4.20 (p. 353) Let T, W be exponential r.v.'s of parameters respectively λ and μ. Let $S = T + W$.

(a) What is the law of S? What is the joint law of T and S?
(b) Compute $\mathrm{E}(T \,|\, S)$.

Recalling the meaning of the conditional expectation as the best approximation in L^2 of T given S, compare with the result of Exercise 2.30 where we computed the regression line, i.e. the best approximation of T by an *affine-linear* function of S.

4.21 (p. 354) Let X, Y be r.v.'s having joint density with respect to the Lebesgue measure

$$f(x, y) = \lambda^2 x e^{-\lambda x(y+1)} \qquad x > 0, y > 0$$

and $f(x, y) = 0$ otherwise.

(a) Compute the densities of X and of Y.
(b) Are the r.v.'s $U = X$ and $V = XY$ independent? What is the density of XY?
(c) Compute the conditional expectation of X given $Y = y$ and the squared L^2 distance $E[(X - E(X|Y))^2]$.

Recall (4.6).

4.22 (p. 356) Let X, Y be independent r.v.'s Gamma(α, λ)- and Gamma(β, λ)-distributed respectively.

(a) What is the density of $X + Y$?
(b) What is the joint density of X and $X + Y$?
(c) What is the conditional density, $\overline{g}(\cdot; z)$, of X given $X + Y = z$?
(d) Compute $E(X|X + Y = z)$ and the regression line of X with respect to $X + Y$.

4.23 (p. 357) Let X, Y be real r.v.'s with joint density

$$f(x, y) = \frac{1}{2\pi\sqrt{1 - r^2}} \exp\left(-\frac{1}{2(1 - r^2)}(x^2 - 2rxy + y^2)\right)$$

where $-1 < r < 1$.

(a) Determine the marginal densities of X and Y.
(b) Compute $E(X|Y)$ and $E(X|X + Y)$.

4.24 (p. 358) Let X be an $N(0, 1)$-distributed r.v. and Y another real r.v. In which of the following situations is the pair (X, Y) Gaussian?

(a) The conditional law of Y given $X = x$ is an $N(\frac{1}{2}x, 1)$ distribution.
(b) The conditional law of Y given $X = x$ is an $N(\frac{1}{2}x^2, 1)$ distribution.
(c) The conditional law of Y given $X = x$ is an $N(0, \frac{1}{4}x^2)$ distribution.

Chapter 5
Martingales

5.1 Stochastic Processes

A stochastic process is a mathematical object that is intended to model a quantity that performs a random motion. It will therefore be something like $(X_n)_n$, where the r.v.'s X_n are defined on some probability space $(\Omega, \mathcal{F}, P)$ and take their values on the same measurable space $(E, \mathcal{E})$. Here n is to be seen as a time. It is also possible to consider families $(X_t)_t$ with $t \in \mathbb{R}^+$, i.e. in continuous time, but we shall only deal with discrete time models.

> A *filtration* is an increasing family $(\mathcal{F}_n)_n$ of sub-σ-algebras of $\mathcal{F}$. A process $(X_n)_n$ is said to be *adapted* to the filtration $(\mathcal{F}_n)_n$ if, for every n, X_n is $\mathcal{F}_n$-measurable.

Given a process $(X_n)_n$, we can always consider the filtration $(\mathcal{G}_n)_n$ defined as $\mathcal{G}_n = \sigma(X_1, \ldots, X_n)$. This is the *natural filtration* of the process and, of course, is the smallest filtration with respect to which the process is adapted.

Just a moment for intuition: $X_1, \ldots, X_n$ are the positions of the process before ($\leq$) time n and therefore are quantities that are known to an observer at time n. The σ-algebra $\mathcal{F}_n$ represents the family of events for which, at time n, it is known whether they have taken place or not.

© The Author(s), under exclusive license to Springer Nature Switzerland AG 2023   205
P. Baldi, *Probability*, Universitext, https://doi.org/10.1007/978-3-031-38492-9_5

5.2 Martingales: Definitions and General Facts

Let $(\Omega, \mathcal{F}, P)$ be a probability space and $(\mathcal{F}_n)_n$ a filtration on it.

Definition 5.1 A *martingale* (resp. a supermartingale, a submartingale) of the filtration $(\mathcal{F}_n)_n$ is a process $(M_n)_n$ adapted to $(\mathcal{F}_n)_n$, such that M_n *is integrable* for every n and, for every $n \geq m$,

$$E(M_n | \mathcal{F}_m) = M_m \qquad (\text{resp. } \leq M_m, \geq M_m) . \qquad (5.1)$$

Martingales are an important tool in probability, appearing in many contexts of the theory. For more information on this subject, in addition to almost all references mentioned at the beginning of Chap. 1, see also [1], [21].

Of course (5.1) is equivalent to requiring that, for every n,

$$E(M_n | \mathcal{F}_{n-1}) = M_{n-1} , \qquad (5.2)$$

as this relation entails, for $m < n$,

$$E(M_n | \mathcal{F}_m) = E\big[E(M_n | \mathcal{F}_{n-1}) | \mathcal{F}_m\big] = E(M_{n-1} | \mathcal{F}_m) = \cdots = E(M_{m+1} | \mathcal{F}_m) = M_m .$$

It is sometimes important to specify with respect to which filtration $(M_n)_n$ is a martingale. Note for now that if $(M_n)_n$ is a martingale with respect to $(\mathcal{F}_n)_n$, then it is a martingale with respect to every smaller filtration (provided it contains the natural filtration). Indeed if $(\mathcal{F}'_n)_n$ is another filtration to which the martingale is adapted and with $\mathcal{F}'_n \subset \mathcal{F}_n$, then

$$E(M_n | \mathcal{F}'_m) = E\big[E(M_n | \mathcal{F}_m) | \mathcal{F}'_m\big] = E(M_m | \mathcal{F}'_m) = M_m .$$

Of course a similar property holds for super- and submartingales. When the filtration is not specified we shall understand it to be the natural filtration.

The following example presents three typical situations giving rise to martingales.

Example 5.2

(a) Let $(Z_k)_k$ be a sequence of real centered independent r.v.'s and let $X_n = Z_1 + \cdots + Z_n$. Then $(X_n)_n$ is a martingale.

> Indeed we have $X_n = X_{n-1} + Z_n$ and, as Z_n is independent of $X_1, \ldots, X_{n-1}$ and therefore of $\mathscr{F}_{n-1} = \sigma(X_1, \ldots, X_{n-1})$,
>
> $$E(X_n | \mathscr{F}_{n-1}) = E(X_{n-1} | \mathscr{F}_{n-1}) + E(Z_n | \mathscr{F}_{n-1}) = X_{n-1} + E(Z_n) = X_{n-1} .$$
>
> (b) Let $(U_k)_k$ be a sequence of real independent r.v.'s such that $E(U_k) = 1$ for every k and let $Y_n = U_1 \cdots U_n$. Then $(Y_n)_n$ is a martingale: with an idea similar to (a)
>
> $$E(Y_n | \mathscr{F}_{n-1}) = E(U_n Y_{n-1} | \mathscr{F}_{n-1}) = Y_{n-1} E(U_n) = Y_{n-1} .$$
>
> A particularly important instance of these martingales appears when the U_n are positive.
> (c) Let X be an integrable r.v. and $(\mathscr{F}_n)_n$ a filtration, then $X_n = E(X | \mathscr{F}_n)$ is a martingale. Indeed, if $n > m$, (Proposition 4.5 (c))
>
> $$E(X_n | \mathscr{F}_m) = E\big[E(X | \mathscr{F}_n) | \mathscr{F}_m\big] = E(X | \mathscr{F}_m) = X_m .$$
>
> We shall see that these martingales may have very different behaviors.

It is clear that linear combinations of martingales are also martingales and linear combinations with positive coefficients of supermartingales (resp. submartingales) are again supermartingales (resp. submartingales). If $(M_n)_n$ is a supermartingale, $(-M_n)_n$ is a submartingale and conversely.

Moreover, if M is a martingale (resp. a submartingale) and $\Phi : \mathbb{R} \to \mathbb{R}$ is a l.s.c. convex function (resp. convex and increasing) such that $\Phi(M_n)$ is also integrable, then $(\Phi(M_n))_n$ is a submartingale with respect to the same filtration. This is a consequence of Jensen's inequality, Proposition 4.6 (d): if M is a martingale we have for $n \geq m$

$$E\big[\Phi(M_n) | \mathscr{F}_m\big] \geq \Phi\big(E[M_n | \mathscr{F}_m]\big) = \Phi(M_m) .$$

In particular, if $(M_n)_n$ is a martingale then $(|M_n|)_n$ is a submartingale.

We say that $(M_n)_n$ is a martingale (resp. supermartingale, submartingale) of L^p, $p \geq 1$, if $M_n \in L^p$ for every n and we shall speak of *square integrable* martingales (resp. supermartingales, submartingales) for $p = 2$. If $(M_n)_n$ is a martingale of L^p, $p \geq 1$, then $(|M_n|^p)_n$ is a submartingale.

Beware of a possible mistake: it is not granted that if M is a submartingale the same is true of $(|M_n|)_n$ or $(M_n^2)_n$ (even if M is square integrable): the functions $x \mapsto |x|$ and $x \mapsto x^2$ are indeed convex but *not increasing*. This statement becomes true if we add the assumption that M is positive: the functions $x \mapsto |x|$ and $x \mapsto x^2$ are increasing when restricted to $\mathbb{R}^+$.

5.3 Doob's Decomposition

A process $(A_n)_n$ is said to be *a predictable increasing process* for the filtration $(\mathscr{F}_n)_n$ if

- $A_0 = 0$,
- for every n, $A_n \leq A_{n+1}$ a.s.,
- for every n, A_{n+1} is $\mathscr{F}_n$-measurable.

Let $(X_n)_n$ be an $(\mathscr{F}_n)_n$-submartingale and recursively define

$$A_0 = 0, \quad A_{n+1} = A_n + \mathrm{E}(X_{n+1}|\mathscr{F}_n) - X_n . \tag{5.3}$$

By construction $(A_n)_n$ is a predictable increasing process. Actually by induction A_{n+1} is $\mathscr{F}_n$-measurable and, as X is a submartingale, $A_{n+1} - A_n = \mathrm{E}(X_{n+1}|\mathscr{F}_n) - X_n \geq 0$.

If $M_n = X_n - A_n$ then

$$\mathrm{E}(M_{n+1}|\mathscr{F}_n) = \mathrm{E}(X_{n+1}|\mathscr{F}_n) - A_{n+1} = X_n - A_n = M_n$$

(we use the fact that A_{n+1} is $\mathscr{F}_n$-measurable). Hence $(M_n)_n$ is a martingale.

Such a decomposition is unique: if $X_n = M'_n + A'_n$ is another decomposition of $(X_n)_n$ into the sum of a martingale M' and of a predictable increasing process A', then $A_0 = A'_0 = 0$ and

$$A'_{n+1} - A'_n = X_{n+1} - X_n - (M'_{n+1} - M'_n).$$

Conditioning with respect to $\mathscr{F}_n$, we obtain $A'_{n+1} - A'_n = \mathrm{E}(X_{n+1}|\mathscr{F}_n) - X_n = A_{n+1} - A_n$; hence $A'_n = A_n$ and $M'_n = M_n$. We have thus obtained that

every submartingale $(X_n)_n$ can be decomposed uniquely into the sum of a martingale $(M_n)_n$ and of a predictable increasing process $(A_n)_n$.

This is *Doob's decomposition*. The process A is the *compensator* of $(X_n)_n$.

If $(M_n)_n$ is a square integrable martingale, then $(M_n^2)_n$ is a submartingale. Its compensator is the *associated increasing process* to the martingale $(M_n)_n$.

5.4 Stopping Times

When dealing with stochastic processes an important technique consists in the investigation of its value when stopped at some random time. This section introduces the right notion in this direction.

Let $(\Omega, \mathcal{F}, P)$ be a probability space and $(\mathcal{F}_n)_n$ a filtration on it. Let $\mathcal{F}_\infty = \sigma(\bigcup_{n \geq 0} \mathcal{F}_n)$.

Definition 5.3

(a) A *stopping time* of the filtration $(\mathcal{F}_n)_n$ is a map $\tau : \Omega \to \mathbb{N} \cup \{+\infty\}$ (the value $+\infty$ is allowed) such that, for every $n \geq 0$,

$$\{\tau \leq n\} \in \mathcal{F}_n .$$

(b) Let

$$\mathcal{F}_\tau = \{A \in \mathcal{F}_\infty; \text{ for every } n \geq 0, \ A \cap \{\tau \leq n\} \in \mathcal{F}_n\} .$$

$\mathcal{F}_\tau$ is the σ-*algebra of events prior to time* τ.

Remark 5.4 In (a) and (b), the conditions $\{\tau \leq n\} \in \mathcal{F}_n$ and $A \cap \{\tau \leq n\} \in \mathcal{F}_n$ are equivalent to requiring that $\{\tau = n\} \in \mathcal{F}_n$ and $A \cap \{\tau = n\} \in \mathcal{F}_n$, respectively, as

$$\{\tau \leq n\} = \bigcup_{k=0}^{n} \{\tau = k\}, \qquad \{\tau = n\} = \{\tau \leq n\} \setminus \{\tau \leq n - 1\}$$

so that if, for instance, $\{\tau = n\} \in \mathcal{F}_n$ for every n then also $\{\tau \leq n\} \in \mathcal{F}_n$ and conversely.

Remark 5.5 Note that a deterministic time $\tau \equiv m$ is a stopping time. Indeed

$$\{\tau \leq n\} = \begin{cases} \emptyset & \text{if } n < m \\ \Omega & \text{if } n \geq m , \end{cases}$$

and in any case $\{\tau \leq n\} \in \mathcal{F}_n$. Not unexpectedly in this case $\mathcal{F}_\tau = \mathcal{F}_m$: if $A \in \mathcal{F}_\infty$ then

$$A \cap \{\tau \leq n\} = \begin{cases} \emptyset & \text{if } n < m \\ A & \text{if } n \geq m \, . \end{cases}$$

Therefore $A \cap \{\tau \leq n\} \in \mathcal{F}_n$ for every n if and only if $A \in \mathcal{F}_m$.

A stopping time is a random time at which a process adapted to the filtration $(\mathcal{F}_n)_n$ is observed or modified. Recalling that intuitively $\mathcal{F}_n$ is the σ-algebra of events that are known at time n, the condition $\{\tau \leq n\} \in \mathcal{F}_n$ imposes the condition that at time n it is known whether τ has already happened or not. A typical example is the first time at which the process takes some values, as in the following example.

Example 5.6 (Passage Times) Let X be a stochastic process with values in $(E, \mathcal{E})$ adapted to the filtration $(\mathcal{F}_n)_n$. Let, for $A \in \mathcal{E}$,

$$\tau_A(\omega) = \inf\{n; \, X_n(\omega) \in A\} \, , \qquad (5.4)$$

with the understanding $\inf \emptyset = +\infty$. Then τ_A is a stopping time as

$$\{\tau_A = n\} = \{X_0 \notin A, X_1 \notin A, \ldots, X_{n-1} \notin A, X_n \in A\} \in \mathcal{F}_n \, .$$

τ_A is the *passage time* at A, i.e. the first time at which X visits the set A. Conversely, let

$$\rho_A(\omega) = \sup\{n; \, X_n(\omega) \in A\}$$

i.e. *the last time* at which the process visits the set A. This *is not* in general a stopping time: in order to know whether $\rho_A \leq n$ you need to know the positions of the process at times after time n.

The following proposition states some important properties of stopping times. They are immediate consequences of the definitions: we advise the reader to try to work out the proofs (without looking at them beforehand...) as an exercise.

Proposition 5.7 *Let $(\mathcal{F}_n)_n$ be a filtration and τ_1, τ_2 stopping times of this filtration. Then the following properties hold.*

(continued)

> **Proposition 5.7** (continued)
> (a) $\tau_1 + \tau_2$, $\tau_1 \vee \tau_2$, $\tau_1 \wedge \tau_2$ *are stopping times with respect to the same filtration.*
> (b) *If* $\tau_1 \leq \tau_2$, *then* $\mathscr{F}_{\tau_1} \subset \mathscr{F}_{\tau_2}$.
> (c) $\mathscr{F}_{\tau_1 \wedge \tau_2} = \mathscr{F}_{\tau_1} \cap \mathscr{F}_{\tau_2}$.
> (d) *Both events* $\{\tau_1 < \tau_2\}$ *and* $\{\tau_1 = \tau_2\}$ *belong to* $\mathscr{F}_{\tau_1} \cap \mathscr{F}_{\tau_2}$.

Proof

(a) The statement follows from the relations

$$\{\tau_1 + \tau_2 \leq n\} = \bigcup_{k=0}^{n} \{\tau_1 = k, \tau_2 \leq n - k\} \in \mathscr{F}_n ,$$

$$\{\tau_1 \wedge \tau_2 \leq n\} = \{\tau_1 \leq n\} \cup \{\tau_2 \leq n\} \in \mathscr{F}_n ,$$

$$\{\tau_1 \vee \tau_2 \leq n\} = \{\tau_1 \leq n\} \cap \{\tau_2 \leq n\} \in \mathscr{F}_n .$$

(b) Let $A \in \mathscr{F}_{\tau_1}$, i.e. such that $A \cap \{\tau_1 \leq n\} \in \mathscr{F}_n$ for every n; we must prove that also $A \cap \{\tau_2 \leq n\} \in \mathscr{F}_n$ for every n. As $\tau_1 \leq \tau_2$, we have $\{\tau_2 \leq n\} \subset \{\tau_1 \leq n\}$ and therefore

$$A \cap \{\tau_2 \leq n\} = \underbrace{A \cap \{\tau_1 \leq n\}}_{\in \mathscr{F}_n} \cap \{\tau_2 \leq n\} \in \mathscr{F}_n .$$

(c) Thanks to (b) $\mathscr{F}_{\tau_1 \wedge \tau_2} \subset \mathscr{F}_{\tau_1}$ and $\mathscr{F}_{\tau_1 \wedge \tau_2} \subset \mathscr{F}_{\tau_2}$, hence $\mathscr{F}_{\tau_1 \wedge \tau_2} \subset \mathscr{F}_{\tau_1} \cap \mathscr{F}_{\tau_2}$. Conversely, let $A \in \mathscr{F}_{\tau_1} \cap \mathscr{F}_{\tau_2}$. Then, for every n, we have $A \cap \{\tau_1 \leq n\} \in \mathscr{F}_n$ and $A \cap \{\tau_2 \leq n\} \in \mathscr{F}_n$. Taking the union we find that

$$\big(A \cap \{\tau_1 \leq n\}\big) \cup \big(A \cap \{\tau_2 \leq n\}\big) = A \cap \big(\{\tau_1 \leq n\} \cup \{\tau_2 \leq n\}\big) = A \cap \{\tau_1 \wedge \tau_2 \leq n\}$$

so that $A \cap \{\tau_1 \wedge \tau_2 \leq n\} \in \mathscr{F}_n$, hence the opposite inclusion $\mathscr{F}_{\tau_1 \wedge \tau_2} \supset \mathscr{F}_{\tau_1} \cap \mathscr{F}_{\tau_2}$.

(d) Let us prove that $\{\tau_1 < \tau_2\} \in \mathscr{F}_{\tau_1}$: we must show that $\{\tau_1 < \tau_2\} \cap \{\tau_1 \leq n\} \in \mathscr{F}_n$. We have

$$\{\tau_1 < \tau_2\} \cap \{\tau_1 \leq n\} = \bigcup_{k=0}^{n} \big(\{\tau_1 = k\} \cap \{\tau_2 > k\}\big) .$$

This event belongs to $\mathcal{F}_n$, as $\{\tau_2 > k\} = \{\tau_2 \leq k\}^c \in \mathcal{F}_k \subset \mathcal{F}_n$. Therefore $\{\tau_1 < \tau_2\} \in \mathcal{F}_{\tau_1}$. Similarly

$$\{\tau_1 < \tau_2\} \cap \{\tau_2 \leq n\} = \bigcup_{k=0}^{n} \Big(\{\tau_2 = k\} \cap \{\tau_1 < k\} \Big)$$

and again we find that $\{\tau_1 < \tau_2\} \cap \{\tau_2 \leq n\} \in \mathcal{F}_n$. Therefore $\{\tau_1 < \tau_2\} \in \mathcal{F}_{\tau_1} \cap \mathcal{F}_{\tau_2}$. Finally note that

$$\{\tau_1 = \tau_2\} = \{\tau_1 < \tau_2\}^c \cap \{\tau_2 < \tau_1\}^c \in \mathcal{F}_{\tau_1} \cap \mathcal{F}_{\tau_2}.$$

∎

For a given filtration $(\mathcal{F}_n)_n$ let X be an adapted process and τ a *finite* stopping time. Then we can define its position at time τ:

$$X_\tau = X_n \qquad \text{on } \{\tau = n\}, \ n \in \mathbb{N}.$$

Note that X_τ is $\mathcal{F}_\tau$-measurable as

$$\{X_\tau \in A\} \cap \{\tau = n\} = \{X_n \in A\} \cap \{\tau = n\} \in \mathcal{F}_n.$$

A typical operation that is applied to a process is stopping: if $(X_n)_n$ is adapted to the filtration $(\mathcal{F}_n)_n$ and τ is a stopping time for this filtration, the process stopped at time τ is $(X_{\tau \wedge n})_n$, i.e. a process that moves as $(X_n)_n$ up to time τ and then stays fixed at the position X_τ (at least if $\tau < +\infty$).

Also the stopped process is adapted to the filtration $(\mathcal{F}_n)_n$, as $\tau \wedge n$ is a stopping time which is $\leq n$ so that $X_{\tau \wedge n}$ is $\mathcal{F}_{\tau \wedge n}$-measurable and by Proposition 5.7 (b) $\mathcal{F}_{\tau \wedge n} \subset \mathcal{F}_n$.

The following remark states that a stopped martingale is also a martingale.

Remark 5.8 If $(X_n)_n$ is an $(\mathcal{F}_n)_n$-martingale (resp. supermartingale, submartingale), the same is true for the stopped process $X_n^\tau = X_{n \wedge \tau}$, where τ is a stopping time of the filtration $(\mathcal{F}_n)_n$.

Actually $X_{(n+1) \wedge \tau} = X_{n \wedge \tau}$ on $\{\tau \leq n\}$ and therefore

$$\mathrm{E}(X_{n+1}^\tau - X_n^\tau | \mathcal{F}_n) = \mathrm{E}[(X_{n+1} - X_n) 1_{\{\tau \geq n+1\}} | \mathcal{F}_n].$$

By the definition of stopping time $\{\tau \geq n+1\} = \{\tau \leq n\}^c \in \mathcal{F}_n$ and therefore

$$\mathrm{E}(X_{n+1}^\tau - X_n^\tau | \mathcal{F}_n) = 1_{\{\tau \geq n+1\}} \mathrm{E}(X_{n+1} - X_n | \mathcal{F}_n) = 0. \tag{5.5}$$

Note that the stopped process is a martingale *with respect to the same filtration*, $(\mathcal{F}_n)_n$, which may be larger than the natural filtration of the stopped process.

Remark 5.9 Let $M = (M_n)_n$ be a square integrable martingale with respect to the filtration $(\mathcal{F}_n)_n$, τ a stopping time for this filtration and M^τ the stopped martingale, which is of course also square integrable as $|M_{n \wedge \tau}| \leq \sum_{k=0}^n |M_k|$. Let A be the associated increasing process of M and A' the associated increasing process of M^τ. Is it true that $A'_n = A_{n \wedge \tau}$?

Note first that $(A_{n \wedge \tau})_n$ is an increasing predictable process. Actually it is obviously increasing and, as

$$A_{(n+1) \wedge \tau} = \sum_{k=1}^n A_k 1_{\{\tau = k\}} + A_{n+1} 1_{\{\tau > n\}}$$

$A_{(n+1) \wedge \tau}$ is the sum of $\mathcal{F}_n$-measurable r.v.'s hence $\mathcal{F}_n$-measurable itself.

Finally, by definition $(M_n^2 - A_n)_n$ is a martingale, hence so is $(M_{n \wedge \tau}^2 - A_{n \wedge \tau})_n$ (stopping a martingale always gives rise to a martingale). Therefore $A'_n = A_{n \wedge \tau}$, as the associated increasing process is unique.

5.5 The Stopping Theorem

The following result is the key tool in the proof of many properties of martingales appearing in the sequel.

Theorem 5.10 (The Stopping Theorem) Let $X = (\Omega, \mathcal{F}, (\mathcal{F}_n)_n, (X_n)_n, P)$ be a supermartingale and τ_1, τ_2 stopping times of the filtration $(\mathcal{F}_n)_n$, a.s. bounded and such that $\tau_1 \leq \tau_2$ a.s. Then the r.v.'s X_{τ_1} and X_{τ_2} are integrable and

$$E(X_{\tau_2} | \mathcal{F}_{\tau_1}) \leq X_{\tau_1}. \tag{5.6}$$

In particular, $E(X_{\tau_2}) \leq E(X_{\tau_1})$.

Proof The integrability of X_{τ_1} and X_{τ_2} is immediate, as, for $i = 1, 2$ and denoting by k a number larger than τ_2, $|X_{\tau_i}| \leq \sum_{j=1}^k |X_j|$.

In order to prove (5.6) let us first assume $\tau_2 \equiv k \in \mathbb{N}$ and let $A \in \mathcal{F}_{\tau_1}$. As $A \cap \{\tau_1 = j\} \in \mathcal{F}_j$, we have, for $j \leq k$,

$$E(X_{\tau_1} 1_{A \cap \{\tau_1 = j\}}) = E(X_j 1_{A \cap \{\tau_1 = j\}}) \geq E(X_k 1_{A \cap \{\tau_1 = j\}}),$$

where the inequality holds because $(X_n)_n$ is a supermartingale and $A \cap \{\tau_1 = j\} \in \mathcal{F}_j$. Taking the sum with respect to j, $0 \leq j \leq k$,

$$E\big(X_{\tau_1} 1_A\big) = \sum_{j=0}^{k} E\big(X_j 1_{A \cap \{\tau_1 = j\}}\big) \geq \sum_{j=0}^{k} E\big(X_k 1_{A \cap \{\tau_1 = j\}}\big) = E\big(X_{\tau_2} 1_A\big) \,,$$

which proves the theorem if τ_2 is a constant stopping time. Let us now assume more generally $\tau_2 \leq k$. If we apply the result of the first part of the proof to the stopped martingale $(X_n^{\tau_2})_n$ (recall that $X_n^{\tau_2} = X_{n \wedge \tau_2}$) and to the stopping times τ_1 and k, we have

$$E\big(X_{\tau_1} 1_A\big) = E\big(X_{\tau_1}^{\tau_2} 1_A\big) \geq E\big(X_k^{\tau_2} 1_A\big) = E\big(X_{\tau_2} 1_A\big) \,,$$

which concludes the proof.   ∎

Theorem 5.10 applied to X and $-X$ gives

Corollary 5.11 *Under the assumptions of Theorem 5.10, if moreover X is a martingale,*

$$E(X_{\tau_2} | \mathcal{F}_{\tau_1}) = X_{\tau_1} \,.$$

In some sense the stopping theorem states that the martingale (resp. supermartingale, submartingale) relation (5.1) still holds if the times m, n are replaced by bounded stopping times.

If X is a martingale, applying Corollary 5.11 to the stopping times $\tau_1 = 0$ and $\tau_2 = \tau$ we find that the mean $E(X_\tau)$ is constant as τ ranges among bounded stopping times.

Beware: these stopping times must be *bounded*, i.e. a number k must exist such that $\tau_2(\omega) \leq k$ for every ω a.s. A finite stopping time is not necessarily bounded.

Very often however we shall need to apply the relation (5.6) to unbounded stopping times: as we shall see, this can often be done in a simple way by approximating the unbounded stopping times with bounded ones.

The following is a first application of the stopping theorem.

Theorem 5.12 (Maximal Inequalities) *Let X be a supermartingale and $\lambda > 0$. Then*

$$\lambda P\left(\sup_{0 \le n \le k} X_n \ge \lambda\right) \le E(X_0) + E(X_k^-), \tag{5.7}$$

$$\lambda P\left(\inf_{0 \le n \le k} X_n \le -\lambda\right) \le E\left(X_k 1_{\{\inf_{0 \le n \le k} X_n \le -\lambda\}}\right). \tag{5.8}$$

Proof Let

$$\tau(\omega) = \begin{cases} \inf\{n; n \le k, X_n(\omega) \ge \lambda\} \\ k \text{ if } \{\} = \emptyset. \end{cases}$$

τ is a bounded stopping time and, by (5.6) applied to the stopping times $\tau_2 = \tau$, $\tau_1 = 0$,

$$E(X_0) \ge E(X_\tau) = E\left(X_\tau 1_{\{\sup_{0 \le n \le k} X_n \ge \lambda\}}\right) + E\left(X_k 1_{\{\sup_{0 \le n \le k} X_n < \lambda\}}\right)$$

and now just note that, as $X_\tau \ge \lambda$ on $\{\sup_{0 \le n \le k} X_n \ge \lambda\}$,

$$E\left(X_\tau 1_{\{\sup_{0 \le n \le k} X_n \ge \lambda\}}\right) \ge \lambda P\left(\sup_{1 \le n \le k} X_n \ge \lambda\right),$$

$$E\left(X_k 1_{\{\sup_{0 \le n \le k} X_n < \lambda\}}\right) \ge -E\left(X_k^- 1_{\{\sup_{0 \le n \le k} X_n < \lambda\}}\right) \ge -E(X_k^-),$$

which gives (5.7). As for (5.8), if

$$\tau(\omega) = \begin{cases} \inf\{n; n \le k, X_n(\omega) \le -\lambda\} \\ k \text{ if } \{\} = \emptyset, \end{cases}$$

τ is again a bounded stopping time and now Theorem 5.10 applied to the stopping times $\tau_2 = k$ and $\tau_1 = \tau$ gives

$$E(X_k) \le E(X_\tau) = E\left(X_\tau 1_{\{\inf_{0 \le n \le k} X_n \le -\lambda\}}\right) + E\left(X_k 1_{\{\inf_{0 \le n \le k} X_n > -\lambda\}}\right)$$

$$\le -\lambda P\left(\inf_{0 \le n \le k} X_n \le -\lambda\right) + E\left(X_k 1_{\{\inf_{0 \le n \le k} X_n > -\lambda\}}\right)$$

and therefore

$$\lambda P\left(\inf_{0\le n\le k} X_n \le -\lambda\right) \le E\left(X_k 1_{\{\inf_{0\le n\le k} X_n > -\lambda\}}\right) - E(X_k)$$

$$= -E\left(X_k 1_{\{\inf_{0\le n\le k} X_n \le -\lambda\}}\right),$$

i.e. (5.8). ∎

Note that (5.7) implies that if a supermartingale is such that $\sup_{k\ge 0} E(X_k^-) < +\infty$ (in particular if it is a positive supermartingale) then the r.v. $\sup_{n\ge 0} X_n$ is finite a.s. Indeed, by (5.7),

$$\lambda P\left(\sup_{n\ge 0} X_n \ge \lambda\right) = \lim_{k\to\infty} \lambda P\left(\sup_{0\le n\le k} X_n \ge \lambda\right) \le E(X_0) + \sup_{k\ge 0} E(X_k^-) < +\infty,$$

from which

$$\lim_{\lambda\to +\infty} P\left(\sup_{n\ge 0} X_n \ge \lambda\right) = 0,$$

i.e. the r.v. $\sup_{n>0} X_n$ is a.s. finite. This will become more clear in the next section.

5.6   Almost Sure Convergence

One of the reasons for the importance of martingales is the result of this section: it guarantees that, under assumptions that are quite weak and easy to check, a martingale converges a.s.

Let $[a, b] \subset \mathbb{R}$, $a < b$, be an interval and

$$\gamma_{a,b}^k(\omega) = \text{how many times the path } (X_n(\omega))_{n\le k} \text{ crosses ascending } [a, b].$$

We say that $(X_n(\omega))_n$ crosses ascending the interval $[a, b]$ once in the time interval $[i, j]$ if

$$X_i(\omega) < a,$$

$$X_m(\omega) \le b \text{ for } m = i+1, \ldots, j-1,$$

$$X_j(\omega) > b.$$

When this happens we say that the process $(X_n)_n$ has performed one *upcrossing* on the interval $[a, b]$ (see Fig. 5.1). $\gamma_{a,b}^k(\omega)$ is therefore the number of upcrossings on the interval $[a, b]$ of the path $(X_n(\omega))_n$ up to time k.

The proof of the convergence theorem has some technical points, but the baseline is quite simple: in order to prove that a sequence converges we must prove first of

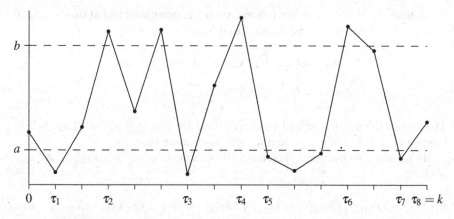

Fig. 5.1 Here $\gamma_{a,b}^k = 3$

all that it does not oscillate too much. Hence the following proposition, which states that a supermartingale cannot make too many upcrossings, is the key tool.

Proposition 5.13 *If X is a supermartingale, then*

$$(b - a)\,\mathrm{E}(\gamma_{a,b}^k) \le \mathrm{E}[(X_k - a)^-]\,. \tag{5.9}$$

Proof Let us consider the following sequence of stopping times

$$\tau_1(\omega) = \begin{cases} \inf\{n;\, n \le k,\, X_n(\omega) < a\} \\ k \text{ if } \{\,\} = \emptyset\,, \end{cases}$$

$$\tau_2(\omega) = \begin{cases} \inf\{n;\, \tau_1(\omega) < n \le k,\, X_n(\omega) > b\} \\ k \text{ if } \{\,\} = \emptyset\,, \end{cases}$$

$$\cdots$$

$$\tau_{2m-1}(\omega) = \begin{cases} \inf\{n;\, \tau_{2m-2}(\omega) < n \le k,\, X_n(\omega) < a\} \\ k \text{ if } \{\,\} = \emptyset\,, \end{cases}$$

$$\tau_{2m}(\omega) = \begin{cases} \inf\{n;\, \tau_{2m-1}(\omega) < n \le k,\, X_n(\omega) > b\} \\ k \text{ if } \{\,\} = \emptyset\,, \end{cases}$$

i.e. at time τ_{2i}, if $X_{\tau_{2i}} > b$, the i-th upcrossing is completed and at time τ_{2i-1}, if $X_{\tau_{2i-1}} < a$, the i-th upcrossing is initialized. Let

$$A_{2m} = \{\tau_{2m} \leq k, X_{\tau_{2m}} > b\} = \{\gamma_{a,b}^{k} \geq m\},$$

$$A_{2m-1} = \{\gamma_{a,b}^{k} \geq m - 1, X_{\tau_{2m-1}} < a\}.$$

The idea of the proof is to find an upper bound for $P(\gamma_{a,b}^{k} \geq m) = P(A_{2m})$. It is immediate that $A_i \in \mathscr{F}_{\tau_i}$, as τ_i and X_{τ_i} are $\mathscr{F}_{\tau_i}$-measurable r.v.'s.

By the stopping theorem, Theorem 5.10, with the stopping times τ_{2m-1} and τ_{2m} we have

$$E[(X_{\tau_{2m}} - a)1_{A_{2m-1}} | \mathscr{F}_{\tau_{2m-1}}] = 1_{A_{2m-1}} E(X_{\tau_{2m}} - a | \mathscr{F}_{\tau_{2m-1}}) \leq 1_{A_{2m-1}}(X_{\tau_{2m-1}} - a).$$

As $X_{\tau_{2m-1}} < a$ on A_{2m-1}, taking the expectation we have

$$0 \geq E[(X_{\tau_{2m-1}} - a)1_{A_{2m-1}}] \geq E[(X_{\tau_{2m}} - a)1_{A_{2m-1}}]. \tag{5.10}$$

Obviously $A_{2m-1} = A_{2m} \cup (A_{2m-1} \setminus A_{2m})$ and

$$X_{\tau_{2m}} \geq b \qquad \text{on } A_{2m}$$

$$X_{\tau_{2m}} = X_k \qquad \text{on } A_{2m-1} \setminus A_{2m}$$

so that (5.10) gives

$$0 \geq E[(X_{\tau_{2m}} - a)1_{A_{2m}}] + E[(X_{\tau_{2m}} - a)1_{A_{2m-1} \setminus A_{2m}}]$$

$$\geq (b - a)P(A_{2m}) + \int_{A_{2m-1} \setminus A_{2m}} (X_k - a)\, dP,$$

from which we deduce

$$(b - a)P(\gamma_{a,b}^{k} \geq m) \leq -\int_{A_{2m-1} \setminus A_{2m}} (X_k - a)\, dP \leq \int_{A_{2m-1} \setminus A_{2m}} (X_k - a)^{-}\, dP. \tag{5.11}$$

The events $A_{2m-1} \setminus A_{2m}$ are pairwise disjoint as m ranges over $\mathbb{N}$ so that, taking the sum in m in (5.11),

$$(b - a) \sum_{m=1}^{\infty} P(\gamma_{a,b}^{k} \geq m) \leq E[(X_k - a)^{-}]$$

and the result follows recalling that $\sum_{m=1}^{\infty} P(\gamma_{a,b}^{k} \geq m) = E(\gamma_{a,b}^{k})$ (Remark 2.1). ∎

Theorem 5.14 *Let X be a supermartingale such that*

$$\sup_{n\geq 0} E(X_n^-) < +\infty \,. \tag{5.12}$$

Then it converges a.s. to a finite limit.

Proof For fixed $a < b$ let $\gamma_{a,b}(\omega)$ denote the number of upcrossings on the interval $[a, b]$ of the whole path $(X_n(\omega))_n$. As $(X_n - a)^- \leq a^+ + X_n^-$, by Proposition 5.13,

$$\begin{aligned}
E(\gamma_{a,b}) &= \lim_{k\to\infty} E(\gamma_{a,b}^k) \leq \frac{1}{b-a} \sup_{n\geq 0} E[(X_n - a)^-] \\
&\leq \frac{1}{b-a} \left(a^+ + \sup_{n\geq 0} E(X_n^-) \right) < +\infty \,.
\end{aligned} \tag{5.13}$$

In particular $\gamma_{a,b} < +\infty$ a.s., i.e. there exists a negligible event $N_{a,b}$ such that $\gamma_{a,b}(\omega) < +\infty$ for $\omega \notin N_{a,b}$; taking the union, N, of the sets $N_{a,b}$ as a, b range in $\mathbb{Q}$ with $a < b$, we can assume that, outside the negligible event N, we have $\gamma_{a,b} < +\infty$ for every $a, b \in \mathbb{R}$.

Let us show that for $\omega \notin N$ the sequence $(X_n(\omega))_n$ converges: otherwise, if $a = \underline{\lim}_{n\to\infty} X_n(\omega) < \overline{\lim}_{n\to\infty} X_n(\omega) = b$, the sequence $(X_n(\omega))_n$ would take values close to a infinitely many times and also values close to b infinitely many times. Hence, for every $\alpha, \beta \in \mathbb{R}$ with $a < \alpha < \beta < b$, the path $(X_n(\omega))_n$ would cross the interval $[\alpha, \beta]$ infinitely many times and we would have $\gamma_{\alpha,\beta}(\omega) = +\infty$.

The limit is moreover finite: thanks to (5.13)

$$\lim_{b\to+\infty} E(\gamma_{a,b}) = 0$$

but $\gamma_{a,b}(\omega)$ is decreasing in b and therefore

$$\lim_{b\to+\infty} \gamma_{a,b}(\omega) = 0 \qquad \text{a.s.}$$

As $\gamma_{a,b}$ can only take integer values, $\gamma_{a,b}(\omega) = 0$ for b large enough and $(X_n(\omega))_n$ is therefore bounded from above a.s. In the same way we see that it is bounded from below. ∎

The assumptions of Theorem 5.14 are in particular satisfied by all positive supermartingales.

Remark 5.15 Note that, if X is a martingale, condition (5.12) of Theorem 5.14 is equivalent to boundedness in L^1. Indeed, of course if $(X_n)_n$ is bounded in L^1 then (5.12) is satisfied. Conversely, taking into account the decomposition $X_n = X_n^+ - X_n^-$, as $n \mapsto \mathrm{E}(X_n) := c$ is constant, we have $\mathrm{E}(X_n^+) = c + \mathrm{E}(X_n^-)$, hence

$$\mathrm{E}(|X_n|) = \mathrm{E}(X_n^+) + \mathrm{E}(X_n^-) = c + 2\mathrm{E}(X_n^-).$$

Example 5.16 As a first application of the a.s. convergence Theorem 5.14, let us consider the process $S_n = X_1 + \cdots + X_n$, where the r.v.'s X_i are such that $P(X_i = \pm 1) = \frac{1}{2}$. $(S_n)_n$ is a martingale (it is an instance of Example 5.2 (a)). $(S_n)_n$ is a model of a random motion that starts at 0 and, at each iteration, makes a step to the left or to the right with probability $\frac{1}{2}$.

Let $k \in \mathbb{Z}$. What is the probability of visiting k or, to be precise, if $\tau = \inf\{n; S_n = k\}$ is the passage time at k, what is the value of $P(\tau < +\infty)$?

Assume, for simplicity, that $k < 0$ and consider the stopped martingale $(S_{n \wedge \tau})_n$. This martingale is bounded from below as $S_{n \wedge \tau} \geq k$, hence $S_{n \wedge \tau}^- \leq -k$ and condition (5.12) is verified. Hence $(S_{n \wedge \tau})_n$ converges a.s. But on $\{\tau = +\infty\}$ convergence cannot take place as $|S_{n+1} - S_n| = 1$, so that $(S_{n \wedge \tau})_n$ cannot be a Cauchy sequence on $\{\tau = +\infty\}$. Hence $P(\tau = +\infty) = 0$ and $(S_n)_n$ visits every integer $k \in \mathbb{Z}$ with probability 1.

A process $(S_n)_n$ of the form $S_n = X_1 + \cdots + X_n$ where the X_n are i.i.d. integer-valued is a *random walk* on $\mathbb{Z}$. The instance of this exercise is a *simple* (because X_n takes the values ± 1 only) random walk. It is a model of random motion where at every step a displacement of one unit is made to the right or to the left.

Martingales are an important tool in the investigation of random walks, as will be revealed in many of the examples and exercises below. Actually martingales are a critical tool in the investigation of any kind of stochastic processes.

Example 5.17 Let $(X_n)_n$ and $(S_n)_n$ be a random walk as in the previous example. Let a, b be positive integers and let $\tau = \inf\{n; X_n \geq b \text{ or } X_n \leq -a\}$ be the exit time of S from the interval $]-a, b[$. We know, thanks to Example 5.16, that $\tau < +\infty$ with probability 1. Therefore we can define the r.v. S_τ, which is the position of $(S_n)_n$ at the exit from the interval $]-a, b[$. Of course, S_τ can only take the values $-a$ or b. What is the value of $P(S_\tau = b)$?

Let us assume for a moment that we can apply Theorem 5.10, the stopping theorem, to the stopping times $\tau_2 = \tau$ and $\tau_1 = 0$ (we are not allowed to do so because τ is finite but we do not know whether it is bounded, and actually it is not), then we would have

$$0 = E(S_0) = E(S_\tau) . \tag{5.14}$$

From this relation, as $P(S_\tau = -a) = 1 - P(S_\tau = b)$, we deduce that

$$0 = E(S_\tau) = b\,P(S_\tau = b) - a\,P(S_\tau = -a) = b\,P(S_\tau = b) - a\,(1 - P(S_\tau = b)) ,$$

i.e.

$$P(S_\tau = b) = \frac{a}{a+b} .$$

The problem is therefore solved if (5.14) holds. Actually this is easy to prove: for every n the stopping time $\tau \wedge n$ is bounded and the stopping theorem gives

$$0 = E(S_{\tau \wedge n}) .$$

Now observe that $\lim_{n \to \infty} S_{\tau \wedge n} = S_\tau$ and that, as $-a \leq S_{\tau \wedge n} \leq b$, the sequence $(S_{\tau \wedge n})_n$ is bounded, so that we can apply Lebesgue's Theorem and obtain (5.14).

This example shows a typical application of the stopping theorem in order to obtain the distribution of a process stopped at some stopping time. In this case the process under consideration is itself a martingale. For a more general process $(S_n)_n$ one can look for a function f such that $M_n = f(S_n)$ is a martingale to which the stopping theorem can be applied. This is the case in Exercise 5.12, for example. Other properties of exit times can also be investigated via the stopping of suitable martingales, as will become clear in the exercises. This kind of problem (investigation of properties of exit times), is very often reduced to the question of "finding the right martingale".

This example also shows how to apply the stopping theorem to stopping times that are not bounded: just apply the stopping theorem to the stopping times $\tau \wedge n$, which are bounded, and then hope to be able to pass to the limit using Lebesgue's Theorem as above or some other statement, such as Beppo Levi's Theorem.

5.7   Doob's Inequality and L^p Convergence, $p > 1$

A martingale M is said to be *bounded in* L^p if

$$\sup_{n \geq 1} E(|M_n|^p) < +\infty .$$

Note that, for $p \geq 1$, $(|M_n|^p)_n$ is a submartingale so that $n \mapsto E(|M_n|^p)$ is increasing.

Theorem 5.18 (Doob's Maximal Inequality) *Let M be a martingale bounded in L^p for $p > 1$. Then if $M^* = \sup_n |M_n|$ (the maximal r.v.), M^* belongs to L^p and*

$$\|M^*\|_p \leq q \sup_{n \geq 1} \|M_n\|_p , \tag{5.15}$$

where $q = \frac{p}{p-1}$ is the exponent conjugated to p.

Theorem 5.18 is a consequence of the following.

Lemma 5.19 *If X is a positive submartingale, then for every $p > 1$ and $n \in \mathbb{N}$,*

$$E\Big(\max_{0 \leq k \leq n} X_k^p \Big) \leq \Big(\frac{p}{p-1} \Big)^p E(X_n^p) .$$

Proof Note that if $X_n \notin L^p$ the term on the right-hand side is equal to $+\infty$ and there is nothing to prove. If instead $X_n \in L^p$, then $X_k \in L^p$ also for $k \leq n$ as $(X_k^p)_{k \leq n}$ is itself a submartingale (see the remarks at the end of Sect. 5.2) and $k \mapsto E(X_k^p)$ is increasing. Hence also $Y := \max_{1 \leq k \leq n} X_k$ belongs to L^p. Let, for $\lambda > 0$,

$$\tau_\lambda(\omega) = \begin{cases} \inf\{k; \ 0 \leq k \leq n, X_k(\omega) > \lambda\} \\ n+1 \text{ if } \{\ \} = \emptyset . \end{cases}$$

We have $\sum_{k=1}^{n} 1_{\{\tau_\lambda = k\}} = 1_{\{Y > \lambda\}}$, so that, as $X_k \geq \lambda$ on $\{\tau_\lambda = k\}$,

$$\lambda 1_{\{Y > \lambda\}} \leq \sum_{k=1}^{n} X_k 1_{\{\tau_\lambda = k\}}$$

and, for every $p > 1$,

$$Y^p = p \int_0^Y \lambda^{p-1} \, d\lambda = p \int_0^{+\infty} \lambda^{p-1} 1_{\{Y > \lambda\}} \, d\lambda$$

$$\leq p \int_0^{+\infty} \lambda^{p-2} \sum_{k=1}^n X_k 1_{\{\tau_\lambda = k\}} \, d\lambda .$$

(5.16)

As $1_{\{\tau_\lambda = k\}}$ is $\mathscr{F}_k$-measurable, $E(X_k 1_{\{\tau_\lambda = k\}}) \leq E(X_n 1_{\{\tau_\lambda = k\}})$ and taking the expectation in (5.16) we have

$$\frac{1}{p} E(Y^p) = \int_0^{+\infty} \lambda^{p-2} E\Big(\sum_{k=1}^n X_k 1_{\{\tau_\lambda = k\}} \Big) \, d\lambda \leq \int_0^{+\infty} E\Big(\lambda^{p-2} X_n \sum_{k=1}^n 1_{\{\tau_\lambda = k\}} \Big) \, d\lambda$$

$$= \frac{1}{p-1} E\Big(X_n \times \underbrace{(p-1) \int_0^{+\infty} \lambda^{p-2} \sum_{k=1}^n 1_{\{\tau_\lambda = k\}} \, d\lambda}_{= Y^{p-1}} \Big) = \frac{1}{p-1} E(Y^{p-1} X_n) .$$

Hölder's inequality now gives

$$E(Y^p) \leq \frac{p}{p-1} E\big[(Y^{p-1})^{\frac{p}{p-1}} \big]^{\frac{p-1}{p}} E(X_n^p)^{\frac{1}{p}} = \frac{p}{p-1} E(Y^p)^{\frac{p-1}{p}} E(X_n^p)^{\frac{1}{p}} .$$

As we know already that $E(Y^p) < +\infty$, we can divide both sides of the equation by $E(Y^p)^{\frac{p-1}{p}}$, which gives

$$E\Big(\max_{0 \leq k \leq n} X_k^p \Big)^{1/p} = E(Y^p)^{1/p} \leq \frac{p}{p-1} E(X_n^p)^{1/p} .$$

∎

Proof of Theorem 5.18. Lemma 5.19 applied to the positive submartingale $(|M_k|)_k$ gives, for every n,

$$E\Big(\max_{0 \leq k \leq n} |M_k|^p \Big) \leq \Big(\frac{p}{p-1} \Big)^p E(|M_n|^p) ,$$

and now we can just note that, as $n \to \infty$,

$$\max_{0 \leq k \leq n} |M_k|^p \uparrow (M^*)^p ,$$

$$E(|M_n|^p) \uparrow \sup_n E(|M_n|^p) .$$

∎

Doob's inequality (5.15) provides simple conditions for the L^p convergence of a martingale if $p > 1$.

Assume that M is bounded in L^p with $p > 1$. Then $\sup_{n \geq 0} M_n^- \leq M^*$. As by Doob's inequality M^* is integrable, condition (5.12) of Theorem 5.14 is satisfied and M converges a.s. to an r.v. M_∞ and, of course, $|M_\infty| \leq M^*$. As $|M_n - M_\infty|^p \leq 2^{p-1}(|M_n|^p + |M_\infty|^p) \leq 2^p M^{*p}$, Lebesgue's Theorem gives

$$\lim_{n \to \infty} E(|M_n - M_\infty|^p) = 0 .$$

Conversely, if $(M_n)_n$ converges in L^p, then it is also bounded in L^p and by the same argument as above it also converges a.s.

Therefore for $p > 1$ the behavior of a martingale bounded in L^p is very simple:

Theorem 5.20 *If $p > 1$ a martingale is bounded in L^p if and only if it converges a.s. and in L^p.*

In the next section we shall see what happens concerning L^1 convergence of a martingale. Things are not so simple (and somehow more interesting).

5.8 L^1 Convergence, Regularity

The key tool for the investigation of the L^1 convergence of martingales is uniform integrability, which was introduced in Sect. 3.6.

Proposition 5.21 *Let $Y \in L^1$. Then the family $\mathcal{H} := \{E(Y \mid \mathcal{G})\}_{\mathcal{G}}$, as $\mathcal{G}$ ranges among all sub-σ-algebras of $\mathcal{F}$, is uniformly integrable.*

Proof We shall prove that the family $\mathcal{H}$ satisfies the criterion of Proposition 3.33. First note that $\mathcal{H}$ is bounded in L^1 as

$$E[|E(Y \mid \mathcal{G})|] \leq E[E(|Y| \mid \mathcal{G})] = E(|Y|)$$

and therefore, by Markov's inequality,

$$P(|E(Y \mid \mathcal{G})| \geq R) \leq \frac{1}{R} E(|Y|) . \tag{5.17}$$

Let us fix $\varepsilon > 0$ and let $\delta > 0$ be such that

$$\int_A |Y|\, d\mathrm{P} < \varepsilon$$

for every $A \in \mathscr{F}$ such that $\mathrm{P}(A) \leq \delta$, as guaranteed by Proposition 3.33, as $\{Y\}$ is a uniformly integrable family. Let now $R > 0$ be such that

$$\mathrm{P}\big(|\mathrm{E}(Y\,|\,\mathscr{G})| > R\big) \leq \frac{1}{R}\,\mathrm{E}(|Y|) < \delta\,.$$

We have then

$$\int_{\{|\mathrm{E}(Y\,|\,\mathscr{G})|>R\}} \big|\mathrm{E}(Y\,|\,\mathscr{G})\big|\, d\mathrm{P} \leq \int_{\{|\mathrm{E}(Y\,|\,\mathscr{G})|>R\}} \mathrm{E}\big(|Y|\,\big|\,\mathscr{G}\big)\, d\mathrm{P}$$

$$= \int_{\{|\mathrm{E}(Y\,|\,\mathscr{G})|>R\}} |Y|\, d\mathrm{P} < \varepsilon\,,$$

where the last equality holds because the event $\{|\mathrm{E}(Y\,|\,\mathscr{G})| > R\}$ is $\mathscr{G}$-measurable. ∎

In particular, recalling Example 5.2 (c), if $(\mathscr{F}_n)_n$ is a filtration on $(\Omega, \mathscr{F}, \mathrm{P})$ and $Y \in L^1$, then $(\mathrm{E}(Y\,|\,\mathscr{F}_n))_n$ is a uniformly integrable martingale. A martingale of this form is called a *regular martingale*.

Conversely, every uniformly integrable martingale $(M_n)_n$ is regular: indeed, as $(M_n)_n$ is bounded in L^1, condition (5.12) holds and $(M_n)_n$ converges a.s. to some r.v. Y. By Theorem 3.34, $Y \in L^1$ and the convergence takes place in L^1. Hence

$$M_m = \mathrm{E}(M_n\,|\,\mathscr{F}_m) \xrightarrow[n\to\infty]{L^1} \mathrm{E}(Y\,|\,\mathscr{F}_m)$$

(recall that the conditional expectation is a continuous operator in L^1, Remark 4.10). We have therefore proved the following characterization of regular martingales.

Theorem 5.22 *A martingale $(M_n)_n$ is uniformly integrable if and only if it is regular, i.e. of the form $M_n = \mathrm{E}(Y\,|\,\mathscr{F}_n)$ for some $Y \in L^1$, and if and only if it converges a.s. and in L^1.*

The following statement specifies the limit of a regular martingale.

Proposition 5.23 *Let* $Y \in L^1(\Omega, \mathcal{F}, P)$, $(\mathcal{F}_n)_n$ *a filtration on* $(\Omega, \mathcal{F})$ *and* $\mathcal{F}_\infty = \sigma\left(\bigcup_{n=1}^\infty \mathcal{F}_n\right)$, *the σ-algebra generated by the $\mathcal{F}_n$'s. Then*

$$\lim_{n \to \infty} E(Y | \mathcal{F}_n) = E(Y | \mathcal{F}_\infty) \quad \text{a.s. and in } L^1.$$

Proof If $Z = \lim_{n \to \infty} E(Y | \mathcal{F}_n)$ a.s. then Z is $\mathcal{F}_\infty$-measurable, being the limit of $\mathcal{F}_\infty$-measurable r.v.'s (recall Remark 1.15 if you are worried about the a.s.). In order to prove that $Z = E(Y | \mathcal{F}_\infty)$ a.s. we must check that

$$E(Z 1_A) = E(Y 1_A) \qquad \text{for every } A \in \mathcal{F}_\infty . \tag{5.18}$$

The class $\mathscr{C} = \bigcup_n \mathcal{F}_n$ is stable with respect to finite intersections, generates $\mathcal{F}_\infty$ and contains Ω. If $A \in \mathcal{F}_m$ for some m then as soon as $n \geq m$ we have $E\big[E(Y | \mathcal{F}_n) 1_A\big] = E\big[E(1_A Y | \mathcal{F}_n)\big] = E(Y 1_A)$, as also $A \in \mathcal{F}_n$. Therefore

$$E(Z 1_A) = \lim_{n \to \infty} E\big[E(Y | \mathcal{F}_n) 1_A\big] = E(Y 1_A) .$$

Hence (5.18) holds for every $A \in \mathscr{C}$, and, by Remark 4.3, also for every $A \in \mathcal{F}_\infty$. ∎

Remark 5.24 (Regularity of Positive Martingales) In the case of a positive martingale $(M_n)_n$ the following ideas may be useful in order to check regularity (or non-regularity). Sometimes it is important to establish this feature (see Exercise 5.24, for example).

(a) Regularity is easily established when the a.s. limit $M_\infty = \lim_{n \to \infty} M_n$ is known. We have $E(M_\infty) \leq \lim_{n \to \infty} E(M_n)$ by Fatou's Lemma. If this inequality is strict, then the martingale cannot be regular, as L^1 convergence entails convergence of the expectations. Conversely, if $E(M_\infty) = \lim_{n \to \infty} E(M_n)$ then the martingale is regular, as for positive r.v.'s a.s. convergence in addition to convergence of the expectations entails L^1 convergence (Scheffé's theorem, Theorem 3.25).

(b) (Kakutani's trick) If the limit M_∞ is not known, a possible approach in order to investigate the regularity of $(M_n)_n$ is to compute

$$\lim_{n \to \infty} E(\sqrt{M_n}) . \tag{5.19}$$

If this limit is equal to 0 then necessarily $M_\infty = 0$. Actually by Fatou's Lemma

$$E(\sqrt{M_\infty}) \leq \lim_{n \to \infty} E(\sqrt{M_n}) = 0 ,$$

so that the positive r.v. $\sqrt{M_\infty}$, having expectation equal to 0, is $= 0$ a.s. In this case regularity is not possible (barring trivial situations).

(c) A particular case is martingales of the form

$$M_n = U_1 \cdots U_n \tag{5.20}$$

where the r.v.'s U_k are independent, positive and such that $E(U_k) = 1$, see Example 5.2 (b). In this case we have

$$\lim_{n \to \infty} E(\sqrt{M_n}) = \prod_{k=1}^{\infty} E(\sqrt{U_k}) \tag{5.21}$$

so that if the infinite product above is equal to 0, then $(M_n)_n$ is not regular. Note that in order to determine the behavior of the infinite product Proposition 3.4 may be useful.

By Jensen's inequality

$$E(\sqrt{U_k}) \leq \sqrt{E(U_k)} \leq 1$$

and the inequality is strict unless $U_k \equiv 1$, as the square root is strictly concave so that $E(\sqrt{U_k}) < 1$. In particular, if the U_k are also identically distributed, we have $E(\sqrt{M_n}) = E(\sqrt{U_1})^n \to_{n \to \infty} 0$ and $(M_n)_n$ is not a regular.

Hence a martingale of the form (5.20), if in addition the U_n are i.i.d., cannot be regular (besides the trivial case $M_n \equiv 1$).

The next result states what happens when the infinite product in (5.21) does not vanish.

Proposition 5.25 *Let $(U_n)_n$ be a sequence of independent positive r.v.'s with $E(U_n) = 1$ for every n and let $M_n = U_1 \cdots U_n$. Then if*

$$\prod_{n=1}^{\infty} E(\sqrt{U_n}) > 0$$

$(M_n)_n$ is regular.

Proof Let us prove first that $(\sqrt{M_n})_n$ is a Cauchy sequence in L^2. We have, for $n \geq m$,

$$
\begin{aligned}
\mathrm{E}\big[(\sqrt{M_n} - \sqrt{M_m})^2\big] &= \mathrm{E}\big(M_n + M_m - 2\sqrt{M_n M_m}\big) \\
&= 2\big(1 - \mathrm{E}(\sqrt{M_n M_m})\big) .
\end{aligned}
\tag{5.22}
$$

Now

$$
\mathrm{E}\big(\sqrt{M_n M_m}\big) = \mathrm{E}(U_1 \cdots U_m) \prod_{k=m+1}^{n} \mathrm{E}(\sqrt{U_k}) = \prod_{k=m+1}^{n} \mathrm{E}(\sqrt{U_k}) .
\tag{5.23}
$$

As $\mathrm{E}(\sqrt{U_k}) \leq 1$, it follows that

$$
\prod_{k=m+1}^{n} \mathrm{E}(\sqrt{U_k}) \geq \prod_{k=m+1}^{\infty} \mathrm{E}(\sqrt{U_k}) = \frac{\prod_{k=1}^{\infty} \mathrm{E}(\sqrt{U_k})}{\prod_{k=1}^{m} \mathrm{E}(\sqrt{U_k})}
$$

and, as by hypothesis $\prod_{k=1}^{\infty} \mathrm{E}(\sqrt{U_k}) > 0$, we obtain

$$
\lim_{m\to\infty} \prod_{k=m+1}^{\infty} \mathrm{E}(\sqrt{U_k}) = \lim_{m\to\infty} \frac{\prod_{k=1}^{\infty} \mathrm{E}(\sqrt{U_k})}{\prod_{k=1}^{m} \mathrm{E}(\sqrt{U_k})} = 1 .
$$

Therefore going back to (5.23), for every $\varepsilon > 0$, for n_0 large enough and $n, m \geq n_0$,

$$
\mathrm{E}\big(\sqrt{M_n M_m}\big) \geq 1 - \varepsilon
$$

and by (5.22) $(\sqrt{M_n})_n$ is a Cauchy sequence in L^2 and converges in L^2. This implies that $(M_n)_n$ converges in L^1 (see Exercise 3.1 (b)) and is regular. ∎

Remark 5.26 (Backward Martingales) Let $(\mathscr{B}_n)_n$ be a *decreasing* sequence of σ-algebras. A family $(Z_n)_n$ of integrable r.v.'s is a backward (or reverse) martingale if

$$
\mathrm{E}(Z_n \,|\, \mathscr{B}_{n+1}) = Z_{n+1} .
$$

Backward supermartingales and submartingales are defined similarly.

The behavior of a backward martingale is easily traced back to the behavior of martingales by setting, for every N and $n \leq N$,

$$
Y_n = Z_{N-n} , \qquad \mathscr{F}_n = \mathscr{B}_{N-n} .
$$

As

$$E(Y_{n+1}|\mathcal{F}_n) = E(Z_{N-n-1}|\mathcal{B}_{N-n}) = Z_{N-n} = Y_n \, ,$$

$(Y_n)_{n \leq N}$ is a martingale with respect to the filtration $(\mathcal{F}_n)_{n \leq N}$.

Let us note first that a backward martingale is automatically uniformly integrable thanks to the criterion of Proposition 5.21:

$$Z_n = E(Z_{n-1}|\mathcal{B}_n) = E\big[E(Z_{n-2}|\mathcal{B}_{n-1})|\mathcal{B}_n\big]$$
$$= E(Z_{n-2}|\mathcal{B}_n) = \cdots = E(Z_1|\mathcal{B}_n) \, .$$

In particular, $(Z_n)_n$ is bounded in L^1.

Also, by Proposition 5.13 (the upcrossings) applied to the reversed backward martingale $(Y_n)_{n \leq N}$, a bound similar to (5.9) is proved to hold for $(Z_n)_n$ and this allows us to reproduce the argument of Theorem 5.14, proving a.s. convergence.

In conclusion, the behavior of a backward martingale is very simple: it converges a.s. and in L^1.

For more details and complements, see [21] p. 115, [12] p. 264 or [6] p. 203.

Example 5.27 Let $(\Omega, \mathcal{F}, P)$ be a probability space and let us assume that $\mathcal{F}$ is countably generated. This is an assumption that is very often satisfied (recall Exercise 1.1). In this example we give a proof of the Radon-Nikodym theorem (Theorem 1.29 p. 26) using martingales. The appearance of martingales in this context should not come as a surprise: martingales appear in a natural way in connection with changes of probability (see Exercises 5.23–5.26).

Let Q be a probability on $(\Omega, \mathcal{F})$ such that $Q \ll P$. Let $(F_n)_n \subset \mathcal{F}$ be a sequence of events such that $\mathcal{F} = \sigma(F_n, n = 1, 2, \ldots)$ and let

$$\mathcal{F}_n = \sigma(F_1, \ldots, F_n) \, .$$

For every n let us consider all possible intersections of the $F_k, k = 1, \ldots, n$. Let $G_{n,k}, k = 1, \ldots, N_n$ be the atoms, i.e. the elements among these intersections that do not contain other intersections. Then every event in $\mathcal{F}_n$ is the finite disjoint union of the $G_{n,k}$'s.

Let, for every n,

$$X_n = \sum_{k=1}^{N_n} \frac{Q(G_{n,k})}{P(G_{n,k})} 1_{G_{n,k}} .$$ (5.24)

As $Q \ll P$ if $P(G_{n,k}) = 0$ then also $Q(G_{n,k}) = 0$ and in (5.24) we shall consider the sum as extended only to the indices k such that $P(G_{n,k}) > 0$.

Let us check that $(X_n)_n$ is an $(\mathscr{F}_n)_n$-martingale. If $A \in \mathscr{F}_n$, then A is the finite (disjoint) union of the $G_{n,k}$ for k ranging in some set of indices $\mathscr{I}$. We have therefore

$$E(X_n 1_A) = E\left(X_n \sum_{k \in \mathscr{I}} 1_{G_{n,k}}\right) = \sum_{k \in \mathscr{I}} E(X_n 1_{G_{n,k}}) = \sum_{k \in \mathscr{I}} P(G_{n,k}) \frac{Q(G_{n,k})}{P(G_{n,k})}$$

$$= \sum_{k \in \mathscr{I}} Q(G_{n,k}) = Q(A) .$$

If $A \in \mathscr{F}_n$, then obviously also $A \in \mathscr{F}_{n+1}$ so that

$$E(X_{n+1} 1_A) = Q(A) = E(X_n 1_A) ,$$ (5.25)

hence $E(X_{n+1}|\mathscr{F}_n) = X_n$. Moreover, the previous relations for $A = \Omega$ give $E(X_n) = 1$.

Being a positive martingale, $(X_n)_n$ converges a.s. to some positive r.v. X. Let us prove that $(X_n)_n$ is also uniformly integrable. Thanks to (5.25) with $A = \{X_n \geq R\}$, we have, for every n,

$$E(X_n 1_{\{X_n \geq R\}}) = Q(X_n \geq R)$$ (5.26)

and also, by Markov's inequality, $P(X_n \geq R) \leq R^{-1} E(X_n) = R^{-1}$. By Exercise 3.28, for every $\varepsilon > 0$ there exists a $\delta > 0$ such that if $P(A) \leq \delta$ then $Q(A) \leq \varepsilon$. If R is such that $R^{-1} \leq \delta$ then $P(X_n \geq R) \leq \delta$ and (5.26) gives

$$E(X_n 1_{\{X_n \geq R\}}) = Q(X_n \geq R) \leq \varepsilon \qquad \text{for every } n$$

and the sequence $(X_n)_n$ is uniformly integrable and converges to X also in L^1. It is now immediate that X is a density of Q with respect to P: this is actually Exercise 5.24 below.

Of course this proof can immediately be adapted to the case of finite measures instead of probabilities.

Note however that the Radon-Nikodym Theorem holds even without assuming that $\mathscr{F}$ is countably generated.

Exercises

5.1 (p. 358) Let $(X_n)_n$ be a supermartingale such that, moreover, $E(X_n) = $ const. Then $(X_n)_n$ is a martingale.

5.2 (p. 359) Let M be a *positive* martingale. Prove that, for $m < n$, $\{M_m = 0\} \subset \{M_n = 0\}$ a.s. (i.e. the set of zeros of a positive martingale increases).

5.3 (p. 359) (Product of independent martingales) Let $(M_n)_n$, $(N_n)_n$ be martingales on the same probability space $(\Omega, \mathcal{F}, P)$, with respect to the filtrations $(\mathcal{F}_n)_n$ and $(\mathcal{G}_n)_n$, respectively. Assume moreover that $(\mathcal{F}_n)_n$ and $(\mathcal{G}_n)_n$ are independent (in particular the martingales are themselves independent). Then the product $(M_n N_n)_n$ is a martingale for the filtration $(\mathcal{H}_n)_n$ with $\mathcal{H}_n = \mathcal{F}_n \vee \mathcal{G}_n$.

5.4 (p. 359) Let $(X_n)_n$ be a sequence of independent r.v.'s with mean 0 and variance σ^2 and let $\mathcal{F}_n = \sigma(X_k, k \le n)$. Let $M_n = X_1 + \cdots + X_n$ and let $(Z_n)_n$ be a square integrable process predictable with respect to $(\mathcal{F}_n)_n$.

(a) Prove that

$$Y_n = \sum_{k=1}^{n} Z_k X_k$$

is a square integrable martingale.

(b) Prove that $E(Y_n) = 0$ and that

$$E(Y_n^2) = \sigma^2 \sum_{k=1}^{n} E(Z_k^2) .$$

(c) What is the associated increasing process of $(M_n)_n$? And of $(Y_n)_n$?

5.5 (p. 360) (Martingales with independent increments) Let $M = (\Omega, \mathcal{F}, (\mathcal{F}_n)_n, (M_n)_n, P)$ be a square integrable martingale.

(a) Prove that $E[(M_n - M_m)^2] = E(M_n^2 - M_m^2)$.
(b) M is said to be with *independent increments* if, for every $n \ge m$, $M_n - M_m$ is independent of $\mathcal{F}_m$. Prove that, in this case, the associated increasing process is $A_n = E(M_n^2) - E(M_0^2) = E[(M_n - M_0)^2]$ and is therefore deterministic.
(c) Let $(M_n)_n$ be *a Gaussian martingale* (i.e. such that, for every n, the vector $(M_0, \ldots, M_n)$ is Gaussian). Show that $(M_n)_n$ has independent increments with respect to its natural filtration $(\mathcal{G}_n)_n$.

5.6 (p. 361) Let $(Y_n)_{n \geq 0}$ be a sequence of i.i.d. r.v.'s such that $P(Y_k = \pm 1) = \frac{1}{2}$. Let $\mathcal{F}_0 = \{\emptyset, \Omega\}$, $\mathcal{F}_n = \sigma(Y_k, k \leq n)$ and $S_0 = 0$, $S_n = Y_1 + \cdots + Y_n$, $n \geq 1$. Let $M_0 = 0$ and

$$M_n = \sum_{k=1}^{n} \text{sign}(S_{k-1})Y_k, \quad n = 1, 2, \ldots$$

where

$$\text{sign}(x) = \begin{cases} 1 & \text{if } x > 0 \\ 0 & \text{if } x = 0 \\ -1 & \text{if } x < 0. \end{cases}$$

(a) What is the associated increasing process of the martingale $(S_n)_n$?
(b) Show that $(M_n)_{n \geq 0}$ is a square integrable martingale with respect to $(\mathcal{F}_n)_n$ and compute its associated increasing process.
(c1) Prove that

$$E[(|S_{n+1}| - |S_n|)1_{\{S_n > 0\}}|\mathcal{F}_n] = 0$$

$$E[(|S_{n+1}| - |S_n|)1_{\{S_n < 0\}}|\mathcal{F}_n] = 0$$

and deduce the compensator $(\widetilde{A}_n)_n$ of the submartingale $(|S_n|)_n$.
(c2) Let $N_n = |S_n| - \widetilde{A}_n$. Show that $N_n = M_n$ and that $(M_n)_n$ is adapted to $\mathcal{G}_n = \sigma(|S_1|, \ldots, |S_n|)$ and is a martingale also with respect to this filtration.

5.7 (p. 363) Let $(\xi_n)_n$ be a sequence of i.i.d. r.v.'s with an exponential law of parameter λ and let $\mathcal{F}_n = \sigma(\xi_k, k \leq n)$. Let $Z_0 = 0$ and

$$Z_n = \max_{k \leq n} \xi_k.$$

(a) Show that $(Z_n)_n$ is an $(\mathcal{F}_n)_n$-submartingale.
(b) Compute its compensator $(A_n)_n$.

5.8 (p. 364) Let $(M_n)_n$ be a martingale such that $E(e^{M_n}) < +\infty$ for every n and let $(\mathcal{F}_n)_n$ be its natural filtration $\mathcal{F}_n = \sigma(M_k, k \leq n)$.

(a) Prove that

$$\log E(e^{M_n}|\mathcal{F}_{n-1}) \geq M_{n-1}.$$ (5.27)

(b) Prove that there exists an increasing predictable process $(A_n)_n$ such that

$$X_n = e^{M_n - A_n}$$

is a martingale.

(c) Explicitly compute $(A_n)_n$ in the following instances.

(c1) $M_n = W_1 + \cdots + W_n$ where $(W_n)_n$ is a sequence of i.i.d. centered r.v.'s such that $E(e^{W_i}) < +\infty$.

(c2) $M_n = \sum_{k=1}^n Z_k W_k$ where the r.v.'s W_k are i.i.d., centered, and have a Laplace transform L that is finite on the whole of $\mathbb{R}$ and $(Z_n)_n$ is a bounded predictable process (i.e. such that Z_n is $\mathscr{F}_{n-1}$-measurable for every n).

5.9 (p. 365) Let $(\mathscr{F}_n)_n$ be a filtration, X an integrable r.v. and τ an a.s. finite stopping time. Let $X_n = E(X | \mathscr{F}_n)$; then

$$E(X | \mathscr{F}_\tau) = X_\tau .$$

5.10 (p. 365) Prove that a process $X = (\Omega, \mathscr{F}, (\mathscr{F}_n)_n, (X_n)_n, P)$ is a martingale if and only if, for every bounded $(\mathscr{F}_n)_n$-stopping time τ, $E(X_\tau) = E(X_0)$.

• This is a useful criterion.

5.11 (p. 366) Let $(\Omega, \mathscr{F}, P)$ be a probability space, $(\mathscr{F}_n)_n \subset \mathscr{F}$ a filtration and $(M_n)_n$ an $(\mathscr{F}_n)_n$-martingale.

(a) Let $\mathscr{G}$ be a σ-algebra independent of $(\mathscr{F}_n)_n$ and $\widetilde{\mathscr{F}}_n = \sigma(\mathscr{F}_n, \mathscr{G})$. Prove that $(M_n)_n$ is a martingale also with respect to $(\widetilde{\mathscr{F}}_n)_n$.

(b) Let $\tau : \Omega \to \mathbb{N} \cup \{+\infty\}$ be an r.v. independent of $(\mathscr{F}_n)_n$. Prove that $(M_{n \wedge \tau})_n$ is also a martingale with respect to some filtration to be determined.

5.12 (p. 366) Let $(Y_n)_n$ be a sequence of i.i.d. r.v.'s such that $P(Y_i = 1) = p$, $P(Y_i = -1) = q = 1 - p$ with $q > p$. Let $S_n = Y_1 + \cdots + Y_n$.

(a) Prove that $\lim_{n \to \infty} S_n = -\infty$ a.s.

(b) Prove that

$$Z_n = \left(\frac{q}{p}\right)^{S_n}$$

is a martingale.

(c) Let a, b be strictly positive integers and let $\tau = \tau_{-a,b} = \inf\{n, S_n = b \text{ or } S_n = -a\}$ be the exit time from $] - a, b[$. What is the value of $E(Z_{n \wedge \tau})$? Of $E(Z_\tau)$?

(d1) Compute $P(S_\tau = b)$ (i.e. the probability for the random walk $(S_n)_n$ to exit from the interval $] - a, b[$ at b). How does this quantity behave as $a \to +\infty$?

(d2) Let, for $b > 0$, $\tau_b = \inf\{n; S_n = b\}$ be the passage time of $(S_n)_n$ at b. Note that $\{\tau_b < n\} \subset \{S_{\tau_{-n,b}} = b\}$ and deduce that $P(\tau_b < +\infty) < 1$, i.e. with strictly positive probability the process $(S_n)_n$ never visits b. This was to be expected, as $q > p$ and the process has a preference to make displacements to the left.

(d3) Compute $P(\tau_{-a} < +\infty)$.

5.13 (p. 368) (Wald's identity) Let $(X_n)_n$ be a sequence of i.i.d. integrable real r.v.'s with $E(X_1) = x$. Let $\mathcal{F}_0 = \{\Omega, \emptyset\}$, $\mathcal{F}_n = \sigma(X_k, k \leq n)$, $S_0 = 0$ and, for $n \geq 1$, $S_n = X_1 + \cdots + X_n$. Let τ be an *integrable* stopping time of $(\mathcal{F}_n)_n$.

(a) Let $Z_n = S_n - nx$. Show that $(Z_n)_n$ is an $(\mathcal{F}_n)_n$-martingale.
(b1) Show that, for every n, $E(S_{n \wedge \tau}) = x E(n \wedge \tau)$.
(b2) Show that S_τ is integrable and that $E(S_\tau) = x E(\tau)$, first assuming $X_1 \geq 0$ a.s. and then in the general case.
(c) Assume that, for every n, $P(X_n = \pm 1) = \frac{1}{2}$ and $\tau = \tau_b = \inf\{n; S_n \geq b\}$, where $b > 0$ is an integer. Show that τ_b is not integrable. (Recall that we already know, Example 5.16, that $\tau_b < +\infty$ a.s.)

• Note that no requirement concerning independence of τ and the X_n is made.

5.14 (p. 369) Let $(X_n)_n$ be a sequence of i.i.d. r.v.'s such that $P(X_n = \pm 1) = \frac{1}{2}$ and let $\mathcal{F}_0 = \{\emptyset, \Omega\}$, $\mathcal{F}_n = \sigma(X_k, k \leq n)$, and $S_0 = 0$, $S_n = X_1 + \cdots + X_n$, $n \geq 1$.

(a) Show that $W_n = S_n^2 - n$ is an $(\mathcal{F}_n)_n$-martingale.
(b) Let a, b be strictly positive integers and let $\tau_{a,b}$ be the exit time of $(S_n)_n$ from $]-a, b[$.
(b1) Compute $E(\tau_{a,b})$.
(b2) Let $\tau_b = \inf\{n; X_n \geq b\}$ be the exit time of $(X_n)_n$ from the half-line $]-\infty, b[$. We already know (Example 5.16) that $\tau_b < +\infty$ a.s. Prove that $E(\tau_b) = +\infty$ (already proved in a different way in Exercise 5.13 (c)).

Recall that we already know (Example 5.17 (a)) that $P(S_{\tau_{a,b}} = -a) = \frac{b}{a+b}$, $P(S_{\tau_{a,b}} = b) = \frac{a}{a+b}$.

5.15 (p. 369) Let $(X_n)_n$ be a sequence of i.i.d. r.v.'s with $P(X = \pm 1) = \frac{1}{2}$ and let $S_n = X_1 + \cdots + X_n$ and $Z_n = S_n^3 - 3nS_n$. Let τ be the exit time of $(S_n)_n$ from the interval $]-a, b[$, $a, b > 0$. Recall that we already know that τ is integrable and that $P(S_\tau = -a) = \frac{b}{a+b}$, $P(S_\tau = b) = \frac{a}{a+b}$.

(a) Prove that Z is a martingale.
(b1) Compute $\text{Cov}(S_\tau, \tau)$ and deduce that if $a \neq b$ then τ and S_τ are not independent.
(b2) Assume that $b = a$. Prove that the r.v.'s (S_n, τ) and $(-S_n, \tau)$ have the same joint distributions and deduce that S_τ and τ are independent.

5.16 (p. 371) Let $(X_n)_n$ be a sequence of i.i.d. r.v.'s such that $P(X_i = \pm 1) = \frac{1}{2}$ and let $S_0 = 0$, $S_n = X_1 + \cdots + X_n$, $\mathcal{F}_0 = \{\Omega, \emptyset\}$ and $\mathcal{F}_n = \sigma(X_k, k \leq n)$. Let a be a strictly positive integer and $\tau = \inf\{n \geq 0; S_n = a\}$ be the first passage time of $(S_n)_n$ at a. In this exercise and in the next one we continue to gather information about the passage times of the simple symmetric random walk.

(a) Show that, for every $\theta \in \mathbb{R}$,

$$Z_n^\theta = \frac{e^{\theta S_n}}{(\cosh \theta)^n}$$

is an $(\mathcal{F}_n)_n$-martingale and that if $\theta \geq 0$ then $(Z_{n\wedge\tau}^\theta)_n$ is bounded.

(b1) Show that, for every $\theta \geq 0$, $(Z_{n\wedge\tau}^\theta)_n$ converges a.s. and in L^2 to the r.v.

$$W^\theta = \frac{e^{\theta a}}{(\cosh \theta)^\tau} 1_{\{\tau < +\infty\}}. \tag{5.28}$$

(b2) Compute $\lim_{\theta\to 0+} E(W^\theta)$ and deduce that $P(\tau < +\infty) = 1$ (which we already know from Example 5.16) and that, for every $\theta \geq 0$,

$$E[(\cosh \theta)^{-\tau}] = e^{-\theta a}. \tag{5.29}$$

(b3) Determine the Laplace transform of τ and its convergence abscissas.
   Might be useful: the inverse of $\cosh : [0, +\infty[\to [1, +\infty[$ is $x \mapsto \log\left(x + \sqrt{x^2 - 1}\right)$.

5.17 (p. 372) Let, as in Exercise 5.16, $(X_n)_n$ be a sequence of i.i.d. r.v.'s such that $P(X_n = \pm 1) = \frac{1}{2}$, $S_0 = 0$, $S_n = X_1 + \cdots + X_n$, $\mathcal{F}_0 = \{\Omega, \emptyset\}$ and $\mathcal{F}_n = \sigma(X_k, k \leq n)$. Let $a > 1$ be a positive integer and let $\tau = \inf\{n \geq 0; |S_n| = a\}$ be the exit time of $(S_n)_n$ from $]-a, a[$. In this exercise we investigate the Laplace transform and the existence of moments of τ.

   Let $\lambda \in \mathbb{R}$ be such that $0 < \lambda < \frac{\pi}{2a}$. Note that, as $a > 1$, $0 < \cos \frac{\pi}{2a} < \cos \lambda < 1$ (see Fig. 5.2).

(a) Show that $Z_n = (\cos \lambda)^{-n} \cos(\lambda S_n)$ is an $(\mathcal{F}_n)_n$-martingale.
(b) Show that

$$1 = E(Z_{n\wedge\tau}) \geq \cos(\lambda a) E[(\cos \lambda)^{-n\wedge\tau}].$$

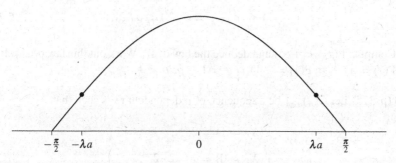

Fig. 5.2 The graph of the cosine function between $-\frac{\pi}{2}$ and $\frac{\pi}{2}$. As $-\lambda a \leq \lambda S_{n\wedge\tau} \leq \lambda a$, $\cos(\lambda S_{n\wedge\tau}) \geq \cos(\lambda a)$

(c) Deduce that $E[(\cos \lambda)^{-\tau}] \le (\cos(\lambda a))^{-1}$ and then that τ is a.s. finite.

(d1) Prove that $E(Z_{n \wedge \tau}) \to_{n \to \infty} E(Z_\tau)$.

(d2) Deduce that the martingale $(Z_{n \wedge \tau})_n$ is regular.

(e) Compute $E[(\cos \lambda)^{-\tau}]$. What are the convergence abscissas of the Laplace transform of τ? For which values of p does $\tau \in L^p$?

5.18 (p. 374) Let $(U_n)_n$ be a positive supermartingale such that $\lim_{n \to \infty} E(U_n) = 0$. Prove that $\lim_{n \to \infty} U_n = 0$ a.s.

5.19 (p. 374) Let $(Y_n)_{n \ge 1}$ be a sequence of $\mathbb{Z}$-valued integrable r.v.'s, i.i.d. and with common law μ. Assume that

* $E(Y_i) = b < 0$,
* $P(Y_i = 1) > 0$ but $P(Y_i \ge 2) = 0$.

   Let $S_0 = 0$, $S_n = Y_1 + \cdots + Y_n$ and

$$W = \sup_{n \ge 0} S_n .$$

The goal of this problem is to determine the law of W. Intuitively, by the Law of Large Numbers, $S_n \to_{n \to \infty} -\infty$ a.s., being sums of independent r.v.'s with a strictly negative expectation. But, before sinking down, $(S_n)_n$ may take an excursion on the positive side. How large?

(a) Prove that $W < +\infty$ a.s.

(b) Recall (Exercise 2.42) that for a real r.v. X, both its Laplace transform and its logarithm are convex functions. Let $L(\lambda) = E(e^{\lambda Y_1})$ and $\psi(\lambda) = \log L(\lambda)$. Prove that $\psi(\lambda) < +\infty$ for every $\lambda \ge 0$. What is the value of $\psi'(0+)$? Prove that $\psi(\lambda) \to +\infty$ as $\lambda \to +\infty$ and that there exists a unique $\lambda_0 > 0$ such that $\psi(\lambda_0) = 0$.

(c) Let λ_0 be as in b). Prove that $Z_n = e^{\lambda_0 S_n}$ is a martingale and that $\lim_{n \to \infty} Z_n = 0$ a.s.

(d) Let $K \in \mathbb{N}$, $K \ge 1$ and let $\tau_K = \inf\{n; S_n \ge K\}$ be the passage time of $(S_n)_n$ at K. Prove that

$$\lim_{n \to \infty} Z_{n \wedge \tau_K} = e^{\lambda_0 K} 1_{\{\tau_K < +\infty\}}. \tag{5.30}$$

(e) Compute $P(\tau_K < +\infty)$ and deduce the law of W. Work out this law precisely if $P(Y_i = 1) = p$, $P(Y_i = -1) = q = 1 - p$, $p < \frac{1}{2}$.

5.20 (p. 375) Let $(X_n)_{n \ge 1}$ be a sequence of independent r.v.'s such that

$$P(X_k = 1) = 2^{-k},$$

$$P(X_k = 0) = 1 - 2 \cdot 2^{-k},$$

$$P(X_k = -1) = 2^{-k}$$

and let $S_n = X_1 + \cdots + X_n$, $\mathscr{F}_n = \sigma(S_k, k \leq n)$.

(a) Prove that $(S_n)_n$ is an $(\mathscr{F}_n)_n$-martingale.
(b) Prove that $(S_n)_n$ is square integrable and compute its associated increasing process.
(c) Does $(S_n)_n$ converge a.s.? In L^1? In L^2?

5.21 (p. 376) Let p, q be probabilities on a countable set E such that $p \neq q$ and $q(x) > 0$ for every $x \in E$. Let $(X_n)_{n \geq 1}$ be a sequence of i.i.d. E-valued r.v.'s having law q. Show that

$$Y_n = \prod_{k=1}^{n} \frac{p(X_k)}{q(X_k)}$$

is a positive martingale converging to 0 a.s. Is it regular?

5.22 (p. 376) Let $(U_n)_n$ be a sequence of real i.i.d. r.v.'s with common density with respect to the Lebesgue measure

$$f(t) = 2(1 - t) \qquad \text{for } 0 \leq t \leq 1$$

and $f(t) = 0$ otherwise (it is a Beta(1, 2) law). Let $\mathscr{F}_0 = \{\emptyset, \Omega\}$ and, for $n \geq 1$, $\mathscr{F}_n = \sigma(U_k, k \leq n)$. For $q \in]0, 1[$ let

$$\begin{aligned} X_0 &= q, \\ X_{n+1} &= \tfrac{1}{2} X_n^2 + \tfrac{1}{2} 1_{[0, X_n]}(U_{n+1}) \quad n \geq 0. \end{aligned} \qquad (5.31)$$

(a) Prove that $X_n \in [0, 1]$ for every $n \geq 0$.
(b) Prove that $(X_n)_n$ is an $(\mathscr{F}_n)_n$-martingale.
(c) Prove that $(X_n)_n$ converges a.s. and in L^2 to an r.v. X_∞ and compute $E(X_\infty)$.
(d) Note that $X_{n+1} - \tfrac{1}{2} X_n^2$ can only take the values 0 or $\tfrac{1}{2}$ and deduce that X_∞ can only take the values 0 or 1 a.s. and has a Bernoulli distribution of parameter q.

5.23 (p. 377) Let P, Q be probabilities on the measurable space $(\Omega, \mathscr{F})$ and let $(\mathscr{F}_n)_n \subset \mathscr{F}$ be a filtration. Assume that, for every $n > 0$, the restriction $Q_{|\mathscr{F}_n}$ of Q to $\mathscr{F}_n$ is absolutely continuous with respect to the restriction, $P_{|\mathscr{F}_n}$, of P to $\mathscr{F}_n$. Let

$$Z_n = \frac{dQ_{|\mathscr{F}_n}}{dP_{|\mathscr{F}_n}}.$$

(a) Prove that $(Z_n)_n$ is a martingale.
(b) Prove that $Z_n > 0$ Q-a.s. and that $(Z_n^{-1})_n$ is a Q-supermartingale.
(c) Prove that if also $Q_{|\mathscr{F}_n} \gg P_{|\mathscr{F}_n}$, then $(Z_n^{-1})_n$ is a Q-martingale.

5.24 (p. 378) Let $(\mathcal{F}_n)_n \subset \mathcal{F}$ be a filtration on the probability space $(\Omega, \mathcal{F}, P)$. Let $(M_n)_n$ be a positive $(\mathcal{F}_n)_n$-martingale such that $E(M_n) = 1$. Let, for every n,

$$dQ_n = M_n \, dP$$

be the probability on $(\Omega, \mathcal{F})$ having density M_n with respect to P.

(a) Assume that $(M_n)_n$ is regular. Prove that there exists a probability Q on $(\Omega, \mathcal{F})$ such that $Q \ll P$ and such that $Q_{|\mathcal{F}_n} = Q_n$.
(b) Conversely, assume that such a probability Q exists. Prove that $(M_n)_n$ is regular.

5.25 (p. 378) Let $(X_n)_n$ be a sequence of $N(0, 1)$-distributed i.i.d. r.v.'s. Let $S_n = X_1 + \cdots + X_n$ and $\mathcal{F}_n = \sigma(X_1, \ldots, X_n)$. Let, for $\theta \in \mathbb{R}$,

$$M_n = e^{\theta S_n - \frac{1}{2} n\theta^2}.$$

(a) Prove that $(M_n)_n$ is an $(\mathcal{F}_n)_n$-martingale and that $E(M_n) = 1$.
(b) For $m > 0$ let Q_m be the probability on $(\Omega, \mathcal{F})$ having density M_m with respect to P.
(b1) What is the law of X_n with respect to Q_m for $n > m$?
(b2) What is the law of X_n with respect to Q_m for $n \le m$?

5.26 (p. 378) Let $(X_n)_n$ be a sequence of independent r.v.'s on $(\Omega, \mathcal{F}, P)$ with $X_n \sim N(0, a_n)$. Let $\mathcal{F}_n = \sigma(X_k, k \le n)$, $S_n = X_1 + \cdots + X_n$, $A_n = a_1 + \cdots + a_n$ and

$$Z_n = e^{S_n - \frac{1}{2} A_n}.$$

(a) Prove that $(Z_n)_n$ is an $(\mathcal{F}_n)_n$-martingale.
(b) Assume that $\lim_{n \to \infty} A_n = +\infty$. Compute $\lim_{n \to \infty} Z_n$ a.s. Is $(Z_n)_n$ regular?
(c1) Assume $\lim_{n \to \infty} A_n = A_\infty < +\infty$. Prove that $(Z_n)_n$ is a regular martingale and determine the law of its limit.
(c2) Let $Z_\infty := \lim_{n \to \infty} Z_n$ and let Q be the probability on $(\Omega, \mathcal{F}, P)$ having density Z_∞ with respect to P. What is the law of X_k under Q? Are the r.v.'s X_n independent also with respect to Q?

5.27 (p. 380) Let $(X_n)_n$ be a sequence of i.i.d. $N(0, 1)$-distributed r.v.'s and $\mathcal{F}_n = \sigma(X_k, k \le n)$.

(a) Determine for which values of $\lambda \in \mathbb{R}$ the r.v. $e^{\lambda X_{n+1} X_n}$ is integrable and compute its expectation.
(b) Let, for $|\lambda| < 1$,

$$Z_n = \lambda \sum_{k=1}^{n} X_{k-1} X_k.$$

Compute

$$\log E(e^{Z_{n+1}} | \mathscr{F}_n)$$

and deduce an increasing predictable process $(A_n)_n$ such that

$$M_n = e^{Z_n - A_n}$$

is a martingale.

(c) Determine $\lim_{n \to \infty} M_n$. Is $(M_n)_n$ regular?

5.28 (p. 381) In this exercise we give a proof of the first part of Kolmogorov's strong law, Theorem 3.12 using backward martingales. Let $(X_n)_n$ be a sequence of i.i.d. integrable r.v.'s with $E(X_k) = b$. Let $S_n = X_1 + \cdots + X_n$, $\overline{X}_n = \frac{1}{n} S_n$ and

$$\mathscr{B}_n = \sigma(S_n, S_{n+1}, \dots) = \sigma(S_n, X_{n+1}, \dots) .$$

(a1) Prove that for $1 \le k \le n$ we have

$$E(X_k | \mathscr{B}_n) = E(X_k | S_n) .$$

(a2) Deduce that, for $k \le n$,

$$E(X_k | \mathscr{B}_n) = \frac{1}{n} S_n .$$

(b1) Prove that $(\overline{X}_n)_n$ is a $(\mathscr{B}_n)_n$-backward martingale.
(b2) Deduce that

$$\overline{X}_n \xrightarrow[n \to \infty]{\text{a.s.}} b .$$

Recall Exercise 4.3.

Chapter 6
Complements

In this chapter we introduce some important notions that might not find their place in a course for lack of time. Section 6.1 will introduce the problem of simulation and the related applications of the Law of Large Numbers. Sections 6.2 and 6.3 will give some hints about deeper properties of the weak convergence of probabilities.

6.1 Random Number Generation, Simulation

In some situations the computation of a probability or of an expectation is not possible analytically and the Law of Large Numbers provides numerical methods of approximation.

Example 6.1 It is sometimes natural to model the subsequent times between events (e.g. failure times) with i.i.d. r.v.'s, Z_i say, having a Weibull law. Recall (see also Exercise 2.9) that the Weibull law of parameters α and λ has density with respect to the Lebesgue measure given by

$$\lambda \alpha t^{\alpha-1} e^{-\lambda t}, \qquad t > 0 .$$

This means that the first failure occurs at time Z_1, the second one at time $Z_1 + Z_2$ and so on. What is the probability of monitoring more than N failures in the time interval $[0, T]$?

This requires the computation of the probability

$$P(Z_1 + \cdots + Z_N \leq T) . \tag{6.1}$$

© The Author(s), under exclusive license to Springer Nature Switzerland AG 2023
P. Baldi, *Probability*, Universitext, https://doi.org/10.1007/978-3-031-38492-9_6

As no simple formulas concerning the d.f. of the sum of i.i.d. Weibull r.v.'s is available, a numerical approach is the following: we ask a computer to simulate n times batches of N i.i.d. Weibull r.v.'s and to keep account of how many times the event $Z_1 + \cdots + Z_N \leq T$ occurs. If we define

$$X_i = \begin{cases} 1 & \text{if } Z_1 + \cdots + Z_N \leq T \text{ for the } i\text{-th simulation} \\ 0 & \text{otherwise} \end{cases}$$

then the X_i are Bernoulli r.v.'s of parameter $p = P(Z_1 + \cdots + Z_N \leq T)$. Hence by the Law of Large Numbers, a.s.,

$$\lim_{n \to \infty} \frac{1}{n}(X_1 + \cdots + X_n) = E(X) = P(Z_1 + \cdots + Z_N \leq T).$$

In other words, an estimate of the probability (6.1) is provided by the proportion of simulations that have given the result $Z_1 + \cdots + Z_N \leq T$ (for n large enough, of course).

In order to effectively take advantage of this technique we must be able to instruct a computer to simulate sequences of r.v.'s with a prescribed distribution.

High level software is available (scilab, matlab, mathematica, python,...), which provides routines that give sequences of "independent" random numbers with the most common distributions, e.g. Weibull.

These software packages are usually interpreted, which is not a suitable feature when dealing with a large number of iterations, for which a compiled program (such as FORTRAN or C, for example) is necessary, being much faster. These compilers usually only provide routines which produce sequences of numbers that can be considered independent and uniformly distributed on $[0, 1]$. In order to produce sequences of random numbers with an exponential law, for instance, it will be necessary to devise an appropriate procedure starting from uniformly distributed ones.

Entire books have been committed to this kind of problem (see e.g. [10, 14, 15]). Useful information has also been gathered in [22]. In this section we shall review some ideas in this direction, mostly in the form of examples.

The first method to produce random numbers with a given distribution is to construct a map Φ such that, if X is uniform on $[0, 1]$, then $\Phi(X)$ has the target distribution.

To be precise the problem is: given an r.v. X uniform on $[0, 1]$ and a discrete or continuous probability μ, find a map Φ such that $\Phi(X)$ has law μ.

If μ is a probability on $\mathbb{R}$ and has a d.f. F which is continuous and strictly increasing, and therefore invertible, then the choice $\Phi = F^{-1}$ solves the problem.

Actually as the d.f. of X is

$$x \mapsto \begin{cases} 0 & \text{if } x < 0 \\ x & \text{if } 0 \le x \le 1 \\ 1 & \text{if } x > 1 \end{cases}$$

and as $0 \le F(t) \le 1$, then, for $0 < t < 1$,

$$P(F^{-1}(X) \le t) = P(X \le F(t)) = F(t)$$

so that the d.f. of the r.v. $F^{-1}(X)$ is indeed F.

Example 6.2 Uniform law on an interval $[a, b]$: its d.f. is

$$F(x) = \begin{cases} 0 & \text{if } x < a \\ \dfrac{x - a}{b - a} & \text{if } a \le x \le b \\ 1 & \text{if } x > b \end{cases}$$

and therefore $F^{-1}(y) = a + (b - a)y$, for $0 \le y \le 1$. Hence if X is uniform on $[0, 1]$, $a + (b - a)X$ is uniform on $[a, b]$.

Example 6.3 Exponential law of parameter λ. Its d.f. is

$$F(t) = \begin{cases} 0 & \text{if } t < 0 \\ 1 - e^{-\lambda t} & \text{if } t \ge 0 . \end{cases}$$

F is therefore invertible $\mathbb{R}^+ \to [0, 1[$ and $F^{-1}(x) = -\frac{1}{\lambda} \log(1 - x)$. Hence if X is uniform on $[0, 1]$ then $-\frac{1}{\lambda} \log(1 - X)$ is exponential of parameter λ.

The method of inverting the d.f. is however useless when the inverse F^{-1} does not have an explicit expression, as is the case with Gaussian laws, or for probabilities on $\mathbb{R}^d$. The following examples provide other approaches to the problem.

Example 6.4 (Gaussian Laws) As always let us begin with an $N(0, 1)$; its d.f. F is only numerically computable and for F^{-1} there is no explicit expression.

A simple algorithm to produce an $N(0, 1)$ law starting from a uniform $[0, 1]$ is provided in Example 2.19: if W and T are independent r.v.'s respectively exponential of parameter $\frac{1}{2}$ and uniform on $[0, 2\pi]$, then the r.v. $X = \sqrt{W} \cos T$ is $N(0, 1)$-distributed. As W and T can be simulated as explained in the previous examples, the $N(0, 1)$ distribution is easily simulated. This is the Box-Müller algorithm.

Other methods for producing Gaussian r.v.'s can be found in the book of Knuth [18]. For simple tasks the fast algorithm of Exercise 3.27 can also be considered.

Starting from the simulation of an $N(0, 1)$ law, every Gaussian law, real or d-dimensional, can easily be obtained using affine-linear transformations.

Example 6.5 How can we simulate an r.v. taking the values $x_1, \dots, x_m$ with probabilities $p_1, \dots, p_m$ respectively?

Let $q_0 = 0$, $q_1 = p_1$, $q_2 = p_1 + p_2, \dots, q_{m-1} = p_1 + \cdots + p_{m-1}$ and $q_m = 1$. The numbers $q_1, \dots, q_m$ split the interval $[0, 1]$ into sub-intervals having amplitude $p_1, \dots, p_m$ respectively. If X is uniform on $[0, 1]$, let

$$Y = x_i \quad \text{if} \quad q_{i-1} \leq X < q_i . \tag{6.2}$$

Obviously Y takes the values $x_1, \dots, x_m$ and

$$P(Y = x_i) = P(q_{i-1} \leq X < q_i) = q_i - q_{i-1} = p_i , \qquad i = 1, \dots, m ,$$

so that Y has the required law.

This method, theoretically, can be used in order to simulate any discrete distribution. Theoretically... see, however, Example 6.8 below.

The following examples suggest simple algorithms for the simulation of some discrete distributions.

Example 6.6 How can we simulate a uniform distribution on the finite set $\{1, 2, \ldots, m\}$?

The idea of the previous Example 6.5 is easily put to work noting that, if X is uniform on $[0, 1]$, then mX is uniform on $[0, m]$, so that

$$Y = \lfloor mX \rfloor + 1 \tag{6.3}$$

is uniform on $\{1, \ldots, m\}$.

Example 6.7 (Binomial Laws) Let $X_1, \ldots, X_n$ be independent numbers uniform on $[0, 1]$ and let, for $0 < p < 1$, $Z_i = 1_{\{X_i \le p\}}$.

Obviously $Z_1, \ldots, Z_n$ are independent and $P(Z_i = 1) = P(X_i \le p) = p$, $P(Z_i = 0) = P(p < X_i \le 1) = 1 - p$.

Therefore $Z_i \sim B(1, p)$ and $Y = Z_1 + \cdots + Z_n \sim B(n, p)$.

Example 6.8 (Simulation of a Permutation) How can we simulate a random deck of 52 cards?

To be precise, we want to simulate a random element in the set E of all permutations of $\{1, \ldots, 52\}$ in such a way that all permutations are equally likely. This is a discrete r.v., but, given the huge cardinality of E ($52! \simeq 8 \cdot 10^{67}$), the method of Example 6.5 is not feasible. What to do?

In general the following algorithm is suitable in order to simulate a permutation on n elements.

(1) Let us denote by x_0 the vector $(1, 2, \ldots, n)$. Let us choose at random a number between 1 and n, with the methods of Example 6.5 or, better, of Example 6.6. If r_0 is this number let us switch in the vector x_0 the coordinates with index r_0 and n. Let us denote by x_1 the resulting vector: x_1 has the number r_0 as its n-th coordinate and n as its r_0-th coordinate.

(2) Let us choose at random a number between 1 and $n - 1$, r_1 say, and let us switch, in the vector x_1, the coordinates with indices r_1 and $n - 1$. Let us denote this new vector by x_2.

(3) Iterating this procedure, starting from a vector x_k, let us choose at random a number r_k in $\{1, \ldots, n-k\}$ and let us switch the coordinates r_k and $n-k$. Let x_{k+1} denote the new vector.

(4) Let us stop when $k = n - 1$. The coordinates of the vector x_{n-1} are now the numbers $\{1, \ldots, n\}$ in a different order, i.e. a permutation of $(1, \ldots, n)$. It is rather immediate that the permutation x_{n-1} can be any permutation of $(1, \ldots, n)$ with a uniform probability.

Example 6.9 (Poisson Laws) The method of Example 6.5 cannot be applied to Poisson r.v.'s, which can take infinitely many possible values.

A possible way of simulating these laws is the following. Let $(Z_n)_n$ be a sequence of i.i.d. exponential r.v.'s of parameter λ and let $X = k$ if k is the largest positive integer such that $Z_1 + \cdots + Z_k \leq 1$, i.e.

$$Z_1 + \cdots + Z_k \leq 1 < Z_1 + \cdots + Z_{k+1} .$$

The d.f. of the r.v. X obtained in this way is

$$P(X \leq k - 1) = P(Z_1 + \cdots + Z_k > 1) = 1 - F_k(1) .$$

The d.f., F_k, of $Z_1 + \cdots + Z_k$, which is Gamma(k, λ)-distributed, is

$$F_k(x) = 1 - e^{-\lambda} \sum_{i=1}^{k-1} \frac{(\lambda x)^i}{i!} ,$$

hence

$$P(X \leq k - 1) = e^{-\lambda} \sum_{i=1}^{k-1} \frac{\lambda^i}{i!} ,$$

which is the d.f. of a Poisson law of parameter λ. This algorithm works for Poisson law, as we know how to sample exponential laws.

However, this method has the drawback that one cannot foresee in advance how many exponential r.v.'s will be needed.

We still do not know how to simulate a Weibull law, which is necessary in order to tackle Example 6.1. This question is addressed in Exercise 6.1 a).

The following proposition introduces a new idea for producing r.v.'s with a uniform distribution on a subset of $\mathbb{R}^d$.

Proposition 6.10 (The Rejection Method) *Let $R \in \mathcal{B}(\mathbb{R}^d)$ and $(Z_n)_n$ a sequence of i.i.d. r.v.'s with values in R and $D \subset R$ a Borel set such that $P(Z_i \in D) = p > 0$. Let τ be the first index i such that $Z_i \in D$, i.e. $\tau = \inf\{i; Z_i \in D\}$, and let*

$$X = \begin{cases} Z_k & \text{if } \tau = k \\ \text{any } x_0 \in D & \text{if } \tau = +\infty. \end{cases}$$

Then, if $A \subset D$,

$$P(X \in A) = \frac{P(Z_1 \in A)}{P(Z_1 \in D)}.$$

In particular, if Z_i is uniform on R then X is uniform on D.

Proof First note that τ has a geometric law of parameter p so that $\tau < +\infty$ a.s. If $A \subset D$ then, noting that $X = Z_k$ if $\tau = k$,

$$P(X \in A) = \sum_{k=1}^{\infty} P(X \in A, \tau = k)$$

$$= \sum_{k=1}^{\infty} P(Z_k \in A, \tau = k) = \sum_{k=1}^{\infty} P(Z_k \in A, Z_1 \notin D, \ldots, Z_{k-1} \notin D)$$

$$= \sum_{k=1}^{\infty} P(Z_k \in A)P(Z_1 \notin D) \ldots P(Z_{k-1} \notin D) = \sum_{k=1}^{\infty} P(Z_k \in A)(1-p)^{k-1}$$

$$= P(Z_1 \in A) \sum_{k=1}^{\infty} (1-p)^{k-1} = P(Z_1 \in A) \times \frac{1}{p} = \frac{P(Z_1 \in A)}{P(Z_1 \in D)}.$$

∎

Example 6.11 How can we simulate an r.v. that is uniform on a domain $D \subset$ $\mathbb{R}^2$? This construction is easily adapted to uniform r.v.'s on domains of $\mathbb{R}^d$.

Note beforehand that this is easy if D is a rectangle $[a, b] \times [c, d]$. Indeed we know how to simulate independent r.v.'s X and Y that are uniform respectively on $[a, b]$ and $[c, d]$ (as explained in Example 6.2). It is clear therefore that the pair (X, Y) is uniform on $[a, b] \times [c, d]$: indeed the densities of X and Y with respect to the Lebesgue measure are respectively

$$
f_X(x) = \begin{cases} \dfrac{1}{b-a} & \text{if } a \leq x \leq b \\ 0 & \text{otherwise} \end{cases} \qquad f_Y(y) = \begin{cases} \dfrac{1}{d-c} & \text{if } c \leq y \leq d \\ 0 & \text{otherwise} \end{cases}
$$

so that the density of (X, Y) is

$$
f(x, y) = \begin{cases} \dfrac{1}{(b-a)(d-c)} & \text{if } (x, y) \in [a, b] \times [c, d] \\ 0 & \text{otherwise,} \end{cases}
$$

which is the density of an r.v. which is uniform on the rectangle $[a, b] \times [c, d]$.

If, in general, $D \subset \mathbb{R}^2$ is a bounded domain, Proposition 6.10 allows us to solve the problem with the following algorithm: if R is a rectangle containing D,

(1) simulate first an r.v. (X, Y) uniform on R as above;
(2) let us check whether $(X, Y) \in D$. If $(X, Y) \notin D$ go back to (1); if $(X, Y) \in D$ then the r.v. (X, Y) is uniform on D.

For instance, in order to simulate a uniform distribution on the unit ball of $\mathbb{R}^2$, the steps to perform are the following:

(1) first simulate r.v.'s X_1, X_2 uniform on $[0, 1]$ and independent; then let $Y_1 := 2X_1 - 1, Y_2 := 2X_2 - 1$, so that Y_1 and Y_2 are uniform on $[-1, 1]$ and independent; (Y_1, Y_2) is therefore uniform on the square $[-1, 1] \times [-1, 1]$;
(2) check whether (Y_1, Y_2) belongs to the unit ball $\{x^2 + y^2 \leq 1\}$. In order to do this just compute $W = Y_1^2 + Y_2^2$; if $W > 1$ we go back to (1) for two new values X_1, X_2; if $W \leq 1$ instead, then (Y_1, Y_2) is uniform on the unit ball.

Example 6.12 (Monte Carlo Methods) Let $f : [0, 1] \to \mathbb{R}$ be a bounded Borel function and $(X_n)_n$ a sequence of i.i.d. r.v.'s uniform on $[0, 1]$. Then $(f(X_n))_n$ is also a sequence of independent r.v.'s, each of them having mean $E[f(X_1)]$; by the Law of Large Numbers therefore

$$\frac{1}{n} \sum_{k=1}^{n} f(X_k) \xrightarrow[n \to \infty]{\text{a.s.}} E[f(X_1)] . \tag{6.4}$$

But we have also

$$E[f(X_1)] = \int_0^1 f(x) \, dx .$$

These remarks suggest a method of numerical computation of the integral of f: just simulate n random numbers $X_1, X_2, \ldots$ uniformly distributed on $[0, 1]$ and then compute

$$\frac{1}{n} \sum_{k=1}^{n} f(X_k) .$$

This quantity for n large is an approximation of

$$\int_0^1 f(x) \, dx .$$

More generally, if f is a bounded Borel function on a bounded Borel set $D \subset \mathbb{R}^d$, then its integral can be approximated numerically in a similar way: if $X_1, X_2, \ldots$ are i.i.d. r.v.'s uniform on D, then

$$\frac{1}{n} \sum_{k=1}^{n} f(X_k) \xrightarrow[n \to \infty]{\text{a.s.}} \frac{1}{|D|} \int_D f(x) \, dx .$$

This algorithm of computation of integrals is a typical example of a *Monte Carlo method*. These methods are in general much slower than the classical algorithms of numerical integration, but they are much simpler to implement and are particularly useful in dimension larger than 1, where numerical methods become very complicated or downright unfeasible. Let us be more precise: let

$$\bar{I}_n := \frac{1}{n} \sum_{k=1}^{n} f(X_k) ,$$

$$I := \frac{1}{|D|} \int_D f(x) \, dx ,$$

$$\sigma^2 := \operatorname{Var}(f(X_n)) < +\infty ,$$

then the Central Limit Theorem states that, weakly,

$$\frac{1}{\sigma \sqrt{n}} (\overline{I}_n - I) \xrightarrow[n \to \infty]{} N(0, 1) .$$

If we denote by ϕ_β the quantile of order β of a $N(0, 1)$ distribution, from this relation it is easy to derive that, for large n, $\left[\overline{I}_n - \frac{\sigma}{\sqrt{n}} \phi_{1-\alpha/2}, \overline{I}_n + \frac{\sigma}{\sqrt{n}} \phi_{1-\alpha/2} \right]$ is a confidence interval for I of level α. This gives the appreciation that the error of $\overline{I}_n$ as an estimate of the integral I is of order $\frac{1}{\sqrt{n}}$.

This is rather slow, *but independent of the dimension d.*

Example 6.13 (On the Rejection Method)   Assume that we are interested in the simulation of a law on $\mathbb{R}$ having density f with respect to the Lebesgue measure. Let us now present a method that does not require a tractable d.f. We shall restrict ourselves to the case of a bounded function f ($f \le M$ say) having its support contained in a bounded interval $[a, b]$.

The region below the graph of f is contained in the rectangle $[a, b] \times [0, M]$. By the method of Example 6.11 let us produce a 2-dimensional r.v. $W = (X, Y)$ uniform in the subgraph $A = \{(x, y); a \le x \le b, 0 \le y \le f(x)\}$: then X has density f. Indeed

$$P(X \le t) = P((X, Y) \in A_t) = \lambda(A_t) = \int_a^t f(s)\, ds ,$$

where A_t is the intersection (shaded in Fig. 6.1) of the subgraph A of f and of the half plane $\{x \le t\}$.

So far we have been mostly concerned with real-valued r.v.'s. The next example considers a more complicated target space. See also Exercise 6.2.

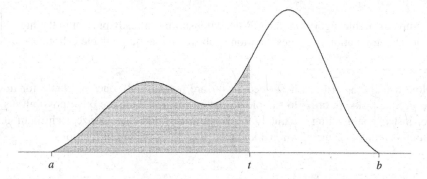

Fig. 6.1 The area of the shaded region is equal to the d.f. of X computed at t

Example 6.14 (Sampling of an Orthogonal Matrix) Sometimes applications require elements in a compact group to be chosen randomly "uniformly". How can we rigorously define this notion?

Given a locally compact topological group G there always exists on $(G, \mathscr{B}(G))$ a Borel measure μ that is invariant under translations, i.e. such that, for every $A \in \mathscr{B}(G)$ and $g \in G$,

$$\mu(gA) = \mu(A) \,,$$

where $gA = \{\ell \in G;\ \ell = gh \text{ for some } h \in A\}$ is "the set A translated by the action of g". This is a *Haar measure* of G (see e.g. [16]). If G is compact it is possible to choose μ so that it is a probability and, with this constraint, such a μ is unique. To sample an element of G with this distribution is a way of choosing an element with a "uniform distribution" on G.

Let us investigate closely how to simulate the random choice of an element of the group of rotations in d dimensions, $O(d)$, with the Haar distribution.

The starting point of the forthcoming algorithm is the QR decomposition: every $d \times d$ matrix M can be decomposed in the form $M = QR$, where Q is orthogonal and R is an upper triangular matrix. This decomposition is unique under the constraint that the entries on the diagonal of R are positive.

The algorithm is very simple: generate a $d \times d$ matrix M with i.i.d. $N(0, 1)$-distributed entries and let $M = QR$ be its QR decomposition. Then Q has the Haar distribution of $O(d)$.

This follows from the fact that if $g \in O(d)$, then the two matrix-valued r.v.'s M and gM have the same distribution, owing to the invariance of the Gaussian laws under orthogonal transformations: if QR is the QR decomposition of M, then gQR is the QR decomposition of gM. Therefore $gQR \sim QR$ and $gQ \sim Q$ by the uniqueness of the QR decomposition. This provides an easily

implementable algorithm: the QR decomposition is already present in the high
level computation packages mentioned above and in the available C libraries.

Note that the algorithms described so far are not the only ones available for the
respective tasks. In order to sample a random rotation there are other possibilities,
for instance simulating separately the Euler angles that characterize each rotation.
But this requires some additional knowledge on the structure of rotations.

6.2 Tightness and the Topology of Weak Convergence

In the investigation that follows we consider probabilities on a *Polish space*, i.e. a
metric space that is complete and separable, or, to be precise, a metric separable
space whose topology is defined by a complete metric. This means that Polish-ness
is a topological property.

Definition 6.15 A family $\mathcal{T}$ of probabilities on $(E, \mathcal{B}(E))$ is *tight* if for every
$\varepsilon > 0$ there exists a compact set K such that $\mu(K) \geq 1 - \varepsilon$ for every $\mu \in \mathcal{T}$.

A family of probabilities $\mathcal{K}$ on $(E, \mathcal{B}(E))$ is said to be *relatively compact* if for
every sequence $(\mu_n)_n \subset \mathcal{K}$ there exists a subsequence converging weakly to some
probability μ on $(E, \mathcal{B}(E))$.

Theorem 6.16 *Suppose that E is separable and complete (i.e. Polish). If a
family $\mathcal{K}$ of probabilities on $(E, \mathcal{B}(E))$ is relatively compact, then it is tight.*

Proof Recall that in a complete metric space relative compactness is equivalent to
total boundedness: a set is totally bounded if and only if, for every $\varepsilon > 0$, it can be
covered by a finite number of open sets having diameter smaller that ε.

Let $(G_n)_n$ be a sequence of open sets increasing to E and let us prove first that
for every $\varepsilon > 0$ there exists an n such that $\mu(G_n) > 1 - \varepsilon$ for all $\mu \in \mathcal{K}$. Otherwise,
for each n we would have $\mu_n(G_n) \leq 1 - \varepsilon$ for some $\mu_n \in \mathcal{K}$. By the assumed
relative compactness of $\mathcal{K}$, there would be a subsequence $(\mu_{n_k})_k \subset \mathcal{K}$ such that
$\mu_{n_k} \to_{k \to \infty} \mu$, for some probability μ on $(E, \mathcal{B}(E))$. But this is not possible: by

the portmanteau theorem, see (3.19), we would have, for every n,

$$\mu(G_n) \leq \varliminf_{k\to\infty} \mu_{n_k}(G_n) \leq \varliminf_{k\to\infty} \mu_{n_k}(G_{n_k}) \leq 1 - \varepsilon$$

from which

$$\mu(E) = \lim_{n\to\infty} \mu(G_n) \leq 1 - \varepsilon ,$$

so that μ cannot be a probability.

As E is separable there is, for each k, a sequence $U_{k,1}, U_{k,2}, \ldots$ of open balls of radius $\frac{1}{k}$ covering E. Let n_k be large enough so that the open set $G_{n_k} = \bigcup_{j=1}^{n_k} U_{k,j}$ is such that

$$\mu(G_{n_k}) \geq 1 - \varepsilon 2^{-k} \qquad \text{for every } \mu \in \mathcal{K} .$$

The set

$$A = \bigcap_{k=1}^{\infty} G_{n_k} = \bigcap_{k=1}^{\infty} \bigcup_{j=1}^{n_k} U_{n,j}$$

is totally bounded hence relatively compact. As, for every $\mu \in \mathcal{K}$,

$$\mu(A^c) = \mu\left(\bigcup_{k=1}^{\infty} G_{n_k}^c\right) \leq \sum_{k=1}^{\infty} \mu(G_{n_k}^c) \leq \varepsilon \sum_{k=1}^{\infty} 2^{-k} = \varepsilon$$

we have $\mu(A) \geq 1 - \varepsilon$. The closure of A is a compact set K satisfying the requirement. ∎

Note that a Polish space need not be locally compact.

The following (almost) converse to Theorem 6.16 is especially important.

Theorem 6.17 (Prohorov's Theorem) *If E is a metric separable space and $\mathcal{T}$ is a tight family of probabilities on $(E, \mathcal{B}(E))$ then it is also relatively compact.*

We shall skip the proof of Prohorov's theorem (see [2], Theorem 5.1, p. 59). Note that it holds under weaker assumptions than those made in Theorem 6.16 (no completeness assumptions).

Let us denote by $\mathscr{P}$ the family of probabilities on the Polish space $(E, \mathscr{B}(E))$. Let us define, for $\mu, \nu \in \mathscr{P}$, $\rho(\mu, \nu)$ as

$$\rho(\mu, \nu) = \inf\{\varepsilon; \ \mu(A^\varepsilon) \le \nu(A) + \varepsilon \text{ and } \mu(A^\varepsilon) \le \nu(A) + \varepsilon \text{ for every } A \in \mathscr{B}(E)\},$$

where $A^\varepsilon = \{x \in E; d(x, A) \le \varepsilon\}$ is the neighborhood of radius ε of A.

Theorem 6.18 *Let $(E, \mathscr{B}(E))$ be a Polish space, then*

- *ρ is a distance on $\mathscr{P}$, the Prohorov distance,*
- *$\mathscr{P}$ endowed with this distance is also a Polish space,*
- *$\mu_n \to_{n\to\infty} \mu$ weakly if and only if $\rho(\mu_n, \mu) \to_{n\to\infty} 0$.*

See again [2] Theorem 6.8, p. 83 for a proof.

Therefore we can speak of "the topology of weak convergence", which makes $\mathscr{P}$ a metric space and Prohorov's Theorem 6.17 gives a characterization of the relatively compact sets for this topology.

Example 6.19 (Convergence of the Empirical Distributions) Let $(X_n)_n$ be an i.i.d. sequence of r.v.'s with values in the Polish space $(E, \mathscr{B}(E))$ having common law μ. Then the maps $\omega \to \delta_{X_n(\omega)}$ are r.v.'s $\Omega \to \mathscr{P}$, being the composition of the measurable maps $X_n : \Omega \to E$ and $x \to \delta_x$, which is continuous $E \to \mathscr{P}$, (see Example 3.24 a)).

Let

$$\overline{\mu}_n = \frac{1}{n} \sum_{k=1}^{n} \delta_{X_k},$$

which is a sequence of r.v.'s with values in $\mathscr{P}$. For every bounded measurable function $f : E \to \mathbb{R}$ we have

$$\int_E f \, d\overline{\mu}_n = \frac{1}{n} \sum_{k=1}^{n} f(X_k)$$

and by the Law of Large Numbers

$$\lim_{n\to\infty} \frac{1}{n} \sum_{k=1}^{n} f(X_k) = \mathrm{E}[f(X_1)] = \int_E f \, d\mu \qquad \text{a.s.}$$

Assuming, in addition, f continuous, this gives that the sequence of random probabilities $(\overline{\mu}_n)_n$ converges a.s., as $n \to \infty$ to the constant r.v. μ in the topology of weak convergence.

Example 6.20 Let μ be a probability on the Polish space $(E, \mathscr{B}(E))$ and let $C = \{\nu \in \mathscr{P}; H(\nu; \mu) \leq M\}$, H denoting the relative entropy (or Kullback-Leibler divergence) defined in Exercise 2.24, p. 105. In this example we see that C is a tight family. Recall that $H(\nu; \mu) = +\infty$ if $\nu \not\ll \mu$ and, noting $\Phi(t) = t \log t$ for $t \geq 0$,

$$H(\nu; \mu) = \int_E \Phi\left(\frac{d\nu}{d\mu}\right) d\mu$$

if $\nu \ll \mu$. As $\lim_{t \to +\infty} \frac{1}{t} \Phi(t) = +\infty$, the family of densities $\mathscr{H} = \{\frac{d\nu}{d\mu}; \nu \in C\}$ is uniformly integrable in $(E, \mathscr{B}(E), \mu)$ by Proposition 3.35. Hence (Proposition 3.33) for every $\varepsilon > 0$ there exists a $\delta > 0$ such that if $\mu(A) \leq \delta$ then $\nu(A) = \int_A \frac{d\nu}{d\mu} d\mu \leq \varepsilon$.

As the family $\{\mu\}$ is tight, for every $\varepsilon > 0$ there exists a compact set $K \subset E$ such that $\mu(K^c) \leq \delta$. Then we have for every probability $\nu \in C$

$$\nu(K^c) = \int_{K^c} \frac{d\nu}{d\mu} d\mu \leq \varepsilon$$

therefore proving that the level sets, C, of the relative entropy are tight.

6.3 Applications

In this section we see two typical applications of Prohorov's theorem.

The first one is the following enhanced version of P. Lévy's Theorem as announced in Chap. 3, p. 132.

Theorem 6.21 (P. Lévy's Revisited) *Let $(\mu_n)_n$ be a sequence of probabilities on $\mathbb{R}^d$. If $(\widehat{\mu}_n)_n$ converges pointwise to a function κ and if κ is continuous at 0, then κ is the characteristic function of a probability μ and $(\mu_n)_n$ converges weakly to μ.*

Proof The idea of the proof is simple: in Proposition 6.23 below we prove that the condition "$(\widehat{\mu}_n)_n$ converges pointwise to a function κ that is continuous at 0" implies that the sequence $(\mu_n)_n$ is tight. By Prohorov's Theorem every subsequence of $(\mu_n)_n$ has a subsequence that converges weakly to a probability μ. Necessarily $\widehat{\mu} = \kappa$, which proves simultaneously that κ is a characteristic function and that $\mu_n \to_{n\to\infty} \mu$. ∎

First, we shall need the following lemma, which states that the regularity of the characteristic function at the origin gives information concerning the behavior of the probability at infinity.

Lemma 6.22 *Let μ be a probability on $\mathbb{R}$. Then, for every $t > 0$,*

$$\mu\left(|x| > \tfrac{1}{t}\right) \le \frac{C}{t} \int_0^t \left(1 - \Re\widehat{\mu}(\theta)\right) d\theta$$

for some constant $C > 0$ independent of μ.

Proof We have

$$\frac{1}{t} \int_0^t \left(1 - \Re\widehat{\mu}(\theta)\right) d\theta = \frac{1}{t} \int_0^t d\theta \int_{-\infty}^{+\infty} (1 - \cos\theta x)\, d\mu(x)$$

$$= \frac{1}{t} \int_{-\infty}^{+\infty} d\mu(x) \int_0^t (1 - \cos\theta x)\, d\theta = \int_{-\infty}^{+\infty} \left(1 - \frac{\sin tx}{tx}\right) d\mu(x).$$

Note that the use of Fubini's Theorem is justified, all integrands being positive. As $1 - \frac{\sin y}{y} \ge 0$, we have

$$\cdots \ge \int_{\{|x| \ge \frac{1}{t}\}} \left(1 - \frac{\sin tx}{tx}\right) d\mu(x) \ge \mu\left(|x| > \tfrac{1}{t}\right) \times \inf_{|y| \ge 1}\left(1 - \frac{\sin y}{y}\right)$$

and the proof is completed with $C = \left(\inf_{|y| \ge 1}(1 - \frac{\sin y}{y})\right)^{-1}$. ∎

Proposition 6.23 *Let $(\mu_n)_n$ be a sequence of probabilities on $\mathbb{R}^d$. If $(\widehat{\mu}_n)_n$ converges pointwise to a function κ and if κ is continuous at 0 then the family $(\mu_n)_n$ is tight.*

Proof Let us assume first $d = 1$. Lemma 6.22 gives

$$\varlimsup_{n \to \infty} \mu_n \left(|x| > \tfrac{1}{t} \right) \leq \frac{C}{t} \varlimsup_{n \to \infty} \int_0^t \left(1 - \Re \widehat{\mu}_n(\theta) \right) d\theta$$

and by Lebesgue's Theorem

$$\varlimsup_{n \to \infty} \mu_n \left(|x| > \tfrac{1}{t} \right) \leq \frac{C}{t} \int_0^t \left(1 - \Re \kappa(\theta) \right) d\theta \, .$$

Let us fix $\varepsilon > 0$ and let $t_0 > 0$ be such that $1 - \Re \kappa(\theta) \leq \frac{\varepsilon}{C}$ for $0 \leq \theta \leq t_0$, which is possible as κ is assumed to be continuous at 0. Setting $R_0 = \frac{1}{t_0}$ we obtain

$$\varlimsup_{n \to \infty} \mu_n \left(|x| > R_0 \right) \leq \varepsilon \, .$$

i.e. $\mu_n(|x| \geq R_0) \leq 2\varepsilon$ for every n larger than some n_0. As the family formed by a single probability μ_k is tight, for every $k = 1, \ldots, n_0$ there are positive numbers $R_1, \ldots, R_{n_0}$ such that

$$\mu_k(|x| \geq R_k) \leq 2\varepsilon$$

and taking $R = \max(R_0, \ldots, R_{n_0})$ we have $\mu_n(|x| \geq R) \leq 2\varepsilon$ for every n.

Let $d > 1$: we have proved that for every j, $1 \leq j \leq d$, there exists a compact set K_j such that $\mu_{n,j}(K_j^c) \leq \varepsilon$ for every n, where we denote by $\mu_{n,j}$ the j-th marginal of μ_n. Now just note that $K := K_1 \times \cdots \times K_d$ is a compact set and

$$\mu_n(K^c) \leq \mu_{n,1}(K_1^c) + \cdots + \mu_{n,d}(K_d^c) \leq d\varepsilon \, .$$

∎

Example 6.24 Let E, G be Polish spaces. Let $(\mu_n)_n$ be a sequence of probabilities on $(E, \mathscr{B}(E))$ converging weakly to some probability μ. Let $(\nu_n)_n$ be a sequence of probabilities on $(G, \mathscr{B}(G))$ converging weakly to some probability ν. Is it true that

$$\mu_n \otimes \nu_n \underset{n \to \infty}{\to} \mu \otimes \nu \, ?$$

We have already met this question when E and G are Euclidean spaces (Exercise 3.14), where characteristic functions allowed us to conclude the result easily.

In this setting we can argue using Prohorov's Theorem (both implications). As the sequence $(\mu_n)_n$ converges weakly, it is tight and, for every $\varepsilon > 0$ there exists a compact set $K_1 \subset E$ such that $\mu_n(K_1) \geq 1 - \varepsilon$. Similarly there exists a compact set $K_2 \subset G$ such that $\nu_n(K_2) \geq 1 - \varepsilon$. Therefore

$$\mu_n \otimes \nu_n(K_1 \times K_2) = \mu_n(K_1)\nu_n(K_2) \geq (1 - \varepsilon)^2 \geq 1 - 2\varepsilon \,.$$

As $K_1 \times K_2 \subset E \times G$ is a compact set, the sequence $(\mu_n \otimes \nu_n)_n$ is tight and for every subsequence there exists a further subsequence $(\mu_{n_k} \otimes \nu_{n_k})_k$ converging to some probability γ on $(E \times G, \mathscr{B}(E \times G))$. Let us prove that necessarily $\gamma = \mu \otimes \nu$. For every pair of bounded continuous functions $f_1 : E \to \mathbb{R}$, $f_2 : G \to \mathbb{R}$ we have

$$\int_{E \times G} f_1(x)f_2(y)\,d\gamma(x, y) = \lim_{k \to \infty} \int_{E \times G} f_1(x)f_2(y)\,d\mu_{n_k}(x)\,d\nu_{n_k}(y)$$

$$= \lim_{k \to \infty} \int_E f_1(x)\,d\mu_{n_k}(x) \int_G f_2(y)\,d\nu_{n_k}(y) = \int_E f_1(x)d\mu(x) \int_G f_2(y)d\nu(y)$$

$$= \int_{E \times G} f_1(x)f_2(y)\,d\mu \otimes \nu(x, y) \,.$$

By Proposition 1.33 necessarily $\gamma = \mu \otimes \nu$ and the result follows thanks to the sub-sub-sequences Criterion 3.8 applied to the sequence $(\mu_n \otimes \nu_n)_n$ in the Polish space $\mathscr{P}(E \times G)$ endowed with the Prohorov metric.

The previous example and the enhanced P. Lévy's theorem are typical applications of tightness and of Prohorov's Theorem: in order to prove weak convergence of a sequence of probabilities, first prove tightness and then devise some argument in order to identify the limit. This is especially useful for convergence of stochastic processes that the reader may encounter in more advanced courses.

Exercises

6.1 (p. 382) Devise a procedure for the simulation of the following probability distributions on $\mathbb{R}$.

(a) A Weibull distribution with parameters α, λ.
(b) A Gamma(α, λ) distribution with α semi-integer, i.e. $\alpha = \frac{k}{2}$ for some $k \in \mathbb{N}$.
(c) A Beta(α, β) distribution with α, β both half-integers.
(d) A Student $t(n)$.
(e) A Laplace distribution of parameter λ.
(f) A geometric law with parameter p.

(a) see Exercise 2.9, (c) see Exercise 2.20(b), (e) see Exercise 2.43, (f) see Exercise 2.12(a).

6.2 (p. 382) (A uniform r.v. on the sphere) Recall (or take it as granted) that the normalized Lebesgue measure of the sphere $\mathbb{S}_{d-1}$ of $\mathbb{R}^d$ is characterized as being the unique probability on $\mathbb{S}_{d-1}$ that is invariant with respect to rotations.

Let X be an $N(0, I)$-distributed d-dimensional r.v. Prove that the law of the r.v.

$$Z = \frac{X}{|X|}$$

is the normalized Lebesgue measure of the sphere.

6.3 (p. 382) For every $\alpha > 0$ let us consider the probability density with respect to the Lebesgue measure

$$f(t) = \frac{\alpha}{(1+t)^{\alpha+1}} \qquad t > 0. \tag{6.5}$$

(a) Determine a function $\Phi :]0, 1[\to \mathbb{R}$ such that if X is an r.v. uniform on $]0, 1[$ then $\Phi(X)$ has density f.
(b) Let Y be a Gamma$(\alpha, 1)$-distributed r.v. and X an r.v. having a conditional law given $Y = y$ that is exponential with parameter y. Determine the law of X and devise another method in order to simulate an r.v. having a law with density (6.5) with respect to the Lebesgue measure.

Chapter 7
Solutions

1.1 Let $D \subset E$ be a dense countable subset and $\mathscr{D}$ the family of open balls with center in D and rational radius. $\mathscr{D}$ is a countable family of open sets. Let $A \subset E$ be an open set. For every $x \in A \cap D$, let B_x be an open ball centered at x and with a rational radius small enough so that $B_x \subset A$. A is then the union (countable, obviously) of these open balls. Hence the σ-algebra generated by $\mathscr{D}$ contains all open sets and therefore also the Borel σ-algebra which is the smallest one enjoying the property of containing the open sets.

1.2 (a) Every open set of $\mathbb{R}$ is a countable union of open intervals (this is also a particular case of Exercise 1.1). Thus the σ-algebra generated by the open intervals, $\widetilde{\mathscr{B}}_1$ say, contains all open sets of $\mathbb{R}$ hence also the Borel σ-algebra $\mathscr{B}(\mathbb{R})$. This concludes the proof, as the opposite inclusion is obvious.

(b) We have, for every $a < b$,

$$]a, b[= \bigcup_{n=1}^{\infty}]a, b - \tfrac{1}{n}] \, .$$

Thus the σ-algebra generated by the half-open intervals, $\widetilde{\mathscr{B}}_2$ say, contains all open intervals, hence also $\mathscr{B}(\mathbb{R})$ thanks to (a). Conversely,

$$]a, b] = \bigcap_{n=1}^{\infty}]a, b + \tfrac{1}{n}[\, .$$

Hence $\mathscr{B}(\mathbb{R})$ contains all half-open intervals and also $\widetilde{\mathscr{B}}_2$.

(c) The σ-algebra generated by the open half-lines $]a, \infty[$, $\widetilde{\mathscr{B}}_3$ say, contains, by complementation, the half lines of the form $] - \infty, b]$ and, by intersection, the half-open intervals $]a, b]$. Thanks to (b), $\widetilde{\mathscr{B}}_3 \supset \mathscr{B}(\mathbb{R})$. The opposite inclusion is obvious.

(d) Just a repetition of the arguments above.

© The Author(s), under exclusive license to Springer Nature Switzerland AG 2023
P. Baldi, *Probability*, Universitext, https://doi.org/10.1007/978-3-031-38492-9_7

1.3 (a) We know (see p. 4) that every real continuous map is measurable with respect to the Borel σ-algebra $\mathscr{B}(E)$. Therefore $\mathscr{B}_0(E)$, which is the smallest σ-algebra enjoying this property, is contained in $\mathscr{B}(E)$.

(b) In a metric space the function "distance from a point" is continuous. Hence, for every $x \in E$ and $r > 0$ the open ball with radius r and centered at x belongs to $\mathscr{B}_0(E)$, being the pullback of the interval $]-\infty, r[$ by the map $y \mapsto d(x, y)$. As every open set of E is a countable union of these balls (see Exercise 1.1), $\mathscr{B}_0(E)$ contains also all open sets and therefore also the Borel σ-algebra $\mathscr{B}(E)$.

1.4 Let us check the three properties of σ-algebras.

(i) $S \in \mathscr{E}_S$ as $S = E \cap S$.
(ii) If $B \in \mathscr{E}_S$ then B is of the form $B = A \cap S$ for some $A \in \mathscr{E}$ and therefore its complement in S is

$$S \setminus B = A^c \cap S.$$

As $A^c \in \mathscr{E}$, the complement set $S \setminus B$ belongs to $\mathscr{E}_S$.
(iii) Finally, if $(B_n)_n \subset \mathscr{E}_S$, then each B_n is of the form $B_n = A_n \cap S$ for some $A_n \in \mathscr{E}$. Hence

$$\bigcup_{n=1}^{\infty} B_n = \bigcup_{n=1}^{\infty} (A_n \cap S) = \left(\bigcup_{n=1}^{\infty} A_n \right) \cap S$$

and, as $\bigcup_n A_n \in \mathscr{E}$, also $\bigcup_n B_n \in \mathscr{E}_S$.

1.5 (a) We have seen already (p. 4) that the functions

$$\varlimsup_{n \to \infty} f_n \quad \text{and} \quad \varliminf_{n \to \infty} f_n$$

are measurable. L is the set where these two functions coincide and is therefore measurable.

(b) If the sequence $(f_n)_n$ takes values in a metric space G, the set of the points x for which the Cauchy condition is satisfied can be written

$$H := \bigcap_{\ell=0}^{\infty} \bigcup_{n=0}^{\infty} \bigcap_{m,k \geq n} \left\{ x \in E; d\big(f_m(x), f_k(x)\big) \leq \frac{1}{\ell} \right\}.$$

The distance function $d : G \times G \to \mathbb{R}$ is continuous, so that all sets appearing in the definition of H are measurable. If G is also complete, then $H = L = \{x; \lim_{n \to \infty} f_n(x) \text{ exists}\}$ is measurable.

1.6 (a) Immediate as $\Phi \circ f = \lim_{n \to \infty} \Phi \circ f_n$ and the functions $\Phi \circ f_n$ are real-valued.

(b) Let $D \subset G$ be a countable dense subset and let us denote by $B_z(r)$ the open ball centered at $z \in D$ and with radius r. Then if $\Phi(x) = d(x, z)$ we have $f^{-1}(B_z(r)) = (\Phi \circ f)^{-1}([0, r[)$. Hence $f^{-1}(B_z(r)) \in \mathscr{E}$. Every open set of (G, d) is the (countable) union of balls $B_z(r)$ with $z \in D$ and radius $r \in \mathbb{Q}$. Hence $f^{-1}(A) \in \mathscr{E}$ for every open set $A \subset G$ and the proof is complete thanks to Remark 1.5.

1.7 (a) This is a rather intuitive inequality as, if the events were disjoint, we would have an equality. A first way of proving this rigorously is to trace back to a sequence of disjoint sets to which σ-additivity can be applied, following the same idea as in Remark 1.10(b). To be precise, recursively define

$$B_1 = A_1, \quad B_2 = A_2 \setminus A_1 \quad, \dots, \quad B_n = A_n \setminus \bigcup_{k=1}^{n-1} A_k, \dots$$

The B_n are pairwise disjoint and $B_1 \cup \cdots \cup B_n = A_1 \cup \cdots \cup A_n$, therefore

$$\bigcup_{n=1}^{\infty} A_n = \bigcup_{n=1}^{\infty} B_n.$$

Moreover $B_n \subset A_n$, so that

$$\mu\left(\bigcup_{n=1}^{\infty} A_n\right) = \mu\left(\bigcup_{n=1}^{\infty} B_n\right) = \sum_{n=1}^{\infty} \mu(B_n) \le \sum_{n=1}^{\infty} \mu(A_n).$$

There is a second method, which is simpler, but uses the integral and Beppo Levi's Theorem. If $A = \bigcup_{n=1}^{\infty} A_n$, then clearly

$$1_A \le \sum_{k=1}^{\infty} 1_{A_k}$$

as the sum on the right-hand side certainly takes a value which is ≥ 1 on A. Now we have, thanks to Corollary 1.22(a),

$$\mu(A) = \int_E 1_A \, d\mu \le \int_E \sum_{k=1}^{\infty} 1_{A_k} \, d\mu = \sum_{k=1}^{\infty} \int_E 1_{A_k} \, d\mu = \sum_{k=1}^{\infty} \mu(A_k).$$

(b) Immediate as, thanks to (a),

$$\mu(A) \le \sum_{n=1}^{\infty} \mu(A_n) = 0.$$

(c) If $A \in \mathcal{A}$ then obviously $A^c \in \mathcal{A}$. If $(A_n)_n \subset \mathcal{A}$ and $\mu(A_n) = 0$ for every n then, thanks to (b), also $\mu(\bigcup_n A_n) = 0$, hence $\bigcup_n A_n \in \mathcal{A}$. Otherwise, if there exists an n_0 such that $\mu(A_{n_0}^c) = 0$, then

$$\mu\left(\left(\bigcup_{n=1}^{\infty} A_n\right)^c\right) \le \mu(A_{n_0}^c) = 0$$

and again $\bigcup_n A_n \in \mathcal{A}$.

1.8 (a) Let $(x_n)_n \subset F$ be a sequence converging to some $x \in E$ and let us prove that $x \in F$. If $r > 0$ then the ball $B_x(r)$ contains at least one of the x_n (actually infinitely many of them). Hence it also contains a ball $B_{x_n}(r')$, for some $r' > 0$. Hence $\mu(B_x(r)) > \mu(B_{x_n}(r')) > 0$, as $x_n \in F$. Hence also $x \in F$.

(b1) Let $D \subset E$ be a dense subset. For every $x \in D \cap F^c$ there exists a neighborhood V_x of x such that $\mu(V_x) = 0$ and that we can assume to be disjoint from F, which is closed. F^c is then the (countable) union of such V_x's for $x \in D$ and is a negligible set, being the countable union of negligible sets (Exercise 1.7(b)).

(b2) If F_1 is a closed set strictly contained in F such that $\mu(F_1^c) = 0$, then there exist $x \in F \setminus F_1$ and $r > 0$ such that $B_x(r) \subset F_1^c$. But then we would have $\mu(B_x(r)) = 0$, in contradiction with the fact that $x \in F$.

1.9 (a) We have, for every $n \in \mathbb{N}$,

$$|f| \ge n 1_{\{|f|=+\infty\}}$$

and therefore

$$\int_E |f| \, d\mu \ge n\mu(|f| = +\infty) \,.$$

As this relation holds for every n, if $\mu(f = +\infty) > 0$ we would have $\int |f| \, d\mu = +\infty$, in contradiction with the integrability of $|f|$.

(b) Let, for every positive integer n, $A_n = \{f \ge \frac{1}{n}\}$. Obviously $f \ge \frac{1}{n} 1_{A_n}$ and therefore

$$\int_E f \, d\mu \ge \int_E \frac{1}{n} 1_{A_n} \, d\mu = \frac{1}{n} \mu(A_n) \,.$$

Hence $\mu(A_n) = 0$ for every n. Now

$$\{f > 0\} = \bigcup_{n=1}^{\infty} \{f \ge \tfrac{1}{n}\} = \bigcup_{n=1}^{\infty} A_n \,,$$

hence $\{f > 0\}$ is negligible, being the countable union of negligible sets (Exercise 1.7(b)).

(c) Let $A_n = \{f \leq -\frac{1}{n}\}$. Then

$$\int_{A_n} f \, d\mu \leq -\frac{1}{n} \mu(A_n) \,.$$

Therefore as we assume that $\int_A f \, d\mu \geq 0$ for every $A \in \mathscr{E}$, necessarily $\mu(A_n) = 0$ for every n. But

$$\{f < 0\} = \bigcup_{n=1}^{\infty} A_n$$

hence again $\{f < 0\}$ is negligible, being the countable union of negligible sets.

1.10 By Beppo Levi's Theorem we have

$$\int_E |f| \, d\mu = \lim_{n \to \infty} \uparrow \int_E |f| \wedge n \, d\mu \,.$$

But, for every n, $|f| \wedge n \leq n \, 1_N$, so that

$$\int_E |f| \wedge n \, d\mu \leq n \, \mu(N) = 0 \,.$$

Taking $n \to \infty$, Beppo Levi's Theorem gives $\int_E |f| \, d\mu = 0$, hence also $\int_E f \, d\mu = 0$.

- In particular the integral of a function taking the value $+\infty$ on a set of measure 0 and vanishing elsewhere is equal to 0.

1.11 (a) Let μ be the measure on $\mathbb{N}$ defined as

$$\mu(n) = w_n \,.$$

With this definition we can write

$$\phi(t) = \int_{\mathbb{N}} e^{-tx} \, d\mu(x) \,.$$

Let us check the conditions of Theorem 1.21 (derivation under the integral sign) for the function $f(t, x) = e^{-tx}$. Let $a > 0$ be such that $I =]a, +\infty[$ is a half-line containing t. Then

$$\left| \frac{\partial f}{\partial t}(t, x) \right| = |x| e^{-tx} \leq |x| e^{-ax} := g(x) \,. \tag{7.1}$$

g is integrable with respect to μ as

$$\int_{\mathbb{N}} g(x)\, d\mu(x) = \sum_{n=1}^{\infty} n w_n \, e^{-an}$$

and the series on the right-hand side is summable. Thanks to Theorem 1.21, for every $a > 0$, ϕ is differentiable in $]a, +\infty[$ and

$$\phi'(t) = \int_{\mathbb{N}} \frac{\partial f}{\partial t}(t, x)\, d\mu(x) = -\sum_{n=1}^{\infty} n w_n \, e^{-tn} .$$

(b) If $w_n^+ = w_n \vee 0$, $w_n^- = -w_n \wedge 0$, then the two sequences $(w_n^+)_n$, $(w_n^-)_n$ are positive and

$$\phi(t) = \sum_{n=1}^{\infty} w_n^+ e^{-tn} - \sum_{n=1}^{\infty} w_n^- e^{-tn} := \phi^+(t) - \phi^-(t)$$

and now both ϕ^+ and ϕ^- are differentiable thanks to (a) above and (1.34) follows.

(c1) Just consider the measure on $\mathbb{N}$

$$\mu(n) = \sqrt{n} .$$

In order to repeat the argument of (a) we just have to check that the function g of (7.1) is integrable with respect to the new measure μ, i.e. that

$$\sum_{n=1}^{\infty} n^{3/2} e^{-an} < +\infty ,$$

which is immediate.

(c2) Again the answer is positive provided that

$$\sum_{n=1}^{\infty} n\, e^{\sqrt{n}} e^{-an} < +\infty . \tag{7.2}$$

Now just write $n\, e^{\sqrt{n}} e^{-an} = n\, e^{\sqrt{n}} e^{-\frac{1}{2}an} \cdot e^{-\frac{1}{2}an}$. As

$$\lim_{n\to\infty} n\, e^{\sqrt{n}} e^{-\frac{1}{2}an} = 0$$

the general term of the series in (7.2) is bounded above, for n large, by $e^{-\frac{1}{2}an}$, which is the general term of a convergent series.

1.12 There are many possible solutions of this exercise, of course.

(a) Let us choose $E = \mathbb{R}$, $\mathscr{E} = \mathscr{B}(\mathbb{R})$ and $\mu =$ Lebesgue's measure. If $A_n = [n, +\infty[$ then $A = \bigcap_n A_n = \emptyset$, so that $\mu(A) = 0$ whereas $\mu(A_n) = +\infty$ for every n.

(b) Let $(E, \mathscr{E}, \mu)$ be as in (a). Let $f_n = -1_{[n,+\infty]}$. We have $f_n \uparrow 0$ as $n \to \infty$, but the integral of the f_n is equal to $-\infty$ for every n.

1.13 Let $A \in \mathscr{G}$ be such that $\widetilde{\mu}(A) = 0$. Then $\mu(\Phi^{-1}(A)) = \widetilde{\mu}(A) = 0$ hence also $\nu(\Phi^{-1}(A)) = 0$, so that $\widetilde{\nu}(A) = \nu(\Phi^{-1}(A)) = 0$.

1.14 (a) If $(A_n)_n \subset \mathscr{B}([0, 1])$ is a sequence of disjoint sets, then

- if $\lambda(A_n) = 0$ for every n then also $\lambda(\bigcup_n A_n) = 0$, therefore

$$\mu\left(\bigcup_{n=1}^{\infty} A_n\right) = 0 \quad \text{and} \quad \sum_{n=1}^{\infty} \mu(A_n) = 0 \,.$$

- If, instead, $\lambda(A_n) > 0$ for some n, then also $\lambda(\bigcup_n A_n) > 0$ and

$$\mu\left(\bigcup_{n=1}^{\infty} A_n\right) = +\infty \quad \text{and} \quad \sum_{n=1}^{\infty} \mu(A_n) = +\infty \,,$$

so that in any case the σ-additivity of μ is satisfied.

(b) Of course if $\mu(A) = 0$ then also $\lambda(A) = 0$ so that $\lambda \ll \mu$. If a density f of λ with respect to μ existed we would have, for every $A \in \mathscr{B}([0, 1])$,

$$\lambda(A) = \int_A f \, d\mu \,.$$

But this is not possible because the integral on the right-hand side can only take the values 0 (if $1_A f = 0$ μ-a.e.) or $+\infty$ (otherwise).

The hypotheses of the Radon-Nikodym theorem are not satisfied here (μ is not σ-finite).

1.15 (a1) Assume, to begin with, $p < +\infty$. Denoting by M an upper bound of the L^p norms of the f_n (the sequence is bounded in L^p), Fatou's Lemma gives

$$\int_E |f|^p \, d\mu \leq \varliminf_{n \to \infty} \int_E |f_n|^p \, d\mu \leq M^p$$

hence $f \in L^p$. The case $p = +\infty$ is rather obvious but, to be precise, let M be again an upper bound of the norms $\|f_n\|_\infty$. This means that if $A_n = \{|f_n| > M\}$ then $\mu(A_n) = 0$ for every n. We obtain immediately that outside $A = \bigcup_n A_n$, which is also negligible, $|f| \leq M$ μ-a.e.

(a2) Counterexample: μ = the Lebesgue measure of $\mathbb{R}$, $f_n = 1_{[n,n+1]}$. Every f_n has, for every p, L^p norm equal to 1 and $(f_n)_n$ converges to 0 a.e. but certainly not in L^p, as $\|f_n\|_p \equiv 1$ and L^p convergence entails convergence of the L^p-norms (Remark 1.30).

(b) We have $g_n \to g$ a.e. as $n \to \infty$. As $|g_n| \leq |g|$ and by the obvious bound $|g - g_n| \leq |g| + |g_n| \leq 2|g|$, we have by Lebesgue's Theorem

$$\int_E |g_n - g|^p \, d\mu \xrightarrow[n \to \infty]{} 0 .$$

1.16 (a1) Let $p < q$. If $|x| \leq 1$, then $|x|^p \leq 1$; if conversely $|x| \geq 1$, then $|x|^p \leq |x|^q$. Hence, in any case, $|x|^p \leq 1 + |x|^q$. If $p \leq q$ and $f \in L^q$, then $|f|^p \leq 1 + |f|^q$ and we have

$$\|f\|_p^p = \int_E |f|^p \, d\mu \leq \int_E (1 + |f|^q) \, d\mu \leq \mu(E) + \|f\|_q^q ,$$

hence $f \in L^p$.

(a2) If $p \to q-$, then $|f|^p \to |f|^q$ a.e. Moreover, thanks to a1), $|f|^p \leq 1 + |f|^q$. As $|f|^q$ and the constant function 1 are integrable (μ is finite), by Lebesgue's Theorem

$$\lim_{p \to q-} \int_E |f|^p \, d\mu = \int_E |f|^q \, d\mu .$$

(a3) Again we have $|f|^p \to |f|^q$ a.e. as $p \to q-$, and by Fatou's Lemma

$$\varliminf_{p \to q-} \int_E |f|^p \, d\mu \geq \int_E |f|^q \, d\mu = +\infty .$$

(a4) (1.37) follows by Fatou's Lemma again. Moreover, if $f \in L^{q_0}$ for some $q_0 > q$, then for $q \leq p \leq q_0$ we have $|f|^p \leq 1 + |f|^{q_0}$ and (1.38) follows by Lebesgue's Theorem.

(a5) Let μ be the Lebesgue measure. The function

$$f(x) = \frac{1}{x \log^2 x} 1_{[0,\frac{1}{2}]}(x)$$

is integrable (a primitive of $x \mapsto (x \log^2 x)^{-1}$ is $x \mapsto (-\log x)^{-1}$). But $|f|^p = (x^p \log^{2p} x)^{-1}$ is not integrable at 0 for any $p > 1$. Therefore, $\|f\|_1 < +\infty$, whereas $\lim_{p \to 1+} \|f\|_p = +\infty$.

(b1) As $|f| \leq \|f\|_\infty$ a.e.

$$\|f\|_p^p = \int_E |f|^p \, d\mu \leq \|f\|_\infty^p \, \mu(E)$$

which gives

$$\varlimsup_{p\to+\infty} \|f\|_p \leq \|f\|_\infty \lim_{p\to+\infty} \mu(E)^{1/p} = \|f\|_\infty .$$

(b2) We have $|f|^p \geq |f|^p 1_{\{|f|\geq M\}} \geq M^p 1_{\{|f|\geq M\}}$. Hence

$$\int_E |f|^p \, d\mu \geq \int_E M^p 1_{\{|f|\geq M\}} \, d\mu = M^p \mu(|f| \geq M) . \tag{7.3}$$

If $M < \|f\|_\infty$, then $\mu(|f| \geq M) > 0$ and by (7.3)

$$\varliminf_{p\to+\infty} \|f\|_p \geq \lim_{p\to+\infty} M \mu(|f| \geq M)^{1/p} = M .$$

By the arbitrariness of $M < \|f\|_\infty$ and (b1)

$$\lim_{p\to+\infty} \|f\|_p = \|f\|_\infty .$$

1.17 An element of ℓ_p is a sequence $(a_n)_n$ such that

$$\sum_{n=1}^{\infty} |a_n|^p < +\infty . \tag{7.4}$$

If $(a_n)_n \in \ell_p$ then necessarily $|a_n| \to_{n\to\infty} 0$, hence $|a_n| \leq 1$ for n larger than some n_0. If $q \geq p$ then $|a_n|^q \leq |a_n|^p$ for $n \geq n_0$ and the series with general term $|a_n|^q$ is bounded above eventually by the series with general term $|a_n|^p$.

1.18 We have

$$\int_0^{+\infty} \frac{1}{x} e^{-tx} \sin x \, dx = \int_0^{+\infty} e^{-tx} \, dx \int_0^1 \cos(xy) \, dy$$

$$= \int_0^1 dy \int_0^{+\infty} \cos(xy) \, e^{-tx} \, dx .$$

Integrating by parts we find

$$\int_0^{+\infty} \cos(xy) \, e^{-tx} \, dx = -\frac{1}{t} e^{-tx} \cos(xy) \Big|_{x=0}^{x=+\infty} - \frac{y}{t} \int_0^{+\infty} \sin(xy) \, e^{-tx} \, dx$$

$$= \frac{1}{t} + \frac{y}{t^2} e^{-tx} \sin(xy) \Big|_{x=0}^{x=+\infty} - \frac{y^2}{t^2} \int_0^{+\infty} \cos(xy) \, e^{-tx} \, dx ,$$

from which

$$\left(1 + \frac{y^2}{t^2}\right) \int_0^{+\infty} \cos(xy)\, e^{-tx}\, dx = \frac{1}{t}$$

and

$$\int_0^{+\infty} \cos(xy)\, e^{-tx}\, dx = \frac{t}{t^2 + y^2} \cdot$$

Therefore, with the change of variable $z = \frac{y}{t}$,

$$\int_0^{+\infty} \frac{1}{x} \sin x\, e^{-tx}\, dx = \int_0^1 \frac{t}{t^2 + y^2}\, dy = \int_0^{1/t} \frac{1}{1 + z^2}\, dz = \arctan \frac{1}{t} \cdot$$

Of course we can apply Fubini's Theorem as $(x, y) \mapsto \cos(xy)\, e^{-tx}$ is integrable on $\mathbb{R}^+ \times [0, 1]$.

As $t \to 0+$ the integral converges to $\frac{\pi}{2}$.

1.19 We must prove that the integral $\int_{\mathbb{R}^d} |f(y)|\,|g(x - y)|\, dy$ is finite for almost every x. Note first that this integral is well defined, the integrand being positive. By Fubini's Theorem 1.34

$$\int_{\mathbb{R}^d} dx \int_{\mathbb{R}^d} |f(y)|\,|g(x - y)|\, dy = \int_{\mathbb{R}^d} |f(y)|\, dy \int_{\mathbb{R}^d} |g(x - y)|\, dx$$

$$= \int_{\mathbb{R}^d} |f(y)|\, dy \int_{\mathbb{R}^d} |g(x)|\, dx = \|f\|_1 \|g\|_1 \,.$$

Hence $(x, y) \mapsto f(y)g(x - y)$ is integrable and, again by Fubini's Theorem (this is (1.30), to be precise)

$$x \mapsto \int_{\mathbb{R}^d} f(y)g(x - y)\, dy$$

is an a.e. finite measurable function of L^1. Moreover

$$\|f * g\|_1 = \int_{\mathbb{R}^d} |f * g(x)|\, dx = \int_{\mathbb{R}^d} dx \left| \int_{\mathbb{R}^d} f(y)g(x - y)\, dy \right|$$

$$\leq \int_{\mathbb{R}^d} dx \int_{\mathbb{R}^d} \left| f(y)g(x - y) \right| dy = \|f\|_1 \|g\|_1 \,.$$

2.1 We have

$$P\left(\bigcap_{n=1}^{\infty} A_n\right) = 1 - P\left(\left(\bigcap_{n=1}^{\infty} A_n\right)^c\right) = 1 - P\left(\bigcup_{n=1}^{\infty} A_n^c\right) = 1$$

as the events A_n^c are negligible and a countable union of negligible events is also negligible (Exercise 1.7).

2.2 Let us denote by D a dense subset of E.

(a) Let us consider the countable set of the balls $B_x(\frac{1}{n})$ centered at $x \in D$ and with radius $\frac{1}{n}$. As the events $\{X \in B_x(\frac{1}{n})\}$ belong to $\mathcal{G}$, their probability can be equal to 0 or to 1 only. As their union is equal to E, for every n there exists at least an $x_n \in D$ such that $P(X \in B_{x_n}(\frac{1}{n})) = 1$.

(b) Let $A_n = B_{x_1}(1) \cap \cdots \cap B_{x_n}(\frac{1}{n})$. $(A_n)_n$ is clearly a decreasing sequence of measurable subsets of E, A_n has diameter $\leq \frac{2}{n}$, as $A_n \subset B_{x_n}(\frac{1}{n})$, and the event $\{X \in A_n\}$ has probability 1, being the intersection of the events $\{X \in B_{x_k}(\frac{1}{k})\}$, $k = 1, \ldots, n$, all of them having probability 1.

(c) The set

$$A = \bigcap_{n=1}^{\infty} A_n$$

has diameter 0 and therefore is formed by a single $x_0 \in E$ or is $= \emptyset$. But, as the sequence $(A_n)_n$ is decreasing,

$$P(X \in A) = \lim_{n \to \infty} P(X \in A_n) = 1 .$$

Hence A is non-void and is formed by a single x_0. We conclude that $X = x_0$ with probability 1.

2.3 (a) We have, for every $k > 0$,

$$\{Z = +\infty\} = \Big\{ \sup_{n \geq 1} X_n = +\infty \Big\} = \Big\{ \sup_{n \geq k} X_n = +\infty \Big\} ,$$

hence the event $\{Z = +\infty\}$ belongs to the tail σ-algebra of the sequence $(X_n)_n$ and by Kolmogorov's 0-1 law, Theorem 2.15, can only have probability 0 or 1. If $P(Z \leq a) > 0$, necessarily $P(Z = +\infty) < 1$ hence $P(Z = +\infty) = 0$.

(b1) Let $a > 0$. As the events $\{\sup_{k \leq n} X_k \leq a\}$ decrease to $\{Z \leq a\}$ as $n \to \infty$, we have

$$P(Z \leq a) = \lim_{n \to \infty} P\Big(\sup_{k \leq n} X_k \leq a \Big) = \lim_{n \to \infty} \prod_{k=1}^{n} P(X_k \leq a) = \prod_{k=1}^{\infty} (1 - e^{-\lambda_k a}) .$$

The infinite product converges to a strictly positive number if and only if the series $\sum_{k=1}^{\infty} e^{-\lambda_k a}$ is convergent (see Proposition 3.4 p. 119, in case this fact was not already known). In this case

$$\sum_{k=1}^{\infty} e^{-\lambda_k a} = \sum_{k=1}^{\infty} \frac{1}{k^a} \cdot$$

If $a > 1$ the series is convergent, hence $P(Z \le a) > 0$ and, thanks to (a), $Z < +\infty$ a.s.

(b2) Let $K > 0$. As $\{\sup_{k \le n} X_k \ge K\} \subset \{Z \ge K\}$, we have, for every $n \ge 1$,

$$P(Z > K) \ge P\left(\sup_{k \le n} X_k > K\right) = 1 - P\left(\sup_{k \le n} X_k \le K\right)$$

$$= 1 - P(X_1 \le K, \dots, X_n \le K) = 1 - P(X_1 \le K)^n = 1 - (1 - e^{-cK})^n .$$

As this holds for every n, $P(Z > K) = 1$ for every $K > 0$ hence $Z = +\infty$ a.s.

2.4 By assumption

$$E[|X + Y|] = \int_{\infty}^{\infty} \int_{\infty}^{\infty} |x + y| \, d\mu_X(x) \, d\mu_Y(y) < +\infty .$$

By Fubini's Theorem for μ_Y-almost every y we have

$$\int_{\infty}^{\infty} |x + y| \, d\mu_X(x) < +\infty ,$$

hence $E(|y + X|) < +\infty$ for at least one $y \in \mathbb{R}$ and X is integrable, being the sum of the integrable r.v.'s $y + X$ and $-y$. By symmetry Y is also integrable.

2.5 (a) For every bounded measurable function $\phi : \mathbb{R}^d \to \mathbb{R}$, we have

$$E[\phi(X + Y)] = \int_{\mathbb{R}^d} \int_{\mathbb{R}^d} \phi(x + y) \, d\mu(x) \, d\nu(y)$$

$$= \int_{\mathbb{R}^d} d\nu(y) \int_{\mathbb{R}^d} \phi(x + y) f(x) \, dx$$

$$= \int_{\mathbb{R}^d} d\nu(y) \int_{\mathbb{R}^d} \phi(z) f(z - y) \, dz = \int_{\mathbb{R}^d} \phi(z) \, dz \underbrace{\int_{\mathbb{R}^d} f(z - y) \, d\nu(y)}_{:=g(z)} ,$$

which means that $X + Y$ has density g with respect to the Lebesgue measure dz.

(b) Let us try to apply the derivation theorem of an integral depending on a parameter, Proposition 1.21. By assumption

$$\left| \frac{\partial f}{\partial z_i}(z - y) \right| < M$$

for some constant M, as we assume boundedness of the partial derivatives of f. The constants being integrable with respect to v, the condition of Proposition 1.21 is satisfied and we deduce that g is also differentiable and

$$\frac{\partial g}{\partial z_i}(z) = \int_{\mathbb{R}^d} \frac{\partial f}{\partial z_i}(z - y)\, dv(y) . \tag{7.5}$$

This proves (b) for $k = 1$. Derivation under the integral sign applied to (7.5) proves (b) for $k = 2$ and iterating this argument the result follows by induction.

- Recalling that the law of $X + Y$ is the convolution $\mu * v$, this exercise shows that "convolution regularizes".

2.6 (a) If $A_n := \{|x| > n\}$ then $\bigcap_{n=1}^{\infty} A_n = \emptyset$, so that $\lim_{n\to\infty} \mu(A_n) = 0$ and $\mu(A_n) < \varepsilon$ for n large.

(b) Let $\varepsilon > 0$. We must prove that there exists an $M > 0$ such that $|g(x)| < \varepsilon$ for $|x| > M$. Let us choose $M = M_1 + M_2$, with M_1 and M_2 as in the statement of the exercise. We have then

$$|g(x)| = \left| \int_{\mathbb{R}^d} f(x - y)\, \mu(dy) \right|$$

$$\leq \int_{\{|y| \leq M_1\}} |f(x - y)|\, \mu(dy) + \int_{\{|y| > M_1\}} |f(x - y)|\, \mu(dy) := I_1 + I_2 .$$

We have $I_2 \leq \|f\|_\infty \mu(\{|y| > M_1\}) \leq \varepsilon \|f\|_\infty$. Moreover, if $|x| \geq M = M_1 + M_2$ and $|y| \leq M_1$ then $|x - y| \geq M_2$ so that $|f(x - y)| \leq \varepsilon$. Putting things together we have, for $|x| > M$,

$$|g(x)| \leq \varepsilon(1 + \|f\|_\infty) ,$$

from which the result follows thanks to the arbitrariness of ε.

2.7 If $X \sim N(0, 1)$,

$$E(e^{tX^2}) = \frac{1}{\sqrt{2\pi}} \int_{-\infty}^{+\infty} e^{tx^2} e^{-x^2/2}\, dx = \frac{1}{\sqrt{2\pi}} \int_{-\infty}^{+\infty} e^{-x^2(\frac{1}{2}-t)}\, dx .$$

The integral clearly diverges if $t \geq \frac{1}{2}$. If $t < \frac{1}{2}$ instead just write

$$\int_{-\infty}^{+\infty} e^{-x^2(\frac{1}{2}-t)}\, dx = \int_{-\infty}^{+\infty} \exp\left(-\frac{x^2}{2(1 - 2t)^{-1}} \right) dx .$$

We recognize in the integrand, but for the constant, the density of a Gaussian law with mean 0 and variance $(1 - 2t)^{-1}$. Hence for $t < \frac{1}{2}$ the integral is equal to $\sqrt{2\pi}(1 - 2t)^{-1/2}$ and $E(e^{tX^2}) = (1 - 2t)^{-1/2}$.

Recalling that if $X \sim N(0, 1)$ then $Z = \sigma X \sim N(0, \sigma^2)$, we have $E(e^{tZ^2}) = E(e^{t\sigma^2 X^2})$ and in conclusion

$$E(e^{tZ^2}) = \begin{cases} +\infty & \text{if } t \geq \frac{1}{2\sigma^2} \\ \dfrac{1}{\sqrt{1 - 2\sigma^2 t}} & \text{if } t < \frac{1}{2\sigma^2}. \end{cases}$$

2.8 Let us assume first $\sigma > 0$. We have, thanks to the integration rule with respect to an image measure, Proposition 1.27,

$$E\big[(xe^{b+\sigma X} - K)^+\big] = \frac{1}{\sqrt{2\pi}} \int_{-\infty}^{+\infty} (xe^{b+\sigma z} - K)^+ e^{-\frac{1}{2} z^2} \, dz \,.$$

The integrand vanishes if $xe^{b+\sigma z} - K < 0$, i.e. if

$$z \leq \zeta := \frac{1}{\sigma} \Big(\log \frac{K}{x} - b \Big) \,,$$

hence, with a few standard changes of variable,

$$\begin{aligned} E\big[(xe^{b+\sigma X} - K)^+\big] &= \frac{1}{\sqrt{2\pi}} \int_{\zeta}^{+\infty} (xe^{b+\sigma z} - K) e^{-\frac{1}{2} z^2} \, dz \\ &= \frac{x}{\sqrt{2\pi}} \int_{\zeta}^{+\infty} e^{b+\sigma z - \frac{1}{2} z^2} \, dz - \frac{K}{\sqrt{2\pi}} \int_{\zeta}^{+\infty} e^{-\frac{1}{2} z^2} \, dz \\ &= \frac{xe^{b+\frac{1}{2}\sigma^2}}{\sqrt{2\pi}} \int_{\zeta}^{+\infty} e^{-\frac{1}{2}(z-\sigma)^2} \, dz - K\big(1 - \Phi(\zeta)\big) \\ &= \frac{xe^{b+\frac{1}{2}\sigma^2}}{\sqrt{2\pi}} \int_{\zeta-\sigma}^{+\infty} e^{-\frac{1}{2} z^2} \, dz - K\Phi(-\zeta) \\ &= xe^{b+\frac{1}{2}\sigma^2} \Phi(-\zeta + \sigma) - K\Phi(-\zeta) \,. \end{aligned}$$

Finally note that as $\sigma X \sim -\sigma X$, $E[(xe^{b+\sigma X} - K)^+] = E[(xe^{b+|\sigma|X} - K)^+]$.

2.9 (a) Let us first compute the d.f. With the change of variable $s^\alpha = u$, $\alpha s^{\alpha-1} \, ds = du$ we find for the d.f. F of f, for $t > 0$,

$$F(t) = \int_0^t \lambda \alpha s^{\alpha-1} e^{-\lambda s^\alpha} \, ds = \int_0^{t^\alpha} \lambda e^{-\lambda u} \, du = 1 - e^{-\lambda t^\alpha} \,. \tag{7.6}$$

As

$$\int_{-\infty}^{+\infty} f(s)\,ds = \lim_{t\to+\infty}\int_{-\infty}^{t} f(s)\,ds = \lim_{t\to+\infty} F(t) = 1\,,$$

f is a probability density with respect to the Lebesgue measure.

(b1) If X is exponential with parameter λ we have, recalling the values of the constants for the Gamma laws,

$$E(X^{\beta}) = \lambda \int_{0}^{+\infty} t^{\beta} e^{-\lambda t}\,dt = \frac{\lambda \Gamma(\beta+1)}{\lambda^{\beta+1}} = \frac{\Gamma(\beta+1)}{\lambda^{\beta}}\,. \tag{7.7}$$

The d.f., G say, of X^{β} is, for $t > 0$,

$$G(t) = P(X^{\beta} \le t) = P(X \le t^{1/\beta}) = 1 - e^{-\lambda t^{1/\beta}}\,,$$

so that, comparing with (7.6), X^{β} is Weibull with parameters λ and $\alpha = \frac{1}{\beta}$.

(b2) Thanks to (b1) a Weibull r.v. Y with parameters α, λ is of the form $X^{1/\alpha}$, where X is exponential with parameter λ; thanks to (7.7), for $\beta = \frac{1}{\alpha}$ and $\beta = \frac{2}{\alpha}$, we have

$$E(Y) = E(X^{1/\alpha}) = \frac{\Gamma(1+\frac{1}{\alpha})}{\lambda^{1/\alpha}}\,,$$

$$E(Y^{2}) = E(X^{2/\alpha}) = \frac{\Gamma(1+\frac{2}{\alpha})}{\lambda^{2/\alpha}}$$

and for the variance

$$\mathrm{Var}(Y) = E(Y^{2}) - E(Y)^{2} = \frac{\Gamma(1+\frac{2}{\alpha}) - \Gamma(1+\frac{1}{\alpha})^{2}}{\lambda^{2/\alpha}}\,.$$

(c) Just note that $\Gamma(1+2t) - \Gamma(1+t)^{2}$ is the variance of a Weibull r.v. with parameters $\lambda = 1$ and $\alpha = \frac{1}{t}$. Hence it is a positive quantity.

2.10 The density of X is obtained from the joint density as explained in Example 2.16:

$$f_X(x) = \int_{-\infty}^{+\infty} f(x,y)\,dy = (\theta+1)\,e^{\theta x} \int_{0}^{+\infty} \frac{e^{\theta y}}{(e^{\theta x} + e^{\theta y} - 1)^{2+\frac{1}{\theta}}}\,dy$$

$$= -(\theta+1)\,e^{\theta x}\,\frac{1}{\theta(1+\frac{1}{\theta})}\,\frac{1}{(e^{\theta x}+e^{\theta y}-1)^{1+\frac{1}{\theta}}}\Bigg|_{y=0}^{y=+\infty}$$

$$= e^{\theta x}\,\frac{1}{(e^{\theta x})^{1+\frac{1}{\theta}}} = e^{-x}\,.$$

Hence X is exponential of parameter 1. By symmetry Y has the same density. Note that the marginals do not depend on θ.

2.11 (a1) We have, for $t \geq 0$,

$$P(-\log X \leq t) = P(X \geq e^{-t}) = 1 - e^{-t} \,,$$

hence $-\log X$ is an exponential Gamma(1, 1)-distributed r.v.

(a2) $W = -\log X - \log Y$ is therefore Gamma(2, 1)-distributed and its d.f. is, again for $t \geq 0$,

$$F_W(t) = 1 - e^{-t} - te^{-t} \,.$$

Hence the d.f. of $XY = e^{-W}$ is, for $0 < s \leq 1$,

$$F(s) = P(e^{-W} \leq s) = P(W \geq -\log s) = 1 - F_W(-\log s) = s - s \log s$$

and, taking the derivative, the density of XY is

$$f(s) = -\log s \qquad \text{for } 0 < s \leq 1 \,.$$

(b) The r.v.'s XY and Z are independent and their joint law has a density with respect to the Lebesgue measure that is the tensor product of their densities. We have, for $z \in [0, 1]$,

$$P(Z^2 \leq z) = P(Z \leq \sqrt{z}) = \sqrt{z}$$

and, taking the derivative, the density of Z^2 is

$$f_{Z^2}(z) = \frac{1}{2\sqrt{z}} \qquad 0 < z \leq 1 \,.$$

The joint density of (XY, Z) is therefore, for $s, z \in]0, 1]$,

$$f(s, z) = -\frac{1}{2\sqrt{z}} \log s \,.$$

The probability $P(XY < Z^2)$ is the integral of f on the region $\{s < z\}$, i.e.

$$P(XY < Z^2) = \int_0^1 -\log s \, ds \int_s^1 \frac{1}{2\sqrt{z}} \, dz = -\int_0^1 \left(1 - \sqrt{s}\right) \log s \, ds \,.$$

Now

$$-\int_0^1 \log s \, ds = s - s \log s \Big|_0^1 = 1$$

whereas, integrating by parts,

$$\int_0^1 \sqrt{s}\,\log s\,ds = \frac{2}{3}s^{3/2}\log s\,\Big|_0^1 - \frac{2}{3}\int_0^1 s^{3/2}\frac{1}{s}\,ds = -\frac{2}{3}\frac{2}{3}s^{3/2}\Big|_0^1 = -\frac{4}{9}\,.$$

Therefore

$$P(XY < Z^2) = 1 - \frac{4}{9} = \frac{5}{9}\,.$$

2.12 (a) Note first that Z_1 is positive integer-valued, whereas Z_2 takes values in $[0, 1]$. Now, recalling the expression of the d.f. F of the exponential laws,

$$P(Z_1 = k) = P(k \le Z < k+1) = F(k+1) - F(k)$$
$$= 1 - e^{\lambda(k+1)} - (1 - e^{-\lambda k})$$
$$= e^{-\lambda k} - e^{\lambda(k+1)} = e^{-\lambda k}(1 - e^{-\lambda})$$

and we recognize a geometric law of parameter $p = 1 - e^{-\lambda}$. Now

$$\{Z_2 \le t\} = \bigcup_{k=1}^{\infty}\{k \le Z \le k+t\}$$

and, for $0 \le t \le 1$,

$$F_2(t) := P(Z_2 \le t) = \sum_{k=0}^{\infty}(e^{-\lambda k} - e^{-\lambda(k+t)}) = (1 - e^{-\lambda t})\sum_{k=0}^{\infty}e^{-\lambda k}$$
$$= \frac{1 - e^{-\lambda t}}{1 - e^{-\lambda}}\,.$$

(b1) We have, for $k \in \mathbb{N}, 0 < a < b < 1$,

$$P(Z_1 = k, Z_2 \in]a, b]) = P(k + a \le Z \le k + b) = e^{-\lambda(k+a)} - e^{-\lambda(k+b)}$$
$$= e^{-\lambda k}(e^{-\lambda a} - e^{-\lambda b}) = e^{-\lambda k}(1 - e^{-\lambda})\frac{e^{-\lambda a} - e^{-\lambda b}}{1 - e^{-\lambda}}$$
$$= P(Z_1 = k)\,P(Z_2 \in [a, b])\,.$$

(b2) The sets $\{k\}\times]a, b]$ form a class that is stable with respect to finite intersections and generate the product σ-algebra $\mathscr{P}(\mathbb{N}) \otimes \mathscr{B}([0, 1])$. Thanks to (b1) the law of (Z_1, Z_2) coincides with the product of the laws of Z_1 and Z_2 on this class, hence, by Proposition 1.11 (Carathéodory's criterion) the two laws coincide and Z_1 and Z_2 are independent.

2.13 (a) Thanks to Remark 2.1

$$\int_0^{+\infty} g(t)\,dt = \frac{1}{b}\int_0^{+\infty} (1 - F(t))\,dt = \frac{1}{b}\int_0^{+\infty} P(X \geq t)\,dt = \frac{1}{b}E(X) = 1$$

and therefore g is a probability density.

(b1) In this case $\overline{F}(t) = e^{-\lambda t}$, $b = \frac{1}{\lambda}$ and $g(t) = \lambda e^{-\lambda t}$. The density g coincides with the density of X.

(b2) Now $\overline{F}(t) = 1 - t$ for $0 \leq t \leq 1$ whereas $b = \frac{1}{2}$. Hence $g(t) = \frac{1}{2}(1 - t)$, for $0 \leq t \leq 1$ and 0 otherwise (i.e. a Beta(1, 2)).

(b3) We have, for $t \geq 0$,

$$F(t) = 1 - \frac{\theta^\alpha}{(\theta + t)^\alpha}$$

and

$$E(X) = \int_0^{+\infty} P(X \geq t)\,dt = \int_0^{+\infty} \frac{\theta^\alpha}{(\theta + t)^\alpha}\,dt$$

$$= \frac{\theta^\alpha}{1 - \alpha} \frac{1}{(\theta + t)^{\alpha-1}}\bigg|_0^{+\infty} = \frac{\theta}{\alpha - 1}$$

and therefore

$$g(t) = \frac{(\alpha - 1)\theta^{\alpha-1}}{(\theta + t)^\alpha}$$

i.e. a Pareto distribution with parameters $\alpha - 1$ and θ.

(c) The d.f. of a Gamma(n, λ) law is, for $t \geq 0$,

$$t \mapsto 1 - e^{-\lambda t}\sum_{k=0}^{n-1} \frac{(\lambda t)^k}{k!}$$

and its mean is equal to $\frac{n}{\lambda}$, hence, for $t > 0$,

$$g(t) = \frac{1}{n}\sum_{k=0}^{n-1} \underbrace{\frac{\lambda^{k+1}}{k!}t^k e^{-\lambda t}}_{\sim\,\text{Gamma}(k+1,\lambda)}$$

(d) We have

$$\int_0^{+\infty} t g(t)\,dt = \frac{1}{b}\int_0^{+\infty} t\,\overline{F}(t)\,dt = \frac{1}{b}\int_0^{+\infty} t\,P(X > t)\,dt$$

and, recalling again Remark 2.1,

$$\cdots = \frac{1}{2b} E(X^2) = \frac{\sigma^2 + b^2}{2b} = \frac{\sigma^2}{2b} + \frac{b}{2} .$$

2.14 We must compute the image, ν say, of the probability

$$d\mu(\theta, \phi) = \frac{1}{4\pi} \sin\theta \, d\theta \, d\phi, \qquad (\theta, \phi) \in [0, \pi] \times [0, 2\pi]$$

under the map $(\theta, \phi) \mapsto \cos\theta$. Let us use the method of the dumb function: let $\psi : [-1, 1] \to \mathbb{R}$ be a bounded measurable function, by the integration formula with respect to an image measure, Proposition 1.27, we have

$$\int \psi(t) \, d\nu(t) = \int \psi(\cos\theta) \, d\mu(\theta, \phi) = \frac{1}{4\pi} \int_0^{2\pi} d\phi \int_0^\pi \psi(\cos\theta) \sin\theta \, d\theta$$

$$= \frac{1}{2} \int_0^\pi \psi(\cos\theta) \sin\theta \, d\theta = \frac{1}{2} \int_{-1}^1 \psi(u) \, du ,$$

i.e. ν is the uniform distribution on $[-1, 1]$. In some sense all points of the interval $[-1, 1]$ are "equally likely".

• One might wonder what the answer to this question would be for the spheres of $\mathbb{R}^d$ for other values of d. Exercise 2.15 gives an answer for $d = 2$ (i.e. the circle).

2.15 First approach: let us compute the d.f. of W: for $-1 \le t \le 1$

$$F_W(t) = P(W \le t) = P(\cos Z \le t) = P(Z \ge \arccos t) = \frac{1}{\pi} (\pi - \arccos t)$$

(recall that arccos is decreasing). Hence

$$f_W(t) = \frac{1}{\pi \sqrt{1 - t^2}}, \qquad -1 \le t \le 1 . \tag{7.8}$$

Second approach: the method of the dumb function: let $\phi : \mathbb{R} \to \mathbb{R}$ be a bounded Borel function, then

$$E[\phi(\cos Z)] = \frac{1}{\pi} \int_0^\pi \phi(\cos\theta) \, d\theta .$$

Let $t = \cos\theta$, so that $\theta = \arccos t$ and $d\theta = -(1 - t^2)^{-1/2} \, dt$. Recall that arccos is the inverse of the cos function restricted to the interval $[0, \pi]$ and therefore taking values in the interval $[-1, 1]$. This gives

$$E[\phi(\cos Z)] = \int_{-1}^1 \phi(t) \frac{1}{\pi \sqrt{1 - t^2}} \, dt$$

i.e. (7.8).

2.16 (a) The integral of f on $\mathbb{R}^2$ must be equal to 1. In polar coordinates and with the change of variable $r^2 = u$, we have

$$1 = \int_{-\infty}^{+\infty} \int_{-\infty}^{+\infty} f(x, y)\, dx\, dy = 2\pi \int_{0}^{+\infty} g(r^2) r\, dr = \pi \int_{0}^{+\infty} g(u)\, du\ .$$

(b1) We know (Example 2.16) that X has density, with respect to the Lebesgue measure,

$$f_X(x) = \int_{-\infty}^{+\infty} g(x^2 + y^2)\, dy \qquad (7.9)$$

and obviously this quantity is equal to the corresponding one for f_Y.

(b2) Thanks to (7.9) the density f_X is an even function, therefore X is symmetric and $E(X) = 0$. Obviously also $E(Y) = 0$.

(b3) We just need to compute $E(XY)$, as we already know that X and Y are centered. We have, again in polar coordinates and recalling that $x = r\cos\theta$, $y = r\sin\theta$,

$$E(XY) = \int_{-\infty}^{+\infty} \int_{-\infty}^{+\infty} xy\, g(x^2 + y^2)\, dx\, dy$$

$$= \underbrace{\int_{0}^{2\pi} \sin\theta \cos\theta\, d\theta}_{=0} \int_{0}^{+\infty} g(r^2) r^3\, dr\ .$$

Note that the integral $\int_{0}^{+\infty} g(r^2) r^3\, dr$ is finite, as it is equal to $\frac{1}{2\pi} E(X^2 + Y^2)$.

If $g(r) = \frac{1}{2\pi} e^{-\frac{1}{2} r}$, then $f(x, y) = \frac{1}{2\pi} e^{-\frac{1}{2}(x^2 + y^2)}$ can be split into the tensor product of a function of x times a function of y, hence X and Y are independent (and are each $N(0, 1)$-distributed).

If $f = \frac{1}{\pi} 1_C$, where C is the ball of radius 1, X and Y are not independent: as can be seen by looking at Fig. 7.1, the marginal densities are both strictly positive on the interval $[-1, 1]$ so that their product gives strictly positive probability to the areas near the corners, which are of probability 0 for the joint distribution.

- It is a classical result of Bernstein that a probability on $\mathbb{R}^d$ which is invariant under rotations and whose components are independent is necessarily Gaussian (see e.g. [7], p. 82).

(c1) For every bounded Borel function $\phi : \mathbb{R} \to \mathbb{R}$ we have

$$E[\phi(\tfrac{X}{Y})] = \int_{-\infty}^{+\infty} dy \int_{-\infty}^{+\infty} \phi(\tfrac{x}{y}) g(x^2 + y^2)\, dx\ .$$

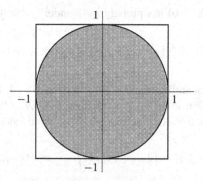

Fig. 7.1 The rounded triangles near the corners have probability 0 for the joint density but strictly positive probability for the product of the marginals

With the change of variable $z = \frac{x}{y}$, $|y|\,dz = dx$ in the inner integral we have

$$\cdots = \int_{-\infty}^{+\infty} dy \int_{-\infty}^{+\infty} \phi(z)g\big(y^2(1+z^2)\big)|y|\,dz$$

$$= \int_{-\infty}^{+\infty} \phi(z)\,dz \int_{-\infty}^{+\infty} g\big(y^2(1+z^2)\big)|y|\,dy$$

$$= \int_{-\infty}^{+\infty} \phi(z)\,dz \int_{0}^{+\infty} 2g\big(y^2(1+z^2)\big)y\,dy\,.$$

Replacing $y\sqrt{1+z^2} = u$, $dy = (1+z^2)^{-1/2}\,du$, we have

$$\cdots = \int_{-\infty}^{+\infty} \phi(z)\,dz \int_{0}^{+\infty} 2g(u^2)\frac{u}{\sqrt{1+z^2}}\frac{du}{\sqrt{1+z^2}}$$

$$= \int_{-\infty}^{+\infty} \phi(z)\frac{1}{1+z^2}\,dz \int_{0}^{+\infty} 2g(u^2)u\,du$$

$$= \int_{-\infty}^{+\infty} \phi(z)\frac{1}{1+z^2}\,dz \int_{0}^{+\infty} g(u)\,du = \int_{-\infty}^{+\infty} \phi(z)\frac{1}{\pi(1+z^2)}\,dz$$

and the result follows.

(c2) Just note that the pair (X, Y) has a density of the type (2.85), so that this is a situation as in (c1) and $\frac{X}{Y}$ has a Cauchy law.

(c3) Just note that in (c2) both $\frac{X}{Y}$ and $\frac{Y}{X}$ have a Cauchy distribution.

2.17 (a) Q is a measure (Theorem 1.28) as X is a density, being positive and integrable. Moreover, $Q(\Omega) = E(X) = 1$ so that Q is a probability.

(b1) As obviously $X1_{\{X=0\}} = 0$, we have $Q(X = 0) = E(X1_{\{X=0\}}) = 0$.

(b2) As the event $\{X > 0\}$ has probability 1 under Q, we have, for every $A \in \mathscr{F}$,

$$\widetilde{P}(A) = E^Q\left(\frac{1}{X}1_A\right) = E^Q\left(\frac{1}{X}1_{A\cap\{X>0\}}\right) = E[1_{A\cap\{X>0\}}] = P(A \cap \{X > 0\})$$

and therefore $\widetilde{P}$ is a probability if and only if $P(X > 0) = 1$. In this case $\widetilde{P} = P$ and $\frac{dP}{dQ} = \frac{1}{X}$ and $P \ll Q$. Conversely, if $P \ll Q$, then, as $Q(X = 0)$, then also $P(X = 0)$.

(c) For every bounded Borel function $\phi : \mathbb{R} \to \mathbb{R}$ we have

$$E^Q[\phi(X)] = E[X\phi(X)] = \int_{-\infty}^{+\infty} \phi(x)x \, d\mu(x) .$$

Hence, under Q, X has law $d\nu(x) = x \, d\mu(x)$. Note that such a ν is also a probability because

$$\int_{-\infty}^{+\infty} d\nu(x) = \int_{-\infty}^{+\infty} x \, d\mu(x) = E(X) = 1 .$$

If $X \sim \text{Gamma}(\lambda, \lambda)$ then its density f with respect to the Lebesgue measure is

$$f(x) = \frac{\lambda^\lambda}{\Gamma(\lambda)} x^{\lambda-1} e^{-\lambda x}$$

and its density with respect to Q is

$$x \mapsto \frac{\lambda^\lambda}{\Gamma(\lambda)} x^\lambda e^{-\lambda x} = \frac{\lambda^{\lambda+1}}{\Gamma(\lambda+1)} x^\lambda e^{-\lambda x} ,$$

which is a $\text{Gamma}(\lambda + 1, \lambda)$.

(d1) Thanks to Theorem 1.28, $E^Q(Z) = E(XZ) = E(X)E(Z) = E(Z)$.

(d2) As X and Z are independent under P, for every bounded Borel function ψ we have

$$E^Q[\psi(Z)] = E[X\psi(Z)] = E(X)E[\psi(Z)] = E[\psi(Z)] , \tag{7.10}$$

hence the laws of Z with respect to P and to Q coincide.

(d3) For every choice of bounded Borel functions $\phi, \psi : \mathbb{R} \to \mathbb{R}$ we have, thanks to (7.10),

$$E^Q[\phi(X)\psi(Z)] = E[X\phi(X)\psi(Z)] = E[X\phi(X)]E[\psi(Z)]$$
$$= E^Q[\phi(X)]E^Q[\psi(Z)] ,$$

i.e. X and Z are independent also with respect to Q.

2.18 (a) We must only check that $\frac{\lambda}{2}(X + Z)$ is a density, i.e. that it is a positive r.v. whose integral is equal to 1, which is immediate.

(b) As X and Z are independent under P and recalling the expressions of the moments of the exponential laws, $E(X) = \frac{1}{\lambda}$, $E(X^2) = \frac{2}{\lambda^2}$, we have

$$E^Q(XZ) = \frac{\lambda}{2} E[XZ(X + Z)] = \frac{\lambda}{2}(E(X^2 Z) + E(XZ^2))$$
$$= \frac{\lambda}{2}(E(X^2)E(Z) + E(X)E(Z^2)) = \frac{\lambda}{2} \times 2 \frac{2}{\lambda^3} = \frac{2}{\lambda^2}. \tag{7.11}$$

(c1) The method of the dumb function: if $\phi : \mathbb{R}^2 \to \mathbb{R}$ is a bounded Borel function,

$$E^Q[\phi(X, Z)] = \frac{\lambda}{2} E[(X + Z)\phi(X, Z)]$$
$$= \frac{\lambda}{2}\int_{-\infty}^{+\infty}\int_{-\infty}^{+\infty} \phi(x, z)(x + z)\lambda^2 e^{-\lambda(x+z)}\, dx\, dz .$$

Hence, under Q, X and Z have a joint law with density, with respect to the Lebesgue measure,

$$g(x, z) = \frac{\lambda^3}{2}(x + z)\, e^{-\lambda(x+z)} \qquad x, z > 0 .$$

As g does not split into the tensor product of functions of x and z, X and Z are not independent under Q. They are even correlated: we have

$$E^Q(X) = \frac{\lambda}{2} E[X(X + Z)] = \frac{\lambda}{2}(E(X^2) + E(XZ)) = \frac{\lambda}{2}\left(\frac{2}{\lambda^2} + \frac{1}{\lambda^2}\right) = \frac{3}{2\lambda}$$

and, recalling (7.11) ,

$$\mathrm{Cov}_Q(X, Z) = E^Q(XZ) - E^Q(X)E^Q(Z) = \left(2 - \frac{9}{4}\right)\frac{1}{\lambda^2} < 0 .$$

(c2) Computing the marginals of g,

$$g_X(x) = \frac{\lambda^3}{2}\int_0^{+\infty}(x + z)e^{-\lambda(x+z)}\, dz$$
$$= \frac{\lambda^3}{2} e^{-\lambda x}\left(x\int_0^{+\infty} e^{-\lambda z}\, dz + \int_0^{+\infty} z e^{-\lambda z}\, dz\right) = \frac{1}{2}\left(\lambda^2 x + \lambda\right)e^{-\lambda x} ,$$

i.e. a linear combination of an exponential and a Gamma$(2, \lambda)$ density. Of course, by symmetry, $g_Z = g_X$.

2.19 (a) Let us argue as in Proposition 2.18. For every bounded Borel function $\phi : \mathbb{R} \to \mathbb{R}$ we have

$$E[\phi(XY)] = \int_{-\infty}^{+\infty} dx \int_{-\infty}^{+\infty} \phi(xy) f(x, y) \, dy$$

and, with the change of variable $xy = z$, $|x| \, dy = dz$, in the inner integral

$$\ldots = \int_{-\infty}^{+\infty} dx \int_{-\infty}^{+\infty} \phi(z) f(x, \tfrac{z}{x}) |x|^{-1} \, dz$$

$$= \int_{-\infty}^{+\infty} \phi(z) \, dz \int_{-\infty}^{+\infty} f(x, \tfrac{z}{x}) |x|^{-1} \, dx$$

so that the law of XY is $d\mu(z) = h(z) \, dz$ with

$$h(z) = \int_{-\infty}^{+\infty} f(x, \tfrac{z}{x}) |x|^{-1} \, dx \; .$$

In the case of the quotient the argument is the same, but for the remark that the r.v. $\frac{X}{Y}$ is defined except on the event $\{Y = 0\}$, which has probability 0, as Y has a density with respect to the Lebesgue measure. With the change of variable $\frac{x}{y} = z$, i.e. $dx = |y| \, dz$, in the inner integral

$$E[\phi(\tfrac{X}{Y})] = \int_{-\infty}^{+\infty} dy \int_{-\infty}^{+\infty} \phi(\tfrac{x}{y}) f(x, y) \, dx$$

$$= \int_{-\infty}^{+\infty} dy \int_{-\infty}^{+\infty} \phi(z) f(yz, y) |y| \, dz$$

$$= \int_{-\infty}^{+\infty} \phi(z) \, dz \int_{-\infty}^{+\infty} f(yz, y) |y| \, dy$$

and therefore the law of $\frac{Y}{X}$ is $d\nu(z) = g(z) \, dz$ with

$$g(z) = \int_{-\infty}^{+\infty} f(yz, y) |y| \, dy \; . \tag{7.12}$$

(b1) We have

$$f(x, y) = \frac{\lambda^{\alpha+\beta}}{\Gamma(\alpha)\Gamma(\beta)} \, x^{\alpha-1} y^{\beta-1} e^{-\lambda(x+y)} \qquad x, y > 0 \, ,$$

so that (7.12) gives

$$
\begin{aligned}
g(z) &= \frac{\lambda^{\alpha+\beta}}{\Gamma(\alpha)\Gamma(\beta)} \int_0^{+\infty} (yz)^{\alpha-1} y^{\beta-1} e^{-\lambda(zy+y)} \, y \, dy \\
&= \frac{\lambda^{\alpha+\beta} z^{\alpha-1}}{\Gamma(\alpha)\Gamma(\beta)} \int_0^{+\infty} y^{\alpha+\beta-1} e^{-\lambda(1+z)y} \, dy \\
&= \frac{\lambda^{\alpha+\beta} z^{\alpha-1}}{\Gamma(\alpha)\Gamma(\beta)} \frac{\Gamma(\alpha+\beta)}{(\lambda(1+z))^{\alpha+\beta}} \\
&= \frac{\Gamma(\alpha+\beta)}{\Gamma(\alpha)\Gamma(\beta)} \frac{z^{\alpha-1}}{(1+z)^{\alpha+\beta}} .
\end{aligned}
\tag{7.13}
$$

(b2) If $U \sim$ Gamma$(\alpha, 1)$, then $\frac{U}{\lambda} \sim$ Gamma(α, λ) (exercise). Let now U, V be two independent r.v.'s Gamma$(\alpha, 1)$- and Gamma$(\beta, 1)$-distributed respectively, then the r.v.'s $\frac{U}{\lambda}$, $\frac{V}{\lambda}$ have the same joint law as X, Y, therefore their quotient has the same law as $\frac{X}{Y}$. Hence $\frac{X}{Y} = \frac{U}{V}$ and the law of $\frac{U}{V}$ does not depend on λ.

(b3) The moment of order p of W is

$$
\mathrm{E}(W^p) = \int_0^{+\infty} z^p g(z) \, dz = \frac{\Gamma(\alpha+\beta)}{\Gamma(\alpha)\Gamma(\beta)} \int_0^{+\infty} \frac{z^{\alpha+p-1}}{(z+1)^{\alpha+\beta}} \, dz .
\tag{7.14}
$$

The integrand tends to 0 at infinity as $z^{p-\beta-1}$, hence the integral converges if and only if $p < \beta$. If this condition is satisfied, the integral is easily computed recalling that (7.13) is a density: just write

$$
\int_0^{+\infty} \frac{z^{\alpha+p-1}}{(z+1)^{\alpha+\beta}} \, dz = \int_0^{+\infty} \frac{z^{\alpha+p-1}}{(z+1)^{\alpha+p+\beta-p}} \, dz
$$

and therefore, thanks to (7.13) with α replaced by $\alpha + p$ and β by $\beta - p$,

$$
\mathrm{E}(W^p) = \frac{\Gamma(\alpha+\beta)}{\Gamma(\alpha)\Gamma(\beta)} \times \frac{\Gamma(\alpha+p)\Gamma(\beta-p)}{\Gamma(\alpha+\beta)} = \frac{\Gamma(\alpha+p)\Gamma(\beta-p)}{\Gamma(\alpha)\Gamma(\beta)} .
\tag{7.15}
$$

(c1) The r.v.'s X^2 and $Y^2 + Z^2$ are Gamma$(\frac{1}{2}, \frac{1}{2})$- and Gamma$(1, \frac{1}{2})$-distributed respectively and independent. Therefore (7.13) with $\alpha = \frac{1}{2}$ and $\beta = 1$ gives for the density of W_1

$$
f_1(z) = \frac{\Gamma(\frac{3}{2})}{\Gamma(\frac{1}{2})\Gamma(1)} \frac{z^{-\frac{1}{2}}}{(z+1)^{3/2}} = \frac{1}{2} \frac{z^{-\frac{1}{2}}}{(z+1)^{3/2}} .
$$

As $W_2 = \sqrt{W_1}$,

$$
\mathrm{P}(W_2 \le t) = \mathrm{P}(W_1 \le t^2) = F_{W_1}(t^2)
$$

and, taking the derivative, the requested density of W_2 is

$$f_2(t) = 2t f_1(t^2) = \frac{1}{(t^2 + 1)^{3/2}} \qquad t > 0 .$$

(c2) The joint law of X and Y has density, with respect to the Lebesgue measure,

$$f(x, y) = \frac{1}{2\pi} e^{-\frac{1}{2}(x^2 + y^2)} .$$

It is straightforward to deduce from (7.12) that $\frac{X}{Y}$ has a Cauchy density

$$g(z) = \frac{1}{\pi(1 + z^2)}$$

but we have already proved this in Exercise 2.16, as a general fact concerning all joint densities that are rotation invariant.

2.20 (a) We can write $(U, V) = \Psi(X, Y)$, with $\Psi(x, y) = (x + y, \frac{x+y}{x})$. Let us make the change of variable $(u, v) = \Psi(x, y)$. Let us first compute Ψ^{-1}: we must solve

$$\begin{cases} u = x + y \\ v = \dfrac{x + y}{x} \end{cases} .$$

We find $x = \frac{u}{v}$ and then $y = u - \frac{u}{v}$, i.e. $\Psi^{-1}(u, v) = (uv, u - \frac{u}{v})$. Its differential is

$$D\Psi^{-1}(u, v) = \begin{pmatrix} \dfrac{1}{v} & -\dfrac{u}{v^2} \\ 1 - \dfrac{1}{v} & \dfrac{u}{v^2} \end{pmatrix}$$

so that $|\det D\Psi^{-1}(u, v)| = \frac{u}{v^2}$. Denoting by f the joint density of (X, Y), i.e.

$$f(x, y) = \frac{1}{\Gamma(\alpha)\Gamma(\beta)} x^{\alpha-1} y^{\beta-1} e^{-(x+y)}, \qquad x, y > 0 ,$$

the joint density of (U, V) is

$$g(u, v) = f(\tfrac{u}{v}, u - \tfrac{u}{v}) \frac{u}{v^2} .$$

The density f vanishes unless both its arguments are positive, hence $g > 0$ for $u > 0$, $v > 1$. If $u > 0$, $v > 1$ we have

$$
\begin{aligned}
g(u, v) &= \frac{1}{\Gamma(\alpha)\Gamma(\beta)} \left(\frac{u}{v}\right)^{\alpha-1} \left(u - \frac{u}{v}\right)^{\beta-1} e^{-\frac{u}{v} - \left(u - \frac{u}{v}\right)} \frac{u}{v^2} \\
&= \frac{1}{\Gamma(\alpha)\Gamma(\beta)} u^{\alpha+\beta-1} e^{-u} \times \frac{(v-1)^{\beta-1}}{v^{\alpha+\beta}} .
\end{aligned}
\tag{7.16}
$$

As the joint density of (U, V) can be split into the product of a function of u and of a function of v, U and V are independent.

(b) We must compute

$$
g_V(v) := \int_{-\infty}^{+\infty} g(u, v)\, du .
$$

By (7.16) we have $g_V(v) = 0$ for $v \le 1$ and

$$
g_V(v) = \frac{\Gamma(\alpha + \beta)}{\Gamma(\alpha)\Gamma(\beta)} \frac{(v-1)^{\beta-1}}{v^{\alpha+\beta}}
$$

for $v > 1$, as we recognized the integral of a Gamma$(\alpha + \beta, 1)$ density.

Note that $V = 1 + \frac{Y}{X}$ and that the density of the quotient $\frac{Y}{X}$ has already been computed in Exercise 2.19(b), from which the density g_V could also be derived.

As for the law of $\frac{1}{V}$, note first that this r.v. takes its values in the interval $[0, 1]$. For $0 \le t \le 1$ we have

$$
P\left(\frac{1}{V} \le t\right) = P\left(V \ge \frac{1}{t}\right) = 1 - G_V\left(\frac{1}{t}\right) ,
$$

with G_V denoting the d.f. of V. Taking the derivative, $\frac{1}{V}$ has density, with respect to the Lebesgue measure,

$$
t \mapsto \frac{1}{t^2} g_V\left(\frac{1}{t}\right) = \frac{\Gamma(\alpha + \beta)}{\Gamma(\alpha)\Gamma(\beta)} \frac{1}{t^2} \left(\frac{1}{t} - 1\right)^{\beta-1} t^{\alpha+\beta} = \frac{\Gamma(\alpha+\beta)}{\Gamma(\alpha)\Gamma(\beta)} t^{\alpha-1}(1 - t)^{\beta-1} ,
$$

i.e. a Beta(α, β) density.

2.21 (a) For every bounded Borel function $\phi : \mathbb{R}^2 \to \mathbb{R}$

$$
E[\phi(Z, W)] = \int_0^{+\infty} f(t)\, dt \int_0^1 \phi(xt, (1 - x)t)\, dx .
$$

With the change of variable $z = xt$, $dz = t\, dx$, in the inner integral we obtain, after Fubinization,

$$
\cdots = \int_0^{+\infty} f(t)\, dt \int_0^t \frac{1}{t} \phi(z, t - z)\, dz = \int_0^{+\infty} dz \int_z^{+\infty} \phi(z, t - z) \frac{1}{t} f(t)\, dt .
$$

With the further change of variable $w = t - z$ and noting that $w = 0$ when $t = z$, we land on

$$\cdots = \int_0^{+\infty} dz \int_0^{+\infty} \phi(z, w) \frac{1}{z + w} f(z + w) \, dz \, ,$$

so that the requested joint density is

$$g(z, w) := \frac{1}{z + w} f(z + w), \qquad z > 0, w > 0 \, .$$

Note that, g being symmetric, Z and W have the same distribution, a fact which was to be expected.

(b) If

$$f(t) = \lambda^2 t \, e^{-\lambda t} , \qquad t > 0$$

then

$$g(z, w) = \lambda^2 e^{-\lambda(z+w)} = \lambda e^{-\lambda z} \times \lambda e^{-\lambda w} \, .$$

Z and W are i.i.d. with an exponential distribution of parameter λ.

2.22 We have

$$G(x, y) = P(x \le X \le Y \le y) = \int_{Q_{x,y}} f(u, v) \, du \, dv \, ,$$

where $Q_{x,y}$ is the square $[x, y] \times [x, y]$. Keeping in mind that $X \le Y$ a.s., $f(u, v) = 0$ for $u > v$ so that

$$G(x, y) = \int_x^y du \int_u^y f(u, v) \, dv \, .$$

Taking the derivative first with respect to x and then with respect to y we find

$$f(x, y) = -\frac{\partial^2 G}{\partial x \partial y}(x, y) \, .$$

(b1) Denoting by H the common d.f. of Z and W, we have

$$G(x, y) = P(x \le X \le Y \le y) = P(x \le Z \le y, x \le W \le y)$$

$$= (H(y) - H(x))^2 \, ,$$

(7.17)

hence the joint density of X, Y is, for $x \leq y$,

$$f(x, y) = -\frac{\partial^2 G}{\partial x \partial y}(x, y) = 2h(x)h(y)$$

and $f(x, y) = 0$ for $x > y$.

(b2) If Z and W are uniform on $[0, 1]$ then $h = 1_{[0,1]}$ and $f(x, y) = 2\,1_{\{0 \leq x \leq y \leq 1\}}$. Therefore

$$E[|Z - W|] = E\big(\max(Z, W) - \min(Z, W)\big) = E(Y - X)$$

$$= 2 \int_0^1 dy \int_0^y (y - x)\, dx = \int_0^1 y^2\, dy = \frac{1}{3}.$$

2.23 (a) Let $f = 1_A$ with $A \in \mathscr{C}$ and $\mu(A) < +\infty$ and $\phi(x) = x^2$. Then $\phi(1_A) = 1_A$ and (2.86) becomes

$$\mu(A) \geq \mu(A)^2$$

hence $\mu(A) \leq 1$. Let now $(A_n)_n \subset \mathscr{C}$ be an increasing sequence of sets of finite μ-measure and such that $E = \bigcup_n A_n$. As $\mu(A_n) \leq 1$ and μ passes to the limit on increasing sequences, we have also $\mu(E) \leq 1$.

(b) Note that (2.86) implies a similar, reverse, inequality for integrable concave functions hence equality for affine-linear ones. Now for $\phi \equiv 1$, recalling that necessarily μ is finite thanks to (a),

$$\mu(E) = \int \phi(1_E)\, d\mu = \phi\Big(\int 1_E\, d\mu \Big) = 1.$$

2.24 (a1) Let $\phi(x) = x \log x$ if $x > 0$, $\phi(0) = 0$, $\phi(x) = +\infty$ if $x < 0$. For $x > 0$ we have $\phi'(x) = 1 + \log x$, $\phi''(x) = \frac{1}{x}$, therefore ϕ is convex and, as $\lim_{x \to 0} \phi(x) = 0$, also lower semi-continuous. It vanishes at 1 and at 0. By Jensen's inequality

$$H(\nu; \mu) = \int_E \phi\Big(\frac{d\nu}{d\mu}\Big)\, d\mu \geq \phi\Big(\int_E \frac{d\nu}{d\mu}\, d\mu \Big) = \phi(\nu(E)) = 0. \tag{7.18}$$

The convexity relation

$$H\big(\lambda \nu_1 + (1 - \lambda)\nu_2; \mu\big) \leq \lambda H(\nu_1; \mu) + (1 - \lambda)H(\nu_2; \mu) \tag{7.19}$$

is immediate if both ν_1 and ν_2 are $\ll \mu$ thanks to the convexity of ϕ. If one at least among ν_1, ν_2 is not absolutely continuous with respect to μ, then also $\lambda \nu_1 + (1 - \lambda)\nu_2 \not\ll \mu$ and in (7.19) both members are $= +\infty$.

Moreover ϕ is strictly convex as $\phi'' > 0$ for $x > 0$. Therefore the inequality (7.18) is strict, unless $\frac{d\nu}{d\mu}$ is constant. As $\frac{d\nu}{d\mu}$ is a density, this constant can only be equal to 1 so that the inequality is strict unless $\nu = \mu$.

(a2) As $\log 1_A = 0$ on A whereas $1_A = 0$ on A^c, $1_A \log 1_A \equiv 0$ and

$$H(v; \mu) = \frac{1}{\mu(A)} \int_E 1_A \log \frac{1_A}{\mu(A)} \, d\mu = \frac{1}{\mu(A)} \int_A - \log \mu(A) \, d\mu$$
$$= - \log \mu(A) .$$

As $v(A^c) = 0$ whereas $\mu(A^c) = 1 - \mu(A) > 0$, μ is not absolutely continuous with respect to v and $H(\mu; v) = +\infty$.

(b1) We have, for $k = 0, 1, \ldots, n$,

$$\frac{dv}{d\mu}(k) = \frac{q^k (1-q)^{n-k}}{p^k (1-p)^{n-k}} ,$$

i.e.

$$\log \frac{dv}{d\mu}(k) = k \log \frac{q}{p} + (n-k) \log \frac{1-q}{1-p} ,$$

so that

$$H(v; \mu) = \sum_{k=0}^{n} v(k) \log \frac{dv}{d\mu}(k)$$

$$= \sum_{k=0}^{n} \binom{n}{k} q^k (1-q)^{n-k} \left(k \log \frac{q}{p} + (n-k) \log \frac{1-q}{1-p} \right)$$

$$= n \left(q \log \frac{q}{p} + (1-q) \log \frac{1-q}{1-p} \right) .$$

(b2) We have, for $t > 0$,

$$\frac{dv}{d\mu}(t) = \frac{\rho}{\lambda} e^{-(\rho-\lambda)t} ,$$

$$\log \frac{dv}{d\mu}(t) = - \log \frac{\lambda}{\rho} - (\rho - \lambda) t$$

and

$$H(v; \mu) = \int_0^{+\infty} \log \frac{dv}{d\mu}(t) \, dv(t) = - \log \frac{\lambda}{\rho} - (\rho - \lambda)\rho \int_0^{+\infty} t e^{-\rho t} \, dt$$

$$= - \log \frac{\lambda}{\rho} - \frac{\rho - \lambda}{\rho} = \frac{\lambda}{\rho} - 1 - \log \frac{\lambda}{\rho} ,$$

which, of course, is a positive function (Fig. 7.2).

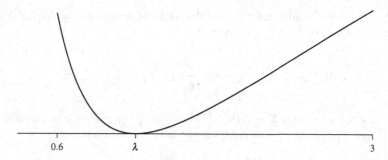

Fig. 7.2 The graph of $\rho \to \frac{\lambda}{\rho} - 1 - \log \frac{\lambda}{\rho}$, for $\lambda = 1.2$

(c) If for one index i, at least, $\nu_i \not\ll \mu_i$, then there exists a set $A_i \in \mathscr{E}_i$ such that $\nu_i(A_i) > 0$ and $\mu_i(A_i) = 0$. Then,

$$\nu(E_1 \times \cdots \times A_i \times \cdots \times E_n) = \nu_i(A_i) > 0 \,,$$

$$\mu(E_1 \times \cdots \times A_i \times \cdots \times E_n) = \mu_i(A_i) = 0 \,,$$

so that also $\nu \not\ll \mu$ and in (2.88) both members are $= +\infty$.

If, instead, $\nu_i \ll \mu_i$ for every i and $f_i := \frac{d\nu_i}{d\mu_i}$, then

$$\frac{d\nu}{d\mu}(x_1, \ldots, x_n) = f_1(x_1) \ldots f(x_n)$$

and, as $\int_{E_i} d\nu_i(x_i) = 1$ for every $i = 1, \ldots, n$,

$$H(\nu; \mu) = \int_{E_1 \times \cdots \times E_n} \log \frac{d\nu}{d\mu} \, d\nu$$

$$= \int_{E_1 \times \cdots \times E_n} \log \big(f_1(x_1) \ldots f_n(x_n) \big) \, d\nu_1(x_1) \ldots d\nu_n(x_n)$$

$$= \int_{E_1 \times \cdots \times E_n} \big(\log f_1(x_1) + \cdots + \log f_n(x_n) \big) \, d\nu_1(x_1) \ldots d\nu_n(x_n)$$

$$= \sum_{i=1}^{n} \int_{E_1 \times \cdots \times E_n} \log f_i(x_i) \, d\nu_1(x_1) \ldots d\nu_n(x_n) = \sum_{i=1}^{n} \int_{E_i} \log f_i(x_i) \, d\nu_i(x_i)$$

$$= \sum_{i=1}^{n} H(\nu_i; \mu_i) \,.$$

- The courageous reader can compute the relative entropy of $v = N(b, \sigma^2)$ with respect to $\mu = N(b_0, \sigma_0^2)$ and find that

$$H(v; \mu) = \frac{1}{2} \left(\frac{\sigma^2}{\sigma_0^2} - \log \left(\frac{\sigma^2}{\sigma_0^2} \right) - 1 \right) + \frac{1}{2\sigma_0^2} (b - b_0)^2 .$$

2.25 (a) We know that if $X \sim N(b, \sigma^2)$ then $Z = X - b \sim N(0, \sigma^2)$, and also that the odd order moments of centered Gaussian laws vanish. Therefore

$$E[(X - b)^3] = E(Z^3) = 0 ,$$

hence $\gamma = 0$. Actually in this computation we have used only the fact that the Gaussian r.v.'s have a law that is symmetric with respect to their mean, i.e. such that $X - b$ and $-(X - b)$ have the same law. For all r.v.'s with a finite third order moment and having this property we have

$$E[(X - b)^3] = E[(-(X - b))^3] = -E[(X - b)^3] ,$$

so that $E[(X - b)^3] = 0$ and $\gamma = 0$.

(b) Recall that if $X \sim \text{Gamma}(\alpha, \lambda)$ its k-th order moment is

$$E(X^k) = \frac{\Gamma(\alpha + k)}{\lambda^k \Gamma(\alpha)} = \frac{(\alpha + k - 1)(\alpha + k - 2) \cdots \alpha}{\lambda^k} ,$$

hence for the first three moments:

$$E(X) = \frac{\alpha}{\lambda} , \qquad E(X^2) = \frac{\alpha(\alpha + 1)}{\lambda^2} , \qquad E(X^3) = \frac{\alpha(\alpha + 1)(\alpha + 2)}{\lambda^3} .$$

With the binomial expansion of the third degree (here $b = \frac{\alpha}{\lambda}$)

$$E[(X - b)^3] = E(X^3) - 3E(X^2)b + 3E(X)b^2 - b^3$$

$$= \frac{1}{\lambda^3} \left(\alpha(\alpha + 1)(\alpha + 2) - 3\alpha^2(\alpha + 1) + 3\alpha^3 - \alpha^3 \right)$$

$$= \frac{\alpha}{\lambda^3} \left(\alpha^2 + 3\alpha + 2 - 3\alpha^2 - 3\alpha + 2\alpha^2 \right) = \frac{2\alpha}{\lambda^3} .$$

On the other hand the variance is equal to $\sigma^2 = \frac{\alpha}{\lambda^2}$, so that

$$\gamma = \frac{\frac{2\alpha}{\lambda^3}}{\frac{\alpha^{3/2}}{\lambda^3}} = 2\alpha^{-1/2} .$$

In particular, the skewness does not depend on λ and for an exponential law is always equal to 2. This fact is not surprising keeping in mind that, as already noted somewhere above, if $X \sim \text{Gamma}(\alpha, 1)$ then $\frac{1}{\lambda} X \sim \text{Gamma}(\alpha, \lambda)$. Hence the moments of order k of a Gamma(α, λ)-distributed r.v. are equal to the same moments of a Gamma$(\alpha, 1)$-distributed r.v. multiplied by λ^{-k} and the λ's in the numerator and in the denominator in (2.89) simplify.

Note also that the skewness of a Gamma law is always positive, which is in agreement with intuition (the graph of the density is always as in Fig. 2.4, at least for $\alpha > 1$).

2.26 By hypothesis, for every $n \geq 1$,

$$\int_{-\infty}^{+\infty} x^n \, d\mu(x) = \int_{-\infty}^{+\infty} x^n \, d\nu(x)$$

and therefore, by linearity, also

$$\int_{-\infty}^{+\infty} P(x) \, d\mu(x) = \int_{-\infty}^{+\infty} P(x) \, d\nu(x) \tag{7.20}$$

for every polynomial P. By Proposition 1.25, the statement follows if we are able to prove that (7.20) holds for every continuous bounded function f (and not just for every polynomial). But if f is a real continuous function then (Weierstrass's Theorem) f is the uniform limit of polynomials on $[-M, M]$. Hence, if P is a polynomial such that $\sup_{-M \leq x \leq M} |f(x) - P(x)| \leq \varepsilon$, then

$$\left| \int_{-\infty}^{+\infty} f(x) \, d\mu(x) - \int_{-\infty}^{+\infty} f(x) \, d\nu(x) \right|$$
$$= \left| \int_{-M}^{M} (f(x) - P(x)) \, d\mu(x) - \int_{-M}^{M} (f(x) - P(x)) \, d\nu(x) \right|$$
$$\leq \int_{-M}^{M} |f(x) - P(x)| \, d\mu(x) + \int_{-M}^{M} |f(x) - P(x)| \, d\nu(x) \leq 2\varepsilon$$

and by the arbitrariness of ε

$$\int_{-\infty}^{+\infty} f(x) \, d\mu(x) = \int_{-\infty}^{+\infty} f(x) \, d\nu(x)$$

for every bounded continuous function f.

2.27 The covariance matrix C is positive definite and therefore is invertible if and only if it is strictly positive definite. Recall (2.33), i.e. for every $\xi \in \mathbb{R}^m$

$$\langle C\xi, \xi \rangle = \mathrm{E}\big(\langle X - \mathrm{E}(X), \xi \rangle^2\big). \tag{7.21}$$

Let us assume that X takes its values in a proper hyperplane of $\mathbb{R}^m$. Such a hyperplane is of the form $\{x; \langle \xi, x \rangle = t\}$ for some $\xi \in \mathbb{R}^m, \xi \neq 0$ and $t \in \mathbb{R}$. Hence

$$\langle \xi, X \rangle = t \quad \text{a.s.}$$

Taking the expectation we have $\langle \xi, E(X) \rangle = t$, so that $\langle \xi, X - E(X) \rangle = 0$ a.s. and by (7.21) $\langle C\xi, \xi \rangle = 0$, so that C cannot be invertible.

Conversely, if C is not invertible there exists a vector $\xi \in \mathbb{R}^m, \xi \neq 0$, such that $\langle C\xi, \xi \rangle = 0$ and by (7.21) $\langle X - E(X), \xi \rangle^2 = 0$ a.s. (the mathematical expectation of a positive r.v. vanishes if and only if the r.v. is a.s. equal to 0, Exercise 1.9). Hence $X \in H$ a.s. where $H = \{x; \langle \xi, x \rangle = \langle \xi, E(X) \rangle\}$.

Let μ denote the law of X. As H has Lebesgue measure equal to 0 whereas $\mu(H) = 1$, μ is not absolutely continuous with respect to the Lebesgue measure.

2.28 Recall the expression of the coefficients a and b, i.e.

$$a = \frac{\text{Cov}(X, Y)}{\text{Var}(X)}, \qquad b = E(Y) - aE(X).$$

(a) As $aX + b = a(X - E(X)) + E(Y)$, we have

$$Y - (aX + b) = Y - E(Y) - a(X - E(X)),$$

which gives $E(Y - (aX + b)) = 0$. Moreover,

$$E\big[(Y - (aX + b))(aX + b)\big]$$
$$= E\big[(Y - E(Y) - a(X - E(X)))(a(X - E(X)) + E(Y))\big]$$
$$= aE\big[(Y - E(Y))(X - E(X))\big] - a^2 E\big[(X - E(X))^2\big]$$
$$= a\text{Cov}(Y, X) - a^2\text{Var}(X) = \frac{\text{Cov}(X, Y)^2}{\text{Var}(X)} - \frac{\text{Cov}(X, Y)^2}{\text{Var}(X)} = 0.$$

(b) As $Y - (aX + b)$ and $aX + b$ are orthogonal in L^2, we have (Pythagoras's theorem)

$$E(Y^2) = E\big[(Y - (aX + b))^2\big] + E\big[(aX + b)^2\big].$$

2.29 (a) We have $\text{Cov}(X, Y) = \text{Cov}(Y + W, Y) = \text{Var}(Y) = 1$, whereas $\text{Var}(X) = \text{Var}(Y) + \text{Var}(W) = 1 + \sigma^2$. As the means vanish, the regression line $x \mapsto ax + b$ of Y with respect to X is given by

$$a = \frac{1}{1 + \sigma^2}, \qquad b = 0.$$

The best approximation of Y by a linear function of X is therefore $\frac{1}{1+\sigma^2} X$ (intuitively one takes the observation X and moves it a bit toward 0, which is the mean of Y). The quadratic error is

$$
E\left[\left(Y - \frac{1}{1+\sigma^2} X\right)^2\right] = \mathrm{Var}(Y) + \frac{1}{(1+\sigma^2)^2} \mathrm{Var}(X) - \frac{2}{1+\sigma^2} \mathrm{Cov}(X, Y)
$$

$$
= 1 + \frac{1}{1+\sigma^2} - \frac{2}{1+\sigma^2} = \frac{\sigma^2}{1+\sigma^2} .
$$

(b) If $X = (X_1, X_2)$, the covariance matrix of X is

$$
C_X = \begin{pmatrix} 1+\sigma^2 & 1 \\ 1 & 1+\sigma^2 \end{pmatrix}
$$

whereas the vector of the covariances of Y and the X_i's is

$$
C_{X,Y} = \begin{pmatrix} 1 \\ 1 \end{pmatrix} .
$$

and the means vanish. We have, with a bit of patience,

$$
C_X^{-1} = \frac{1}{(1+\sigma^2)^2 - 1} \begin{pmatrix} 1+\sigma^2 & -1 \\ -1 & 1+\sigma^2 \end{pmatrix} ,
$$

hence the regression "line" is

$$
\langle C_X^{-1} C_{X,Y}, X \rangle = \frac{1}{2\sigma^2 + \sigma^4} \left\langle \begin{pmatrix} 1+\sigma^2 & -1 \\ -1 & 1+\sigma^2 \end{pmatrix} \begin{pmatrix} 1 \\ 1 \end{pmatrix}, \begin{pmatrix} X_1 \\ X_2 \end{pmatrix} \right\rangle
$$

$$
= \frac{1}{2\sigma^2 + \sigma^4} \left\langle \begin{pmatrix} \sigma^2 \\ \sigma^2 \end{pmatrix}, \begin{pmatrix} X_1 \\ X_2 \end{pmatrix} \right\rangle = \frac{X_1 + X_2}{2 + \sigma^2} .
$$

The quadratic error can be computed as in (a) or in a simpler way, using Exercise 2.28(b),

$$
E\left[\left(Y - \frac{1}{2+\sigma^2} (X_1 + X_2)\right)^2\right] = \mathrm{Var}(Y) - \mathrm{Var}\left(\frac{1}{2+\sigma^2} (X_1 + X_2)\right) .
$$

Now $\mathrm{Var}(X_1 + X_2) = \mathrm{Var}(2Y + W_1 + W_2) = 4 + 2\sigma^2$, therefore

$$
\cdots = 1 - \frac{4 + 2\sigma^2}{(2+\sigma^2)^2} = 1 - \frac{2}{2+\sigma^2} = \frac{\sigma^2}{2+\sigma^2} .
$$

The availability of two independent observations has allowed some reduction of the quadratic error.

2.30 As $\text{Cov}(X, Y) = \text{Cov}(Y, Y) + \text{Cov}(Y, W) = \text{Cov}(Y, Y) = \text{Var}(Y)$, the regression line of Y with respect to X is $y = ax + b$, with the values

$$a = \frac{\text{Cov}(X, Y)}{\text{Var}(X)} = \frac{\text{Var}(Y)}{\text{Var}(Y) + \text{Var}(W)} = \frac{\frac{1}{\lambda^2}}{\frac{1}{\lambda^2} + \frac{1}{\rho^2}} = \frac{\rho^2}{\lambda^2 + \rho^2},$$

$$b = \text{E}(Y) - a\text{E}(X) = \frac{1}{\lambda} - \frac{\rho^2}{\lambda^2 + \rho^2}\left(\frac{1}{\rho} + \frac{1}{\lambda}\right) = \frac{\lambda - \rho}{\lambda^2 + \rho^2}.$$

2.31 Let X be an r.v. having characteristic function ϕ. Then

$$\overline{\phi}(t) = \overline{\text{E}(e^{itX})} = \text{E}(\overline{e^{itX}}) = \text{E}(e^{-itX}) = \phi_{-X}(t),$$

hence $\overline{\phi}$ is the characteristic function of $-X$.

Let Y, Z be independent r.v.'s with characteristic function ϕ. Then the characteristic function of $Y + Z$ is ϕ^2.

Similarly the characteristic function of $Y - Z$ is $\phi \cdot \overline{\phi} = |\phi|^2$.

2.32 (a) The characteristic function of X_1 is

$$\phi_{X_1}(\theta) = \int_{-1/2}^{1/2} e^{i\theta x}\, dx = \frac{1}{i\theta} e^{i\theta x}\Big|_{x=-1/2}^{x=1/2} = \frac{2}{\theta} \sin\frac{\theta}{2}$$

and therefore the characteristic function of $X_1 + X_2$ is

$$\phi_{X_1+X_2}(\theta) = \frac{4}{\theta^2} \sin^2\frac{\theta}{2}.$$

(b) As f is an even function whereas $x \mapsto \sin(\theta x)$ is odd,

$$\phi(\theta) = \int_{-1}^{1} (1 - |x|)\, e^{i\theta x}\, dx = \int_{-1}^{1} (1 - |x|)\cos(\theta x)\, dx$$

$$= 2\int_{0}^{1} (1 - x)\cos(\theta x)\, dx = \frac{2}{\theta}(1 - x)\sin(\theta x)\Big|_{x=0}^{x=1} + \frac{2}{\theta}\int_{0}^{1}\sin(\theta x)\, dx$$

$$= \frac{2}{\theta^2}(1 - \cos\theta) = \frac{4}{\theta^2}\sin^2\frac{\theta}{2}.$$

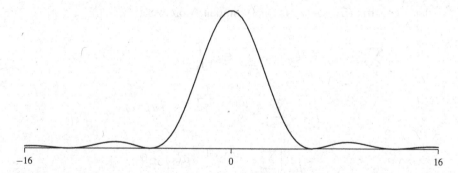

Fig. 7.3 The graph of the density (7.22). Note a typical feature: densities decreasing fast at infinity have very regular characteristic functions and conversely regular densities have characteristic functions decreasing fast at infinity. In this case the density is compactly supported and the characteristic function is very regular. The characteristic function tends to 0 a bit slowly at infinity and the density is not regular

As the probability $f(x)\,dx$ and the law of $X_1 + X_2$ have the same characteristic function, they coincide.

(c) As ϕ is integrable, by the inversion Theorem 2.33,

$$f(x) = \frac{1}{2\pi} \int_{-\infty}^{\infty} \frac{4}{\theta^2} \sin^2 \frac{\theta}{2}\, e^{-i\theta x}\, d\theta\ .$$

Exchanging the roles of x and θ we can write

$$\kappa(\theta) = f(\theta) = \frac{1}{2\pi} \int_{-\infty}^{\infty} \frac{4}{x^2} \sin^2 \frac{x}{2}\, e^{-i\theta x}\, dx\ .$$

As $\kappa(0) = 1$, the positive function

$$g(x) := \frac{2}{\pi x^2} \sin^2 \frac{x}{2} \tag{7.22}$$

is a density, having characteristic function κ. See its graph in Fig. 7.3.

2.33 Let μ be a probability on $\mathbb{R}^d$. We have, for $\theta \in \mathbb{R}^d$,

$$\widehat{\mu}(-\theta) = \overline{\widehat{\mu}(\theta)}\ ,$$

so that the matrix $(\widehat{\mu}(\theta_h - \theta_k))_{h,k}$ is Hermitian. Moreover,

$$\sum_{h,k=1}^{n} \widehat{\mu}(\theta_h - \theta_k)\xi_h\overline{\xi_k} = \sum_{h,k=1}^{n} \xi_h\overline{\xi_k} \int_{\mathbb{R}^d} e^{i\langle\theta_h,x\rangle}e^{-i\langle\theta_k,x\rangle}\,d\mu(x)$$

$$= \int_{\mathbb{R}^d}\left(\sum_{h,k=1}^{n}\xi_h e^{i\langle\theta_h,x\rangle}\,\overline{\xi_k e^{i\langle\theta_k,x\rangle}}\right)d\mu(x)$$

$$= \int_{\mathbb{R}^d}\left|\sum_{h=1}^{n}\xi_h e^{i\langle\theta_h,x\rangle}\right|^2 d\mu(x) \geq 0$$

(the integrand is positive) and therefore $\widehat{\mu}$ is positive definite.

Bochner's Theorem states that the converse is also true: a positive definite function $\mathbb{R}^d \to \mathbb{C}$ taking the value 1 at 0 is the characteristic function of a probability (see e.g. [3], p. 262).

2.34 (a) As $x \mapsto \sin(\theta x)$ is an odd function whereas $x \mapsto \cos(\theta x)$ is even,

$$\widehat{\nu}(\theta) = \frac{1}{2}\int_{-\infty}^{+\infty} e^{-|x|}e^{i\theta x}\,dx = \frac{1}{2}\int_{-\infty}^{+\infty} e^{-|x|}\cos(\theta x)\,dx$$

$$= \int_{0}^{+\infty} e^{-x}\cos(\theta x)\,dx$$

and, twice by parts,

$$\int_{0}^{+\infty} e^{-x}\cos(\theta x)\,dx = -e^{-x}\cos(\theta x)\Big|_{x=0}^{x=+\infty} - \theta\int_{0}^{+\infty} e^{-x}\sin(\theta x)\,dx$$

$$= 1 + \theta e^{-x}\sin(\theta x)\Big|_{x=0}^{x=+\infty} - \theta^2\int_{0}^{+\infty} e^{-x}\cos(\theta x)\,dx$$

$$= 1 - \theta^2\int_{0}^{+\infty} e^{-x}\cos(\theta x)\,dx ,$$

from which

$$(1+\theta^2)\int_{0}^{+\infty} e^{-x}\cos(\theta x)\,dx = 1 ,$$

i.e. (2.90).

(b1) $\theta \mapsto \frac{1}{1+\theta^2}$ is integrable and by the inversion theorem, Theorem 2.33,

$$h(x) = \frac{1}{2}e^{-|x|} = \frac{1}{2\pi}\int_{-\infty}^{+\infty}\frac{e^{-ix\theta}}{1+\theta^2}\,d\theta .$$

Exchanging the roles of x and θ we find

$$\frac{1}{\pi} \int \frac{e^{-ix\theta}}{1+x^2} \, dx = e^{-|\theta|} \, ,$$

hence $\widehat{\mu}(\theta) = e^{-|\theta|}$.

(b2) The characteristic function of $Z = \frac{1}{2}(X+Y)$ is

$$\phi_Z(\theta) = \phi_X(\tfrac{\theta}{2}) \, \phi_Y(\tfrac{\theta}{2}) = e^{-\frac{1}{2}|\theta|} e^{-\frac{1}{2}|\theta|} = e^{-|\theta|} \, .$$

Therefore $\frac{1}{2}(X+Y)$ is also Cauchy-distributed.

- Note that $\widehat{\mu}$ is not differentiable at 0; this is hardly surprising as a Cauchy r.v. does not have finite moment of order 1.

2.35 (a) Yes, $\mu_n = N(\frac{b}{n}, \frac{\sigma^2}{n})$.

(b) Yes, $\mu_n =$ Poiss$(\frac{\lambda}{n})$.

(c) Yes, $\mu_n =$ Gamma$(\frac{1}{n}, \lambda)$.

(d) We have seen in Exercise 2.34 that a Cauchy law μ has characteristic function

$$\widehat{\mu}(\theta) = e^{-|\theta|} \, .$$

Hence if $X_1, \ldots, X_n$ are independent Cauchy r.v.'s, then the characteristic function of $\frac{X_1}{n} + \cdots + \frac{X_n}{n}$ is equal to

$$\widehat{\mu}(\tfrac{\theta}{n})^n = \left(e^{-\frac{|\theta|}{n}} \right)^n = e^{-|\theta|} = \widehat{\mu}(\theta) \, ,$$

hence we can choose μ_n as the law of $\frac{X_1}{n}$, which, by the way, has density $x \mapsto \frac{n}{\pi}(1+n^2 x^2)^{-1}$ with respect to the Lebesgue measure.

2.36 (a) By (2.91), for every $a \in \mathbb{R}$,

$$\mu_\theta(]-\infty, a]) = \mu(H_{\theta,a}) = \nu(H_{\theta,a}) = \nu_\theta(]-\infty, a]) \, .$$

Hence μ_θ and ν_θ have the same d.f. and coincide.

(b) We have

$$\widehat{\mu}(\theta) = \int_{\mathbb{R}^d} e^{i\langle\theta,x\rangle} \, d\mu(x) = \widehat{\mu}_\theta(1) = \widehat{\nu}_\theta(1) = \int_{\mathbb{R}^d} e^{i\langle\theta,x\rangle} \, d\nu(x) = \widehat{\nu}(\theta) \, ,$$

so that μ and ν have the same characteristic function and coincide.

2.37 (a1) Recall that, X being integrable, $\phi'(\theta) = i \, E[Xe^{i\theta X}]$. Hence

$$E^Q(e^{i\theta X}) = E(Xe^{i\theta X}) = -i\phi'(\theta) \tag{7.23}$$

and $-i\phi'$ is therefore the characteristic function of X under Q.

(a2) Going back to (7.23) we have

$$-i\phi'(\theta) = E(Xe^{i\theta X}) = \int_{\mathbb{R}} xe^{i\theta x}\, d\mu(x)\,,$$

i.e. $-i\phi'$ is the characteristic function of the law $d\nu(x) = x\, d\mu(x)$, which is a probability because $\int x\, d\mu(x) = E(X) = 1$.

(a3) If $X \sim \text{Gamma}(\lambda, \lambda)$, then $-i\phi'$ is the characteristic function of the probability having density with respect to the Lebesgue measure given by

$$\frac{\lambda^\lambda}{\Gamma(\lambda)}\, x^\lambda e^{-\lambda x} = \frac{\lambda^{\lambda+1}}{\Gamma(\lambda+1)}\, x^\lambda e^{-\lambda x}\,, \qquad x > 0\,,$$

which is a $\text{Gamma}(\lambda + 1, \lambda)$. If X is geometric of parameter $p = 1$, then $-i\phi'$ is the probability having density with respect to the counting measure of $\mathbb{N}$ given by

$$q_k = kp(1-p)^k\,, \qquad k = 0, 1, \dots$$

i.e. a negative binomial distribution.

(b) Just note that every characteristic function takes the value 1 at 0 and $-i\phi'(0) = E(X)$.

2.38 The problem of establishing whether a given function is a characteristic function is not always a simple one. In this case an r.v. X with characteristic function ϕ would have finite moments of all orders, ϕ being infinitely many times differentiable. Moreover, we have

$$\phi'(\theta) = -4\theta^3 e^{-\theta^4}\,,$$

$$\phi''(\theta) = (16\theta^6 - 12\theta^2)\, e^{-\theta^4}$$

and therefore it would be

$$E(X) = i\phi'(0) = 0\,,$$

$$\text{Var}(X) = E(X^2) - E(X)^2 = -\phi''(0) = 0\,.$$

An r.v. having variance equal to 0 is necessarily a.s. equal to its mean. Therefore such a hypothetical X would be equal to 0 a.s. But then it would have characteristic function equal to the characteristic function of this law, i.e. $\phi \equiv 1$. $\theta \mapsto e^{-\theta^4}$ cannot be a characteristic function.

As further (not needed) evidence, Fig. 7.4 shows the graph, numerically computed using the inversion Theorem 2.33, of what would be the density of an r.v. having this "characteristic function". It is apparent that it is not positive.

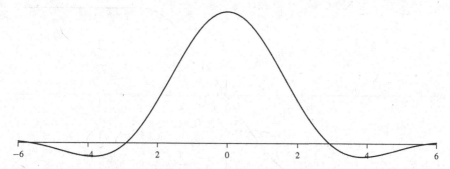

Fig. 7.4 The graph of what the density corresponding to the "characteristic function" $\theta \mapsto e^{-\frac{1}{2}\theta^4}$ would look like. If it was really a characteristic function, this function would have been ≥ 0

2.39 (a) We have, integrating by parts,

$$E\big[Zf(Z)\big] = \frac{1}{\sqrt{2\pi}} \int_{-\infty}^{+\infty} x f(x)\, e^{-x^2/2}\, dx$$

$$= -\frac{1}{\sqrt{2\pi}}\, f(x)\, e^{-x^2/2}\Big|_{-\infty}^{+\infty} + \frac{1}{\sqrt{2\pi}} \int_{-\infty}^{+\infty} f'(x)\, e^{-x^2/2}\, dx$$

$$= E\big[f'(Z)\big]\,.$$

(b1) Let us choose

$$f(x) = \begin{cases} -1 & \text{for } x \leq -1 \\ 1 & \text{for } x \geq 1 \\ \text{connected as in Fig. 7.5} & \text{for } -1 \leq x \leq 1\,. \end{cases}$$

This function belongs to C_b^1. Moreover, $zf(z) \geq 0$ and $zf(z) = |z|$ if $|z| \geq 1$, so that $|Z|1_{\{|Z|\geq 1\}} \leq Zf(Z)$. Hence, as f' is bounded,

$$E(|Z|) \leq 1 + E(|Z|1_{\{|Z|\geq 1\}}) \leq 1 + E[Zf(Z)] = 1 + E[f'(Z)] < +\infty\,,$$

so that $|Z|$ is integrable.

(b2) For $f(x) = e^{i\theta x}$ (2.92) gives

$$E(Ze^{i\theta Z}) = i\theta E(e^{i\theta Z})\,. \tag{7.24}$$

As we know that Z is integrable, its characteristic function ϕ is differentiable and

$$\phi'(\theta) = i\, E(Ze^{i\theta Z}) = -\theta E(e^{i\theta Z}) = -\theta\phi(\theta)$$

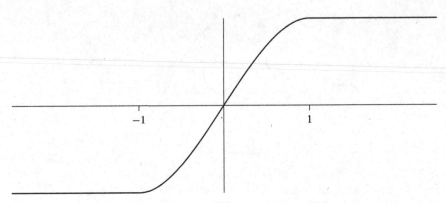

Fig. 7.5 Between -1 and 1 the function is $x \mapsto \frac{1}{2}(3x - x^3)$. Of course other choices of connection are possible in order to obtain a function in C_b^1

and solving this differential equation we obtain $\phi(\theta) = e^{-\theta^2/2}$, hence $Z \sim N(0, 1)$.

2.40 (a) We have

$$\phi(\theta) = \sum_{k=-\infty}^{\infty} P(X = k) e^{i\theta k} . \tag{7.25}$$

As the series converges absolutely, we can (Corollary 1.22) integrate by series and obtain

$$\int_0^{2\pi} \phi(\theta) \, d\theta = \sum_{k=-\infty}^{\infty} P(X = k) \int_0^{2\pi} e^{i\theta k} \, d\theta .$$

All the integrals on the right-hand side above vanish for $k \neq 0$, whereas the one for $k = 0$ is equal to 2π: (2.93) follows.

(b) We have

$$\int_0^{2\pi} e^{-i\theta m} \phi(\theta) \, d\theta = \sum_{k=-\infty}^{\infty} P(X = k) \int_0^{2\pi} e^{-i\theta m} e^{i\theta k} \, d\theta ,$$

and now all the integrals with $k \neq m$ vanish, whereas for $k = m$ the integral is again equal to 2π, i.e.

$$P(X = m) = \frac{1}{2\pi} \int_0^{2\pi} e^{-i\theta m} \phi(\theta) \, d\theta .$$

(c) Of course ϕ cannot be integrable on $\mathbb{R}$, as it is periodic: $\phi(\theta + 2\pi) = \phi(\theta)$. From another point of view, if it was integrable then X would have a density with respect to the Lebesgue measure, thanks to the inversion Theorem 2.33.

2.41 (a) The sets B_n^c are decreasing and their intersection is empty. As probabilities pass to the limit on decreasing sequences of sets,

$$\lim_{n \to \infty} \mu(B_n^c) = 0$$

and therefore $\mu(B_n^c) \leq \eta$ for n large enough.

   (b) We have

$$|e^{i\langle \theta_1, x \rangle} - e^{i\langle \theta_2, x \rangle}| \leq \int_0^1 \left| \frac{d}{dt} e^{i\langle \theta_1 + t(\theta_2 - \theta_1), x \rangle} \right| dt$$

$$= \int_0^1 |\langle \theta_2 - \theta_1, x \rangle| \, dt \leq |x| \, |\theta_2 - \theta_1| \, .$$

   (c) We have

$$|\widehat{\mu}(\theta_1) - \widehat{\mu}(\theta_2)| \leq \int_{\mathbb{R}^d} |e^{i\langle \theta_1, x \rangle} - e^{i\langle \theta_2, x \rangle}| \, d\mu(x)$$

$$= \int_{B_{R_\eta}^c} |e^{i\langle \theta_1, x \rangle} - e^{i\langle \theta_2, x \rangle}| \, d\mu(x) + \int_{B_{R_\eta}} |e^{i\langle \theta_1, x \rangle} - e^{i\langle \theta_2, x \rangle}| \, d\mu(x)$$

$$\leq 2\mu(B_{R_\eta}^c) + |\theta_1 - \theta_2| \int_{B_{R_\eta}} |x| \, d\mu(x) \leq 2\mu(B_{R_\eta}^c) + R_\eta |\theta_1 - \theta_2| \, .$$

Let $\varepsilon > 0$. Choose first $\eta > 0$ so that $2\mu(B_{R_\eta}^c) \leq \frac{\varepsilon}{2}$ and then δ such that $\delta R_\eta < \frac{\varepsilon}{2}$. Then if $|\theta_1 - \theta_2| \leq \delta$ we have $|\widehat{\mu}(\theta_1) - \widehat{\mu}(\theta_2)| \leq \varepsilon$.

2.42 (a) If $0 < \lambda < 1$, by Hölder's inequality with $p = \frac{1}{\lambda}$, $q = \frac{1}{1-\lambda}$, we have, all the integrands being positive,

$$L\big(\lambda s + (1 - \lambda)t\big) = \mathrm{E}\big[(e^{\langle s, X \rangle})^\lambda (e^{\langle t, X \rangle})^{1-\lambda}\big] \leq \mathrm{E}(e^{\langle s, X \rangle})^\lambda \mathrm{E}(e^{\langle t, X \rangle})^{1-\lambda} \tag{7.26}$$

$$= L(s)^\lambda L(t)^{1-\lambda} \, .$$

   (b) Taking logarithms in (7.26) we obtain the convexity of $\log L$. The convexity of L now follows as the exponential function is convex *and* increasing.

2.43 (a) For the Laplace transform we have

$$L(z) = \frac{\lambda}{2} \int_{-\infty}^{+\infty} e^{zt} e^{-\lambda |t|} \, dt \, .$$

The integral does not converge if $\Re z \geq \lambda$ or $\Re z \leq -\lambda$: in the first case the integrand does not vanish at $+\infty$, in the second case it does not vanish at $-\infty$. For real values $-\lambda < t < \lambda$ we have,

$$L(t) = \mathrm{E}(e^{tX}) = \frac{\lambda}{2} \int_0^{+\infty} e^{tx} e^{-\lambda x}\, dx + \frac{\lambda}{2} \int_{-\infty}^0 e^{tx} e^{\lambda x}\, dx$$

$$= \frac{\lambda}{2} \int_0^{+\infty} e^{-(\lambda - t)x}\, dx + \frac{\lambda}{2} \int_{-\infty}^0 e^{(\lambda + t)x}\, dx = \frac{\lambda}{2}\left(\frac{1}{\lambda - t} + \frac{1}{\lambda + t}\right)$$

$$= \frac{\lambda^2}{\lambda^2 - t^2}.$$

As L is holomorphic for $-\lambda < \Re z < \lambda$, by the argument of analytic continuation in the strip we have

$$L(z) = \frac{\lambda^2}{\lambda^2 - z^2}, \qquad -\lambda < \Re z < \lambda.$$

The characteristic function is of course (compare with Exercise 2.34, where $\lambda = 1$)

$$\phi(\theta) = L(i\theta) = \frac{\lambda^2}{\lambda^2 + \theta^2}.$$

(b) The Laplace transform, L_2 say, of Y and W is computed in Example 2.37(c). Its domain is $\mathscr{D} = \{z < \lambda\}$ and, for $z \in \mathscr{D}$,

$$L_2(z) = \frac{\lambda}{\lambda - z}.$$

Then their characteristic function is

$$\phi_2(t) = L_2(it) = \frac{\lambda}{\lambda - it}$$

and the characteristic function of $Y - W$ is

$$\phi_3(t) = \phi_2(t)\overline{\phi_2(t)} = \frac{\lambda}{\lambda - it}\frac{\lambda}{\lambda + it} = \frac{\lambda^2}{\lambda^2 + t^2},$$

i.e. the same as the characteristic function of a Laplace law of parameter λ. Hence $Y - W$ has a Laplace law of parameter λ.

(c1) If $X_1, \ldots, X_n$ and $Y_1, \ldots, Y_n$ are independent and Gamma$(\frac{1}{n}, \lambda)$-distributed, then

$$(X_1 - Y_1) + \cdots + (X_n - Y_n) = \underbrace{(X_1 + \cdots + X_n)}_{\sim \text{Gamma}(1, \lambda)} - \underbrace{(Y_1 + \cdots + Y_n)}_{\sim \text{Gamma}(1, \lambda)} .$$

We have found n i.i.d. r.v.'s whose sum has a Laplace distribution, which is therefore infinitely divisible.

(c2) Recalling the characteristic function of the Gamma$(\frac{1}{n}, \lambda)$ that is computed in Example 2.37(c), if $\lambda = 1$ the r.v.'s $X_k - Y_k$ of (c1) have characteristic function

$$\theta \mapsto \Big(\frac{1}{1 - i\theta}\Big)^{1/n} \Big(\frac{1}{1 + i\theta}\Big)^{1/n} = \frac{1}{(1 + \theta^2)^{1/n}} ,$$

so that ϕ of (2.95) is a characteristic function.

Note that the r.v.'s $X_k - Y_k$ have density with respect to the Lebesgue measure as this is true for both X_k and Y_k, but, for $n \geq 2$, their characteristic function is not integrable so that in this case the inversion theorem does not apply.

2.44 (a) For every $0 < \lambda < x_2$, Markov's inequality gives

$$P(X \geq t) = P(e^{\lambda X} \geq e^{\lambda t}) \leq E(e^{\lambda X}) \cdot e^{-\lambda t} .$$

(b) Let us prove that $E(e^{\lambda' X}) < +\infty$ for every $\lambda' < \lambda$: Remark 2.1 gives

$$E(e^{\lambda' X}) = \int_0^{+\infty} P\big(e^{\lambda' X} \geq s\big)\, ds = \int_0^{t_0} P\big(e^{\lambda' X} \geq s\big)\, ds + \int_{t_0}^{+\infty} P\big(e^{\lambda' X} \geq s\big)\, ds$$

$$\leq t_0 + \int_{t_0}^{+\infty} P\big(X \geq \tfrac{1}{\lambda'} \log s\big)\, ds \leq t_0 + \int_{t_0}^{+\infty} e^{-\frac{\lambda}{\lambda'} \log s}\, ds$$

$$= t_0 + \int_{t_0}^{+\infty} \frac{1}{s^{\lambda/\lambda'}}\, ds < +\infty .$$

Therefore $x_2 \geq \lambda$.

2.45 As we assume that 0 belongs to the convergence strip, the two Laplace transforms, L_μ and L_ν, are holomorphic at 0 (Theorem 2.36), i.e., for z in a neighborhood of 0,

$$L_\mu(z) = \sum_{k=1}^{\infty} \frac{1}{k!} L_\mu^{(k)}(0) z^k , \qquad L_\nu(z) = \sum_{k=1}^{\infty} \frac{1}{k!} L_\nu^{(k)}(0) z^k .$$

By (2.63) we find

$$L_\mu^{(k)}(0) = \int x^k \, d\mu(x) = \int x^k \, dv(x) = L_v^{(k)}(0) \, ,$$

so that the two Laplace transforms coincide in a neighborhood of the origin and, by the uniqueness of the analytic continuation, in the whole convergence strip, hence on the imaginary axis, so that μ and v have the same characteristic function.

2.46 (a) Let us compute the derivatives of ψ:

$$\psi'(t) = \frac{L'(t)}{L(t)} \, , \qquad \psi''(t) = \frac{L(t)L''(t) - L'(t)^2}{L(t)^2} \, .$$

Recalling that $L(0) = 1$, denoting by X any r.v. having Laplace transform L, (2.63) gives

$$\psi'(0) = L'(0) = E(X),$$

$$\psi''(0) = L''(0) - L'(0)^2 = E(X^2) - E(X)^2 = \text{Var}(X) \, .$$

(b1) The integral of $x \mapsto e^{\gamma x} L(\gamma)^{-1}$ with respect to μ is equal to 1 so that $x \mapsto e^{\gamma x} L(\gamma)^{-1}$ is a density with respect to μ and μ_γ is a probability. As for its Laplace transform:

$$\begin{aligned}
L_\gamma(t) &= \int_{-\infty}^{+\infty} e^{tx} \, d\mu_\gamma(x) \, dx \\
&= \frac{1}{L(\gamma)} \int_{-\infty}^{+\infty} e^{(\gamma+t)x} \, d\mu(x) = \frac{L(\gamma + t)}{L(\gamma)} \, .
\end{aligned} \tag{7.27}$$

(b2) Let us compute the mean and variance of μ_γ via the derivatives of $\log L_\gamma$ as seen in (a). As $\log L_\gamma(t) = \log L(\gamma + t) - \log L(\gamma)$ we have

$$\frac{d}{dt} \log L_\gamma(t) = \frac{L'(\gamma + t)}{L(\gamma + t)} \, ,$$

$$\frac{d^2}{dt^2} \log L_\gamma(t) = \frac{L(\gamma + t)L''(\gamma + t) - L'(\gamma + t)^2}{L(\gamma + t)^2}$$

and denoting by Y an r.v. having law μ_γ, for $t = 0$,

$$E(Y) = \frac{L'(\gamma)}{L(\gamma)} = \psi'(\gamma),$$

$$\text{Var}(Y) = \frac{L(\gamma)L''(\gamma) - L'(\gamma)^2}{L(\gamma)^2} = \psi''(\gamma). \tag{7.28}$$

(b3) One of the criteria to establish the convexity of a function is to check that its second order derivative is positive. From the second Eq. (7.28) we have $\psi''(\gamma) = \text{Var}(Y) \geq 0$. Hence ψ is convex. We find again, in a different way, the result of Exercise 2.42. Actually we obtain something more: if X is not a.s. constant then $\psi''(\gamma) = \text{Var}(Y) > 0$, so that ψ is *strictly* convex.

The mean of μ_γ is equal to $\psi'(\gamma)$. As ψ' is increasing (ψ'' is positive), the mean of μ_γ is an increasing function of γ.

(c1) If $\mu \sim N(0, \sigma^2)$ then

$$L(t) = e^{\frac{1}{2}\sigma^2 t^2}, \qquad \psi(t) = \tfrac{1}{2}\sigma^2 t^2.$$

Hence μ_γ has density $x \mapsto e^{\gamma x - \frac{1}{2}\sigma^2\gamma^2}$ with respect to the $N(0, \sigma^2)$ law and therefore its density with respect to the Lebesgue measure is

$$x \mapsto e^{\gamma x - \frac{1}{2}\sigma^2\gamma^2} \frac{1}{\sqrt{2\pi}\,\sigma} e^{-\frac{1}{2\sigma^2}x^2} = \frac{1}{\sqrt{2\pi}\,\sigma} e^{-\frac{1}{2\sigma^2}(x - \sigma^2\gamma)^2}$$

and we recognize an $N(\sigma^2\gamma, \sigma^2)$ law. Alternatively, and possibly in a simpler way, we can compute the Laplace transform of μ_γ using (7.27):

$$L_\gamma(t) = \frac{e^{\frac{1}{2}\sigma^2(t+\gamma)^2}}{e^{\frac{1}{2}\sigma^2\gamma^2}} = e^{\frac{1}{2}\sigma^2(t^2 + 2\gamma t)} = e^{\frac{1}{2}\sigma^2 t^2 + \sigma^2\gamma t},$$

which is the Laplace transform of an $N(\sigma^2\gamma, \sigma^2)$ law.

(c2) Also in this case it is not difficult to compute the density, but the Laplace transform provides the simplest argument: the Laplace transform of a $\Gamma(\alpha, \lambda)$ law is for $t < \lambda$ (Example 2.37(c))

$$L(t) = \left(\frac{\lambda}{\lambda - t}\right)^\alpha.$$

As L is defined only on $]-\infty, \lambda[$, we can consider only $\gamma < \lambda$. The Laplace transform of μ_γ is now

$$L_\gamma(t) = \left(\frac{\lambda}{\lambda - (t+\gamma)}\right)^\alpha \left(\frac{\lambda - \gamma}{\lambda}\right)^\alpha = \left(\frac{\lambda - \gamma}{\lambda - \gamma - t}\right)^\alpha,$$

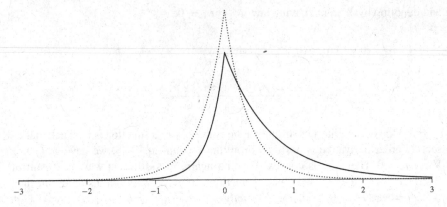

Fig. 7.6 Comparison, for $\lambda = 3$ and $\gamma = 1.5$, of the graphs of the Laplace density f of parameter λ (dots) and of the twisted density f_γ

which is the Laplace transform of a $\Gamma(\alpha, \lambda - \gamma)$ law.

(c3) The Laplace transform of a Laplace law of parameter λ is, for $-\lambda < t < \lambda$, (Exercise 2.43)

$$L(t) = \frac{\lambda^2}{\lambda^2 - t^2} .$$

Hence μ_γ has density

$$x \mapsto \frac{(\lambda^2 - \gamma^2) e^{\gamma x}}{\lambda^2}$$

with respect to μ and density

$$f_\gamma(x) := \frac{\lambda^2 - \gamma^2}{\lambda^2} e^{-\lambda |x| + \gamma x}$$

with respect to the Lebesgue measure (see the graph in Fig. 7.6). Its Laplace transform is

$$L_\gamma(t) = \frac{\lambda^2}{\lambda^2 - (t + \gamma)^2} \frac{\lambda^2 - \gamma^2}{\lambda^2} = \frac{\lambda^2 - \gamma^2}{\lambda^2 - (t + \gamma)^2} .$$

(c4) The Laplace transform of a Binomial $B(n, p)$ law is

$$L(t) = (1 - p + p e^t)^n .$$

Hence μ_γ has density, with respect to the counting measure of $\mathbb{N}$,

$$f_\gamma(k) = \frac{e^{\gamma k}}{(1 - p + p e^\gamma)^n} \binom{n}{k} p^k (1 - p)^{n-k} , \quad k = 0, \ldots, n ,$$

which, with some imagination, can be written

$$f_\gamma(k) = \binom{n}{k}\left(\frac{p\,e^\gamma}{1 - p + p\,e^\gamma}\right)^k \left(1 - \frac{p\,e^\gamma}{1 - p + p\,e^\gamma}\right)^{n-k}, \quad k = 0, \ldots, n,$$

i.e. a binomial law $B(p_\gamma, n)$ with $p_\gamma = p\,e^\gamma(1 - p + p\,e^\gamma)^{-1}$.

(c5) The Laplace transform of a geometric law of parameter p is

$$L(t) = p \sum_{k=0}^\infty (1 - p)^k e^{tk},$$

which is finite for $(1 - p)\,e^t < 1$, i.e. for $t < -\log(1 - p)$ and for these values

$$L(t) = \frac{p}{1 - (1 - p)\,e^t}.$$

Hence μ_γ has density, with respect to the counting measure of $\mathbb{N}$,

$$f_\gamma(k) = e^{\gamma k}\,(1 - (1 - p)\,e^\gamma)(1 - p)^k = (1 - (1 - p)\,e^\gamma)((1 - p)\,e^\gamma)^k,$$

i.e. a geometric law of parameter $p_\gamma = 1 - (1 - p)\,e^\gamma$.

2.47 (a) As the Laplace transforms are holomorphic (Theorem 2.36), by the uniqueness of the analytic continuation they coincide on the whole strip $a < \Re z < b$ of the complex plane, which, under the assumption of (a), contains the imaginary axis. Hence L_μ and L_ν coincide on the imaginary axis, i.e. μ and ν have the same characteristic function and coincide.

(b1) It is immediate that μ_γ, ν_γ are probabilities (see also Exercise 2.46) and that

$$L_{\mu_\gamma}(z) = \frac{L_\mu(z + \gamma)}{L_\mu(\gamma)}, \qquad L_{\nu_\gamma}(z) = \frac{L_\nu(z + \gamma)}{L_\nu(\gamma)}$$

and now L_{μ_γ} and L_{ν_γ} coincide on the interval $]a - \gamma, b - \gamma[$ which, as $b - \gamma > 0$ and $a - \gamma < 0$, contains the origin. Thanks to (a), $\mu_\gamma = \nu_\gamma$.

(b2) Obviously

$$d\mu(x) = L_\mu(\gamma)e^{-\gamma x}\,d\mu_\gamma(x) = L_\nu(\gamma)e^{-\gamma x}\,d\nu_\gamma(x) = d\nu(x).$$

2.48 (a) The method of the distribution function gives, for $x \geq 0$,

$$F_n(x) = P(Z_n \leq x) = P(X_1 \leq x, \ldots, X_n \leq x) = P(X_1 \leq x)^n = (1 - e^{-\lambda x})^n.$$

Taking the derivative we find the density

$$f_n(x) = n\lambda e^{-\lambda x}(1 - e^{-\lambda x})^{n-1}$$

for $x \geq 0$ and $f_n(x) = 0$ for $x < 0$. Noting that $\int_0^{+\infty} x e^{-\lambda x} dx = \lambda^{-2}$, we have

$$E(Z_2) = 2\lambda \int_0^{+\infty} x e^{-\lambda x}(1 - e^{-\lambda x}) dx = 2\lambda \int_0^{+\infty} (x e^{-\lambda x} - x e^{-2\lambda x}) dx$$

$$= 2\lambda \left(\frac{1}{\lambda^2} - \frac{1}{4\lambda^2} \right) = \frac{3}{2} \frac{1}{\lambda} .$$

And also

$$E(Z_3) = 3\lambda \int_0^{+\infty} x e^{-\lambda x}(1 - e^{-\lambda x})^2 dx$$

$$= 3\lambda \int_0^{+\infty} (x e^{-\lambda x} - 2x e^{-2\lambda x} + x e^{-3\lambda x}) dx = 3\lambda \left(\frac{1}{\lambda^2} - \frac{2}{4\lambda^2} + \frac{1}{9\lambda^2} \right)$$

$$= \frac{11}{6} \frac{1}{\lambda} .$$

(b) We have, for $t \in \mathbb{R}$,

$$L_n(t) = n\lambda \int_0^{+\infty} e^{tx} e^{-\lambda x}(1 - e^{-\lambda x})^{n-1} dx .$$

This integral clearly diverges for $t \geq \lambda$, hence the domain of the Laplace transform is $\Re z < \lambda$ for every n. If $t < \lambda$ let $e^{-\lambda x} = u$, $x = -\frac{1}{\lambda} \log u$, i.e. $-\lambda e^{-\lambda x} dx = du$, $e^{tx} = u^{-t/\lambda}$. We obtain

$$L_n(t) = n \int_0^1 u^{-t/\lambda}(1 - u)^{n-1} dt$$

and, recalling from the expression of the Beta laws the relation

$$\int_0^1 u^{\alpha-1}(1 - u)^{\beta-1} du = \frac{\Gamma(\alpha)\Gamma(\beta)}{\Gamma(\alpha + \beta)} ,$$

we have for $\alpha = 1 - \frac{t}{\lambda}$, $\beta = n$,

$$L_n(t) = n \frac{\Gamma(1 - \frac{t}{\lambda})\Gamma(n)}{\Gamma(n + 1 - \frac{t}{\lambda})} .$$

(c) From the basic relation of the Γ function,

$$\Gamma(\alpha + 1) = \alpha \Gamma(\alpha) , \tag{7.29}$$

and taking the derivative we find $\Gamma'(\alpha + 1) = \Gamma(\alpha) + \alpha\Gamma'(\alpha)$ and, dividing both sides by $\Gamma(\alpha + 1)$, (2.98) follows. We can now compute the mean of Z_n by taking the derivative of its Laplace transform at the origin. We have

$$L'_n(t) = n\,\Gamma(n)\,\frac{-\frac{1}{\lambda}\Gamma(n+1-\frac{t}{\lambda})\Gamma'(1-\frac{t}{\lambda}) + \frac{1}{\lambda}\Gamma'(n+1-\frac{t}{\lambda})\Gamma(1-\frac{t}{\lambda})}{\Gamma(n+1-\frac{t}{\lambda})^2}$$

$$= \frac{n\Gamma(n)}{\lambda\Gamma(n+1-\frac{t}{\lambda})}\left(-\Gamma'(1-\tfrac{t}{\lambda}) + \frac{\Gamma'(n+1-\frac{t}{\lambda})\Gamma(1-\frac{t}{\lambda})}{\Gamma(n+1-\frac{t}{\lambda})}\right)$$

and for $t = 0$, as $\Gamma(n+1) = n\Gamma(n)$,

$$L'_n(0) = \frac{1}{\lambda}\left(-\Gamma'(1) + \frac{\Gamma'(n+1)}{\Gamma(n+1)}\right). \tag{7.30}$$

Thanks to (2.98),

$$\frac{\Gamma'(n+1)}{\Gamma(n+1)} = \frac{1}{n} + \frac{\Gamma'(n)}{\Gamma(n)} = \frac{1}{n} + \frac{1}{n-1} + \frac{\Gamma'(n-1)}{\Gamma(n-1)}$$

$$= \cdots = \frac{1}{n} + \frac{1}{n-1} + \cdots + 1 + \Gamma'(1)$$

and replacing in (7.30),

$$E(Z_n) = \frac{1}{\lambda}\left(1 + \frac{1}{2} + \cdots + \frac{1}{n}\right).$$

In particular, $E(Z_n) \sim const \cdot \log n$.

2.49 (a) Immediate as $\langle \xi, X \rangle$ is a linear function of X.

(b1) Taking as ξ the vector having all its components equal to 0 except for the i-th, which is equal to 1, we have that each of the components $X_1, \ldots, X_d$ is Gaussian, hence square integrable.

(b2) We have

$$E(e^{i\theta\langle \xi, X\rangle}) = e^{i\theta b_\xi}e^{-\frac{1}{2}\sigma_\xi^2\theta^2}, \tag{7.31}$$

where by b_ξ, σ_ξ^2 we denote respectively mean and variance of $\langle \xi, X \rangle$. Let b denote the mean of X and C its covariance matrix (we know already that X is square integrable). We have (recalling (2.33))

$$b_\xi = E(\langle \xi, X \rangle) = \langle \xi, b \rangle,$$

$$\sigma_\xi^2 = E(\langle \xi, X - b \rangle^2) = \sum_{i=1}^{d} c_{ij}\xi_i\xi_j = \langle C\xi, \xi \rangle.$$

Now, by (7.31) with $\theta = 1$, the characteristic function of X is

$$\xi \mapsto E(e^{i\langle \xi, X\rangle}) = e^{i\langle \xi, b\rangle} e^{-\frac{1}{2}\langle C\xi, \xi\rangle},$$

which is the characteristic function of a Gaussian law.

2.50 (a) Let us compute the joint density of U and V: we expect to find a function of (u, v) that is the product of a function of u and of a function of v. We have $(U, V) = \Psi(X, Y)$, where

$$\Psi(x, y) = \left(\frac{x}{\sqrt{x^2 + y^2}}, x^2 + y^2 \right).$$

Let us note beforehand that U will be taking values in the interval $[-1, 1]$ whereas V will be positive. In order to determine the inverse Ψ^{-1}, let us solve the system

$$\begin{cases} u = \dfrac{x}{\sqrt{x^2 + y^2}} \\ v = x^2 + y^2. \end{cases}$$

Replacing v in the first equation we find $x = u\sqrt{v}$ and then $y = \sqrt{v}\sqrt{1 - u^2}$, so that

$$\Psi^{-1}(u, v) = \left(u\sqrt{v}, \sqrt{v}\sqrt{1 - u^2} \right).$$

Hence

$$D\Psi^{-1}(u, v) = \begin{pmatrix} \sqrt{v} & \dfrac{u}{2\sqrt{v}} \\ -\dfrac{u\sqrt{v}}{\sqrt{1 - u^2}} & \dfrac{\sqrt{1 - u^2}}{2\sqrt{v}} \end{pmatrix}$$

and

$$\left| \det D\Psi^{-1}(u, v) \right| = \frac{1}{2}\left(\sqrt{1 - u^2} + \frac{u^2}{\sqrt{1 - u^2}} \right) = \frac{1}{2\sqrt{1 - u^2}}.$$

Therefore the joint density of (U, V) is, for $u \in [-1, 1]$, $v \geq 0$,

$$f(\Psi^{-1}(u, v)) \left| \det D\Psi^{-1}(u, v) \right| = \frac{1}{2\pi} e^{-\frac{1}{2}(u^2 v + v(1 - u^2))} \times \frac{1}{2\sqrt{1 - u^2}}$$

$$= \frac{1}{2} e^{-\frac{1}{2}v} \times \frac{1}{\pi} \frac{1}{\sqrt{1 - u^2}}.$$

Hence U and V are independent. We even recognize the product of an exponential Gamma$(1, \frac{1}{2})$ (the law of V, but we knew that beforehand) and of a distribution of density

$$f_U(u) = \frac{1}{\pi}\frac{1}{\sqrt{1-u^2}} \qquad -1 < u < 1$$

with respect to the Lebesgue measure.

(b) Let $X' = X\cos\theta + Y\sin\theta$, $Y' = -X\sin\theta + Y\cos\theta$. X' and Y' are also independent and $N(0, 1)$-distributed as the vector $\left(\begin{smallmatrix}X'\\Y'\end{smallmatrix}\right)$ is obtained from $\left(\begin{smallmatrix}X\\Y\end{smallmatrix}\right)$ through the rotation associated to the matrix

$$\begin{pmatrix} \cos\theta & \sin\theta \\ -\sin\theta & \cos\theta \end{pmatrix}$$

and recalling that the multidimensional $N(0, I)$ distribution is invariant with respect to orthogonal transformations. Now just note that $X^2 + Y^2 = X'^2 + Y'^2$ and

$$U' = \frac{X'}{\sqrt{X'^2 + Y'^2}} \quad \text{and} \quad V' = X'^2 + Y'^2$$

so that $(U', V') \sim (U, V)$, which allows us to conclude that U' and V' are independent, having the same joint law as U and V and $U' \sim U$, $V' \sim V$, for the same reason.

2.51 (a) We have

$$E(e^{\langle AX, X\rangle}) = \frac{1}{(2\pi)^{m/2}} \int_{\mathbb{R}^m} e^{\langle Ax, x\rangle} e^{-\frac{1}{2}|x|^2} dx$$

$$= \frac{1}{(2\pi)^{m/2}} \int_{\mathbb{R}^m} e^{-\frac{1}{2}\langle (I-2A)x, x\rangle} dx .$$

Let us assume that every eigenvalue of A is $< \frac{1}{2}$. Then all the eigenvalues of $I - 2A$ are > 0: indeed if ξ is an eigenvector associated to the eigenvalue λ of A, then $(I-2A)\xi = (1-2\lambda)\xi$ so that ξ is an eigenvector of $I-2A$ associated to the strictly positive eigenvalue $1 - 2\lambda$. Hence the matrix $I - 2A$ is strictly positive definite and we recognize in the integrand, but for the constant, an $N(0, (I-2A)^{-1})$ density. Therefore

$$E(e^{\langle AX, X\rangle}) = \frac{1}{(2\pi)^{m/2}} \frac{(2\pi)^{m/2}}{\sqrt{\det(I-2A)}} = \frac{1}{\sqrt{\det(I-2A)}} .$$

Note that for $m = 1$ we find the result of Exercise 2.7.

(b) A has an eigenvalue that is $\geq \frac{1}{2}$ if and only if $I - 2A$ has an eigenvalue that is ≤ 0. Let C be a diagonal matrix with the eigenvalues of $I - 2A$ on the diagonal and let O be an orthogonal matrix such that $C = O(I - 2A)O^*$. Then, with the change of variable $y = Ox$,

$$E(e^{\langle AX,X \rangle}) = \frac{1}{(2\pi)^{m/2}} \int_{\mathbb{R}^m} e^{-\frac{1}{2} \langle (I-2A)x,x \rangle} \, dx$$

$$= \frac{1}{(2\pi)^{m/2}} \int_{\mathbb{R}^m} e^{-\frac{1}{2} \langle (I-2A)O^*y, O^*y \rangle} \, dy = \frac{1}{(2\pi)^{m/2}} \int_{\mathbb{R}^m} e^{-\frac{1}{2} \langle O(I-2A)O^*y, y \rangle} \, dy$$

$$= \frac{1}{(2\pi)^{m/2}} \int_{\mathbb{R}^m} e^{-\frac{1}{2} \langle Cy, y \rangle} \, dy \, .$$

As the eigenvalues λ_i of $I - 2A$ coincide with those of C and

$$\langle Cy, y \rangle = \sum_{i=1}^{m} \lambda_i y_i^2 \, ,$$

we have (Fubini's Theorem can be applied here because the integrand is positive)

$$\int_{\mathbb{R}^m} E(e^{\langle AX,X \rangle}) = \frac{1}{(2\pi)^{m/2}} \int_{\mathbb{R}^m} e^{-\frac{1}{2} \langle Cy, y \rangle} \, dy$$

$$= \frac{1}{\sqrt{2\pi}} \int_{-\infty}^{+\infty} e^{-\frac{1}{2} \lambda_1 y_1^2} \, dy_1 \cdots \frac{1}{\sqrt{2\pi}} \int_{-\infty}^{+\infty} e^{-\frac{1}{2} \lambda_m y_m^2} \, dy_m$$

and if at least one among the eigenvalues $\lambda_1, \ldots, \lambda_m$ is ≤ 0 the integral diverges.

(c) Just note that

$$\langle Ax, x \rangle = \langle \tilde{A}x, x \rangle$$

where $\tilde{A} = \frac{1}{2}(A + A^*)$ is a symmetric matrix. Hence the results of (a) and (b) still hold with A replaced by $\tilde{A}$.

2.52 (a) We can write $X = Z + \rho$ with $Z \sim N(0, 1)$. Hence, for $t \in \mathbb{R}$, with the usual idea of factoring out a perfect square at the exponent,

$$L(t) = E[e^{t(Z+\rho)^2}] = \frac{1}{\sqrt{2\pi}} \int_{-\infty}^{+\infty} e^{t(x^2 + 2\rho x + \rho^2)} e^{-x^2/2} \, dx$$

$$= e^{\rho^2 t} \times \frac{1}{\sqrt{2\pi}} \int_{-\infty}^{+\infty} \exp\left\{ -\frac{1}{2}(1 - 2t)\left(x^2 - \frac{4\rho t x}{1 - 2t}\right) \right\} dx$$

$$= \exp\left(\frac{2\rho^2 t^2}{1 - 2t} + \rho^2 t \right) \times \frac{1}{\sqrt{2\pi}} \int_{-\infty}^{+\infty} \exp\left\{ -\frac{1}{2}(1 - 2t) \right.$$

$$\left. \times \left(x^2 - \frac{4\rho t x}{1 - 2t} + \frac{4\rho^2 t^2}{(1 - 2t)^2}\right) \right\} dx$$

$$= \exp\left(\frac{\rho^2 t}{1-2t}\right) \times \frac{1}{\sqrt{2\pi}} \int_{-\infty}^{+\infty} \exp\left\{-\frac{1}{2}(1-2t)\left(x - \frac{2\rho t}{(1-2t)}\right)^2\right\} dx .$$

The last integral converges for $t < \frac{1}{2}$ and, under this condition, we recognize in the integrand, but for the constant, an $N(\frac{2\rho t}{1-2t}, \frac{1}{1-2t})$ density. Hence the integral is equal to

$$\frac{\sqrt{2\pi}}{(1-2t)^{1/2}}$$

and, for $\Re z < \frac{1}{2}$ (the domain), the Laplace transform of X is

$$L(z) = \frac{1}{(1-2z)^{1/2}} \exp\left(\frac{\rho^2 z}{1-2z}\right). \tag{7.32}$$

(b1) We can write $X = Z + b$, where $Z \sim N(0, I)$ and $b = (b_1, \ldots, b_m)^*$ is the (column) vector of the means b_i of the marginals. For every orthogonal matrix O we have

$$|X|^2 = |OX|^2 = |OZ + Ob|^2 \sim |Z + Ob|^2 ,$$

where we have taken advantage of the invariance property of the $N(0, I)$ law with respect to rotations (see p. 89). Let us choose the matrix O so that Ob is the vector having all its first $k - 1$ components equal to 0, i.e. $Ob = (0, \ldots, 0, \sqrt{\lambda})$ with $\lambda = b_1^2 + \cdots + b_m^2$. With this choice of O

$$|X|^2 \sim Z_1^2 + \cdots + Z_{m-1}^2 + (Z_m + \sqrt{\lambda})^2 .$$

As Z_i^2 is $\chi^2(1)$-distributed, $E(Z_i^2) = 1$ and

$$E(|X|^2) = m - 1 + E(Z_m^2 + 2Z_m\sqrt{\lambda} + \lambda) = m + \lambda .$$

(b2) The Laplace transform of $|X|^2$ is equal to the product of the Laplace transforms of the r.v.'s $Z_1^2, \ldots, Z_{m-1}^2, (Z_m + \sqrt{\lambda})^2$. Now just apply (7.32), $m - 1$ times with $\rho = 0$ and once with $\rho = \sqrt{\lambda}$.

2.53 (a) We have $|X|^2 = X_1^2 + \cdots + X_m^2$. As $X_i^2 \sim \chi^2(1)$, $|X|^2 \sim \chi^2(m)$.

(b) X has the same law as AZ, where $Z \sim N(0, I)$ and A is a symmetric square matrix such that $A^2 = C$. We know (see p. 87) that A can be chosen of the form $A = OD^{1/2}O^*$, where D is a diagonal matrix having on the diagonal the eigenvalues $\lambda_1, \ldots, \lambda_m$ of C (which are all ≥ 0) and O is an orthogonal matrix. Now

$$|X|^2 = |OD^{1/2}O^*Z|^2 = |D^{1/2}O^*Z|^2 = |D^{1/2}\tilde{Z}|^2 = \sum_{i=1}^{m} \lambda_i \tilde{Z}_i^2 ,$$

where $\widetilde{Z} = O^*Z$ is also $N(0, I)$-distributed as this law is rotationally invariant (p. 89) and we have used the fact that the norm does not change under the action of an orthogonal matrix. Now (2.100) follows as $\widetilde{Z}_i^2 \sim \chi^2(1)$. We have also

$$E(|X|^2) = \sum_{i=1}^{m} \lambda_i E(\widetilde{Z}_i^2) = \sum_{i=1}^{m} \lambda_i = \mathrm{tr} C \ .$$

2.54 The r.v.'s Y_k, $k = 1, \ldots, n$, are jointly Gaussian as $Y = (Y_1, \ldots, Y_n)$ is a linear function of $X = (X_1, \ldots, X_n)$. Hence in order to prove that $Y_1, \ldots, Y_n$ are independent, it is sufficient to check that they are uncorrelated, i.e., as they are centered, that $E(Y_k Y_m) = 0$ for $k \neq m$. Let us assume $k < m$. As $E(X_i X_j) = 0$ for $i \neq j$, we have

$$E(Y_k Y_m) = E\big[(X_1 + \cdots + X_k - kX_{k+1})(X_1 + \cdots + X_m - mX_{m+1})\big]$$
$$= E(X_1^2) + \cdots + E(X_k^2) - kE(X_{k+1}^2) = 0 \ .$$

2.55 (a) Let X, Y be d-dimensional independent r.v.'s, centered and having covariance matrices A and B respectively (take them to be Gaussian, for instance). Let Z be the d-dimensional r.v. defined as $Z_i = X_i Y_i$, $i = 1, \ldots, d$. Z is centered and its covariance matrix is

$$c_{ij} = E(X_i Y_i X_j Y_j) = E(X_i X_j)E(Y_i Y_j) = a_{ij} b_{ij} = g_{ij} \ .$$

G is therefore the covariance matrix of Z and is positive definite, like every covariance matrix.

(b) A closer look at the definition says that f is positive definite if and only if the matrix

$$A_{ij} = f(x_i - x_j)$$

is positive definite for every choice of n and of $x_1, \ldots, x_n \in \mathbb{R}^d$. Thanks to (a), if f and g are positive definite, then the matrix with entries $f(x_i - x_j)g(x_i - x_j)$ is also positive definite and therefore fg is also positive definite.

2.56 (a) Let us compute the d.f. of Z: recalling that $N(0, 1)$-distributed r.v.'s are symmetric,

$$F_Z(z) = P(Z \leq z) = P(Z \leq z, Y = 1) + P(Z \leq z, Y = -1)$$
$$= P(X \leq z, Y = 1) + P(-X \leq z, Y = -1)$$
$$= \frac{1}{2}P(X \leq z) + \frac{1}{2}P(X \geq -z) = P(X \leq z) \ .$$

Therefore Z has the same law as X, i.e. $Z \sim N(0, 1)$.

(b) We have

$$\mathrm{Cov}(X, Z) = \mathrm{E}(XZ) = \mathrm{E}(XZ1_{\{Y=1\}}) + \mathrm{E}(XZ1_{\{Y=-1\}})$$

$$= \mathrm{E}(X^2 1_{\{Y=1\}}) + \mathrm{E}(-X^2 1_{\{Y=-1\}}) = \frac{1}{2}\mathrm{E}(X^2) - \frac{1}{2}\mathrm{E}(X^2) = 0\,,$$

so that X and Z are uncorrelated. If Z and X were independent, they would be jointly Gaussian and their sum would also be Gaussian. So let us postpone this question until we have dealt with (c).

(c) With the same idea as in (a) (splitting according to the values of Y) we have

$$\mathrm{E}\big(e^{i\theta(X+Y)}\big) = \mathrm{E}\big(e^{i\theta(X+Y)}1_{\{Y=1\}}\big) + \mathrm{E}\big(e^{i\theta(X+Y)}1_{\{Y=-1\}}\big)$$

$$= \mathrm{E}\big(e^{2i\theta X}1_{\{Y=1\}}\big) + \mathrm{E}\big(1_{\{Y=-1\}}\big) = \frac{1}{2}e^{2\theta^2} + \frac{1}{2}\,,$$

which is not the characteristic function of a Gaussian r.v. As mentioned above this proves that X and Z are not jointly Gaussian and cannot therefore be independent.

2.57 (a) This is an immediate consequence of Cochran's Theorem 2.42 as $\overline{X}$ and $X_i - \overline{X}$ are the projections of the vector $X = (X_1, \ldots, X_n)$ onto orthogonal subspaces of $\mathbb{R}^n$. Otherwise, directly: the r.v.'s $\overline{X}$ and $X_i - \overline{X}$ are jointly Gaussian, being linear functions of the vector $(X_1, \ldots, X_n)$. We have

$$\mathrm{Cov}(\overline{X}, X_i - \overline{X}) = \mathrm{Cov}(\overline{X}, X_i) - \mathrm{Var}(\overline{X}) = \frac{1}{n}\sum_{k=1}^{n}\mathrm{Cov}(X_k, X_i) - \frac{1}{n}\,.$$

As $\mathrm{Cov}(X_k, X_i) = 0$ for $k \neq i$ and $\mathrm{Cov}(X_i, X_i) = \mathrm{Var}(X_i) = 1$ we find $\mathrm{Cov}(\overline{X}, X_i - \overline{X}) = 0$. $\overline{X}$ and $X_i - \overline{X}$ are uncorrelated, hence independent.

(b) Thanks to (a), $\overline{X}$ is independent of the vector $(X_1 - \overline{X}, \ldots, X_n - \overline{X})$ and now just note that we have also

$$Y = \max_{i=1,\ldots,n}(X_1 - \overline{X}) - \min_{i=1,\ldots,n}(X_i - \overline{X})\,.$$

Y is a function of the vector $(X_1 - \overline{X}, \ldots, X_n - \overline{X})$ and hence independent of $\overline{X}$.

2.58 (a) $\langle a, X\rangle$ and $X - \langle a, X\rangle a$ are jointly Gaussian r.v.'s, as the pair $(\langle a, X\rangle, X - \langle a, X\rangle a)$ is a linear function of X. In order to prove their independence it is therefore sufficient to prove that $\langle a, X\rangle$ is uncorrelated with all the components $X_i - \langle a, X\rangle a_i$.
As $\mathrm{Cov}(X_k, X_i) = \delta_{ki}$, we have, for $i = 1, \ldots, m$,

$$\mathrm{Cov}(\langle a, X\rangle, X_i - \langle a, X\rangle a_i) = \mathrm{Cov}(\langle a, X\rangle, X_i) - \mathrm{Cov}(\langle a, X\rangle, \langle a, X\rangle a_i)$$

$$= \sum_{k=1}^{m}\mathrm{Cov}(a_k X_k, X_i) - a_i \sum_{k=1}^{m}\sum_{j=1}^{m}\mathrm{Cov}(a_k X_k, a_j X_j)$$

$$= a_i - a_i \sum_{k=1}^{m} a_k^2 = 0 \, .$$

(b) Let $P : \mathbb{R}^m \to \mathbb{R}^m$ be the linear map $Px = x - \langle a, x \rangle a$. We have $Px = x$ if x is orthogonal to a, whereas $Px = 0$ if x is of the form $x = ta, t \in \mathbb{R}$. Hence P is the orthogonal projector onto the subspace of $\mathbb{R}^m$ which is orthogonal to a, therefore onto a subspace of dimension $m - 1$. As $X - \langle a, X \rangle a = PX$, by Cochran's Theorem

$$|X - \langle a, X \rangle a|^2 \sim \chi^2(m - 1) \, .$$

3.1 (a) As L^p convergence for $p \geq 1$ implies L^1 convergence,

$$\lim_{n \to \infty} \left| E(X_n) - E(X) \right| = \lim_{n \to \infty} \left| E(X_n - X) \right| \leq \lim_{n \to \infty} E(|X_n - X|) = 0 \, .$$

(b) By the Cauchy-Schwarz inequality

$$E(|X_n Y_n - XY|) = E\big(|X_n Y_n - X_n Y + X_n Y - XY| \big)$$
$$\leq E\big(|X_n||Y_n - Y| + |Y||X_n - X| \big)$$
$$\leq E(X_n^2)^{1/2} E[(Y_n - Y)^2]^{1/2} + E(Y^2)^{1/2} E[(X_n - X)^2]^{1/2}$$

and the result follows since the norms $\|X_n\|_2$ are bounded, as noted in Remark 3.2(b).

(c1) We have $\mathrm{Var}(X_n) = E(X_n^2) - E(X_n)^2$. From (a) we have convergence of the expectations and from Remark 3.2(b) convergence of the second order moments.

(c2) Let us denote by $X_{n,i}$ the i-th component of the random vector X_n. As $X_{n,i} \to_{n \to \infty} X_i$ in L^2 and by (c1) the variances of the $X_{n,i}$ also converge, we obtain the convergence of entries on the diagonal of the covariance matrix. As for the off-diagonal terms, let us prove that, for $i \neq j$,

$$\lim_{n \to \infty} \mathrm{Cov}(X_{n,i}, X_{n,j}) = \mathrm{Cov}(X_i, X_j) \, . \tag{7.33}$$

But $\lim_{n \to \infty} E(X_{n,i} X_{n,j}) = E(X_i X_j)$ thanks to (b) and $\lim_{n \to \infty} E(X_{n,i}) = E(X_i)$ thanks to (a), so that (7.33) follows.

3.2 (a) False. The l.h.s. is the event $\{X_n \geq \delta$ infinitely many times$\}$ and it is possible to have $\overline{\lim}_{n \to \infty} X_n \geq \delta$ with $X_n < \delta$ for every n. The relation becomes true if $=$ is replaced by $\subset$.

(b) True. If $\omega \in \overline{\lim}_{n \to \infty} \{X_n < \delta\}$, then $X_n(\omega) < \delta$ for infinitely many indices n and therefore $\underline{\lim}_{n \to \infty} X_n(\omega) \leq \delta$.

3.3 (a1) $P(A_n) = \frac{1}{n}$, the series therefore diverges.

(a2) Recall that $\overline{\lim}_{n \to \infty} A_n$ is the event of the ω's that belong to A_n for infinitely many indices n. Now if $X(\omega) = x > 0$, $\omega \in A_n$ only for the values of n such that $x \leq \frac{1}{n}$, i.e. only for a finite number of them. Hence $\overline{\lim}_{n \to \infty} A_n = \{X = 0\}$ and

$P(\overline{\lim}_{n\to\infty} A_n) = 0$. Clearly the second half of the Borel-Cantelli Lemma does not apply here as the events $(A_n)_n$ are not independent.

(b1) The events $(B_n)_n$ are now independent and

$$\sum_{n=1}^{\infty} P(B_n) = \sum_{n=1}^{\infty} \frac{1}{n} = +\infty$$

and by the Borel-Cantelli Lemma, second half, $P(\overline{\lim}_{n\to\infty} B_n) = 1$.

(b2) Now instead

$$\sum_{n=1}^{\infty} P(B_n) = \sum_{n=1}^{\infty} \frac{1}{n^2} < +\infty$$

and the Borel-Cantelli Lemma gives $P(\overline{\lim}_{n\to\infty} B_n) = 0$.

3.4 (a) We have

$$\sum_{n=1}^{\infty} P(X_n \geq 1) = \sum_{n=1}^{\infty} e^{-(\log(n+1))^\alpha} = \sum_{n=1}^{\infty} \frac{1}{(n+1)^{\log(n+1)^{\alpha-1}}} \cdot$$

The series converges if $\alpha > 1$ and diverges if $\alpha \leq 1$, hence by the Borel-Cantelli lemma

$$P\left(\overline{\lim_{n\to\infty}} \{X_n \geq 1\} \right) = \begin{cases} 0 & \text{if } \alpha > 1 \\ 1 & \text{if } \alpha \leq 1. \end{cases}$$

(b1) Let $c > 0$. By a repetition of the computation of (a) we have

$$\sum_{n=1}^{\infty} P(X_n \geq c) = \sum_{n=1}^{\infty} e^{-c(\log(n+1))^\alpha} = \sum_{n=1}^{\infty} \frac{1}{(n+1)^{c\log(n+1)^{\alpha-1}}} \cdot \qquad (7.34)$$

• Assume $\alpha > 1$. The series on the right-hand side on (7.34) is convergent for every $c > 0$ so that $P(\overline{\lim}_{n\to\infty}\{X_n \geq c\}) = 0$ and $X_n(\omega) \geq c$ for finitely many indices n only a.s. Therefore there exists a.s. an n_0 such that $X_n < c$ for every $n \geq n_0$, which implies that $\overline{\lim}_{n\to\infty} X_n < c$ and, thanks to the arbitrariness of c,

$$\lim_{n\to\infty} X_n = 0 \qquad \text{a.s.}$$

• Assume $\alpha < 1$ instead. Now, for every $c > 0$, the series on the right-hand side in (7.34) diverges, so that $P(\overline{\lim}_{n\to\infty}\{X_n \geq c\}) = 1$ and $X_n \geq c$ for infinitely many indices n a.s. Hence $\overline{\lim}_{n\to\infty} X_n \geq c$ and, thanks to the arbitrariness of c,

$$\overline{\lim_{n\to\infty}} X_n = +\infty \qquad \text{a.s.}$$

• We are left with the case $\alpha = 1$. We have

$$\sum_{n=1}^{\infty} P(X_n \geq c) = \sum_{n=1}^{\infty} \frac{1}{(n+1)^c} \cdot$$

The series on the right-hand side now converges for $c > 1$ and diverges for $c \leq 1$. Hence if $c \leq 1$, $X_n \geq c$ for infinitely many indices n whereas if $c > 1$ there exists an n_0 such that $X_n \leq c$ for every $n \geq n_0$ a.s. Hence if $\alpha = 1$

$$\overline{\lim_{n\to\infty}} X_n = 1 \qquad \text{a.s.}$$

(b2) For the inferior limit we have, whatever the value of $\alpha > 0$,

$$\sum_{n=1}^{\infty} P(X_n \leq c) = \sum_{n=1}^{\infty} \left(1 - e^{-(c\log(n+1))^\alpha}\right). \tag{7.35}$$

The series on the right-hand side diverges (its general term tends to 1 as $n \to \infty$), therefore, for every $c > 0$, $X_n \leq c$ for infinitely many indices n and $\underline{\lim}_{n\to\infty} X_n = 0$ a.s.

(c) As seen above, $\underline{\lim}_{n\to\infty} X_n = 0$ whatever the value of α. Taking into account the possible values of $\overline{\lim}_{n\to\infty} X_n$ computed in (b) above, the sequence converges only for $\alpha > 1$ and in this case $X_n \to_{n\to\infty} 0$ a.s.

3.5 (a) By Remark 2.1

$$E(Z_1) = \int_0^{+\infty} P(Z_1 \geq s)\,ds = \sum_{n=0}^{\infty} \int_n^{n+1} P(Z_1 \geq s)\,ds$$

and now just note that

$$P(Z_1 \geq n+1) \leq \int_n^{n+1} P(Z_1 \geq s)\,ds \leq P(Z_1 \geq n).$$

(b1) Thanks to (a) the series $\sum_{n=1}^{\infty} P(Z_n \geq n)$ is convergent and by the Borel-Cantelli Lemma the event $\overline{\lim}_{n\to\infty}\{Z_n \geq n\}$ has probability 0 (even if the Z_n were not independent).

(b2) Now the series $\sum_{n=1}^{\infty} P(Z_n \geq n)$ diverges, hence $\overline{\lim}_{n\to\infty}\{Z_n \geq n\}$ has probability 1.

(c1) Assume that $0 < x_2 < +\infty$ and let $0 < \theta < x_2$. Then $E(e^{\theta X_n}) < +\infty$ and thanks to (b1) applied to the r.v.'s $Z_n = e^{\theta X_n}$, $P(\overline{\lim}_{n\to\infty} e^{\theta X_n} \geq n) = 0$ hence $e^{\theta X_n} < n$ eventually, i.e. $X_n < \frac{1}{\theta} \log n$ for n larger than some n_0, so that

$$\overline{\lim_{n\to\infty}} \frac{X_n}{\log n} \leq \frac{1}{\theta} \qquad \text{a.s.} \qquad (7.36)$$

and, by the arbitrariness of $\theta < x_2$,

$$\overline{\lim_{n\to\infty}} \frac{X_n}{\log n} \leq \frac{1}{x_2} \qquad \text{a.s.} \qquad (7.37)$$

Conversely, if $\theta > x_2$ then $e^{\theta X_n}$ is not integrable and by (b2) $P(\overline{\lim}_{n\to\infty} e^{\theta X_n} \geq n) = 1$, hence $X_n > \frac{1}{\theta} \log n$ infinitely many times and

$$\overline{\lim_{n\to\infty}} \frac{X_n}{\log n} \geq \frac{1}{\theta} \qquad \text{a.s.} \qquad (7.38)$$

and, again by the arbitrariness of $\theta > x_2$,

$$\overline{\lim_{n\to\infty}} \frac{X_n}{\log n} \geq \frac{1}{x_2} \qquad \text{a.s.} \qquad (7.39)$$

which together with (7.37) completes the proof. If $x_2 = 0$ then (7.38) gives

$$\overline{\lim_{n\to\infty}} \frac{X_n}{\log n} = +\infty .$$

(c2) We have

$$\overline{\lim_{n\to\infty}} \frac{|X_n|}{\sqrt{\log n}} = \overline{\lim_{n\to\infty}} \sqrt{\frac{X_n^2}{\log n}} \qquad (7.40)$$

and, as $X_n^2 \sim \chi^2(1)$ and for such a distribution $x_2 = \frac{1}{2}$ (Example 2.37(c)), the $\overline{\lim}$ in (7.40) is equal to $\sqrt{2}$ a.s.

- Note that Example 3.5 is a particular case of (c1).

3.6 (a) The r.v. $\overline{\lim}_{n\to\infty} |X_n(\omega)|^{1/n}$, hence also R, is measurable with respect to the tail σ-algebra $\mathscr{B}^\infty$ of the sequence $(X_n)_n$. R is therefore a.s. constant by Kolmogorov's 0-1 law, Theorem 2.15.

(b) As $E(|X_1|) > 0$, there exists an $a > 0$ such that $P(|X_1| > a) > 0$. Then the series $\sum_{n=1}^{\infty} P(|X_n| > a) = \sum_{n=1}^{\infty} P(|X_1| > a)$ is divergent and by the Borel-Cantelli Lemma

$$P\left(\varlimsup_{n \to \infty} \{|X_n| > a\}\right) = 1 .$$

Therefore $|X_n|^{1/n} > a^{1/n}$ infinitely many times and

$$\cdot \varlimsup_{n \to \infty} |X_n|^{1/n} \geq \lim_{n \to \infty} a^{1/n} = 1 \qquad \text{a.s.}$$

i.e. $R \leq 1$ a.s.

(c) By Markov's inequality, for every $b > 1$,

$$P(|X_n| \geq b^n) \leq \frac{E(|X_n|)}{b^n} = \frac{E(|X_1|)}{b^n}$$

hence the series $\sum_{n=1}^{\infty} P(|X_n| \geq b^n)$ is bounded above by a convergent geometric series. By the Borel-Cantelli Lemma $P\left(\varlimsup_{n \to \infty}\{|X_n| \geq b^n\}\right) = 0$, i.e.

$$P\left(|X_n|^{1/n} < b \text{ eventually}\right) = 1$$

and, as this is true for every $b > 1$,

$$\varlimsup_{n \to \infty} |X_n|^{1/n} \leq 1 \qquad \text{a.s.}$$

i.e. $R \geq 1$ a.s. Hence $R = 1$ a.s.

3.7 Assume that $X_n \to_{n \to \infty}^{P} X$. Then for every subsequence of $(X_n)_n$ there exists a further subsequence $(X_{n_k})_k$ such that $X_{n_k} \to_{k \to \infty} X$ a.s., hence also

$$\frac{d(X_{n_k}, X)}{1 + d(X_{n_k}, X)} \xrightarrow[n \to \infty]{\text{a.s.}} 0 .$$

As the r.v.'s appearing on the left-hand side above are bounded, by Lebesgue's Theorem

$$\lim_{k \to \infty} E\left[\frac{d(X_{n_k}, X)}{1 + d(X_{n_k}, X)}\right] = 0 .$$

We have proved that from every subsequence of the quantity on the left-hand side of (3.44) we can extract a further subsequence converging to 0, therefore (3.44) follows by Criterion 3.8.

Conversely, let us assume that (3.44) holds. We have

$$P\left(\frac{d(X_n, X)}{1 + d(X_n, X)} \geq \delta\right) = P\big(d(X_n, X) \geq \delta(1 + d(X_n, X))\big)$$

$$= P\left(d(X_n, X) \geq \frac{\delta}{1 - \delta}\right).$$

Let us fix $\varepsilon > 0$ and let $\delta = \frac{\varepsilon}{1+\varepsilon}$ so that $\varepsilon = \frac{\delta}{1-\delta}$. By Markov's inequality

$$P\big(d(X_n, X) \geq \varepsilon\big) = P\left(\frac{d(X_n, X)}{1 + d(X_n, X)} \geq \delta\right) \leq \frac{1}{\delta} E\left[\frac{d(X_n, X)}{1 + d(X_n, X)}\right],$$

so that $\lim_{n \to \infty} P\big(d(X_n, X) \geq \varepsilon\big) = 0$.

3.8 (a) Let

$$S_n = \sum_{k=1}^{n} X_k.$$

Then we have, for $m < n$,

$$E(|S_n - S_m|) = E\left(\left|\sum_{k=m+1}^{n} X_k\right|\right) \leq \sum_{k=m+1}^{n} E(|X_k|),$$

from which it follows easily that $(S_n)_n$ is a Cauchy sequence in L^1, which implies L^1 convergence.

(b1) As $E(X_k^+) \leq E(|X_k|)$, the argument of (a) gives that, if $S_n^{(1)} = \sum_{k=1}^{n} X_k^+$, the sequence $(S_n^{(1)})_n$ converges in L^1 to some integrable r.v. Z_1. As $(S_n^{(1)})_n$ is increasing, it also converges a.s. to the same r.v. Z_1, as the a.s. and the L^1 limits necessarily coincide.

(b2) By the same argument as in (b1), the sequence $S_n^{(2)} = \sum_{k=1}^{n} X_k^-$ converges a.s. to some integrable r.v. Z_2. We have then

$$\lim_{n \to \infty} S_n = \lim_{n \to \infty} S_n^{(1)} - \lim_{n \to \infty} S_n^{(2)} = Z_1 - Z_2 \qquad \text{a.s.}$$

and there is no danger of encountering a $+\infty - \infty$ form as both Z_1 and Z_2 are finite a.s.

3.9 For every subsequence of $(X_n)_n$ there exists a further subsequence $(X_{n_k})_k$ such that $X_{n_k} \to_{k \to \infty} X$ a.s. By Lebesgue's Theorem

$$\lim_{k \to \infty} E(X_{n_k}) = E(X), \tag{7.41}$$

hence for every subsequence of $(E[X_n])_n$ there exists a further subsequence that converges to $E(X)$, and, by the sub-sub-sequences criterion, $\lim_{n\to\infty} E(X_n) = E(X)$.

3.10 (a1) We have, for $t > 0$,

$$P(U_n > t) = P(X_1 > t, \dots, X_n > t) = P(X_1 > t) \dots P(X_n > t) = e^{-nt}.$$

Hence the d.f. of U_n is $F_n(t) = 1 - e^{-nt}$, $t > 0$, and U_n is exponential of parameter n.

   (a2) We have

$$\lim_{n\to\infty} F_n(t) = \begin{cases} 0 & \text{if } x \le 0 \\ 1 & \text{if } x > 0. \end{cases}$$

The limit coincides with the d.f. F of an r.v. that takes only the value 0 with probability 1 except for its value at 0, which however is not a continuity point of F. Hence (Proposition 3.23) $(U_n)_n$ converges in law (and in probability) to the Dirac mass δ_0.

   (b) For every $\delta > 0$ we have

$$\sum_{n=1}^{\infty} P(U_n > \varepsilon) = \sum_{n=1}^{\infty} e^{-n\varepsilon} < +\infty,$$

hence by Remark 3.7, as $U_n > 0$ for every n, $U_n \to_{n\to\infty} 0$ a.s.

   In a much simpler way, just note that $\lim_{n\to\infty} U_n$ exists certainly, the sequence $U_n(\omega)$ being decreasing for every ω. Therefore $(U_n)_n$ converges a.s. and, by (a), it converges in probability to 0. The result then follows, as the a.s. limit and the limit in probability coincide. No need for Borel-Cantelli...

   (c) We have

$$P\left(V_n > \frac{1}{n^\beta}\right) = P\left(U_n > \frac{1}{n^{\beta/\alpha}}\right) = e^{-n^{1-\beta/\alpha}}.$$

As $1 - \frac{\beta}{\alpha} > 0$,

$$\sum_{n=1}^{\infty} P\left(V_n > \frac{1}{n^\beta}\right) < +\infty$$

and by the Borel-Cantelli Lemma $V_n > \frac{1}{n^\beta}$ for a finite number of indices n only a.s. Hence for n large $V_n \le \frac{1}{n^\beta}$, which is the general term of a convergent series.

3.11 (a1) As X_1 and X_2 are independent and integrable, their product $X_1 X_2$ is also integrable and $E(X_1 X_2) = E(X_1)E(X_2) = 0$ (Corollary 2.10).

Similarly, X_1^2 and X_2^2 are integrable (X_1 and X_2 have finite variance) independent r.v.'s, hence $X_1^2 X_2^2$ is integrable, and $E(X_1^2 X_2^2) = E(X_1^2)E(X_2^2) = \text{Var}(X_1)\text{Var}(X_2) = \sigma^4$. As $X_1 X_2$ is centered, $\text{Var}(X_1 X_2) = E(X_1^2 X_2^2) = \sigma^4$.

(a2) We have $Y_k Y_m = X_k X_{k+1} X_m X_{m+1}$. Let us assume, to fix the ideas, $m > k$: then the r.v.'s X_k, $X_{k+1} X_m$, X_{m+1} are independent and integrable. Hence $Y_k Y_m$ is also integrable and

$$E(Y_k Y_m) = E(X_k)E(X_{k+1} X_m)E(X_{m+1}) = 0$$

(note that, possibly, $k + 1 = m$).

(b) The r.v.'s Y_n are uncorrelated and have common variance σ^4. By Rajchman's strong law

$$\frac{1}{n}\left(X_1 X_2 + X_2 X_3 + \cdots + X_n X_{n+1}\right) = \frac{1}{n}\left(Y_1 + \cdots + Y_n\right) \overset{\text{a.s.}}{\underset{n\to\infty}{\to}} E(Y_1) = 0 \,.$$

3.12 $(X_n^4)_n$ is a sequence of i.i.d. r.v.'s having a common finite variance, as the Laplace laws have finite moments of all orders. Hence by Rajchman's strong law

$$\frac{1}{n}\left(X_1^4 + X_2^4 + \cdots + X_n^4\right) \overset{\text{a.s.}}{\underset{n\to\infty}{\to}} E(X_1^4) \,.$$

Let us compute $E(X_1^4)$: tracing back to the integrals of the Gamma laws,

$$E(X_1^4) = \frac{\lambda}{2}\int_{-\infty}^{+\infty} x^4 e^{-\lambda|x|}\,dx = \lambda \int_0^{+\infty} x^4 e^{-\lambda x}\,dx = \frac{\Gamma(5)}{\lambda^4} = \frac{24}{\lambda^4}\,.$$

Again thanks to Rajchman's strong law

$$\frac{1}{n}\sum_{k=1}^n X_k^2 \overset{\text{a.s.}}{\underset{n\to\infty}{\to}} E(X_1^2) = \frac{2}{\lambda^2}\,,$$

hence

$$\lim_{n\to\infty} \frac{X_1^2 + X_2^2 + \cdots + X_n^2}{X_1^4 + X_2^4 + \cdots + X_n^4} = \lim_{n\to\infty} \frac{\frac{1}{n}\sum_{k=1}^n X_k^2}{\frac{1}{n}\sum_{k=1}^n X_k^4} = \frac{E(X_1^2)}{E(X_1^4)} = \frac{\lambda^2}{12} \qquad \text{a.s.}$$

3.13 (a) We have

$$S_n^2 = \frac{1}{n}\sum_{k=1}^n (X_k^2 - 2X_k \overline{X}_n + \overline{X}_n^2)$$

$$= \frac{1}{n}\sum_{k=1}^n X_k^2 - 2\overline{X}_n \frac{1}{n}\sum_{k=1}^n X_k + \overline{X}_n^2 = \frac{1}{n}\sum_{k=1}^n X_k^2 - \overline{X}_n^2 \,. \qquad (7.42)$$

By Kolmogorov's strong law, Theorem 3.12, applied to the sequence $(X_n^2)_n$,

$$\frac{1}{n}\sum_{k=1}^{n} X_k^2 \underset{n\to\infty}{\overset{\text{a.s.}}{\longrightarrow}} E(X_1^2)$$

and again by Kolmogorov's (or Rajchman's) strong law for the sequence $(X_n)_n$

$$\overline{X}_n^2 \underset{n\to\infty}{\overset{\text{a.s.}}{\longrightarrow}} E(X_1)^2 .$$

In conclusion

$$S_n^2 \underset{n\to\infty}{\overset{\text{a.s.}}{\longrightarrow}} E(X_1^2) - E(X_1)^2 = \sigma^2 .$$

(b) Thinking of (7.42) we have

$$E\Big(\frac{1}{n}\sum_{k=1}^{n} X_k^2\Big) = E(X_1^2)$$

whereas

$$E(\overline{X}_n^2) = \text{Var}(\overline{X}_n) + E(\overline{X}_n)^2 = \frac{1}{n}\sigma^2 + E(X_1)^2$$

and putting things together

$$E(S_n^2) = \underbrace{E(X_1^2) - E(X_1)^2)}_{=\sigma^2} - \frac{1}{n}\sigma^2 = \frac{n-1}{n}\sigma^2 .$$

Therefore $S_n^2 \to_{n\to\infty} \sigma^2$ but, in the average, S_n^2 is always a bit smaller than σ^2.

3.14 (a) For every $\theta_1 \in \mathbb{R}^d$, $\theta_2 \in \mathbb{R}^m$ the weak convergence of the two sequences implies, for their characteristic functions, that

$$\widehat{\mu}_n(\theta_1) \underset{n\to\infty}{\to} \widehat{\mu}(\theta_1), \qquad \widehat{\nu}_n(\theta_2) \underset{n\to\infty}{\to} \widehat{\nu}(\theta_2) .$$

Hence, denoting by ϕ_n, ϕ the characteristic functions of $\mu_n \otimes \nu_n$ and of $\mu \otimes \nu$ respectively, by Proposition 2.35 we have

$$\lim_{n\to\infty} \phi_n(\theta_1, \theta_2) = \lim_{n\to\infty} \widehat{\mu}_n(\theta_1)\widehat{\nu}_n(\theta_2) = \widehat{\mu}(\theta_1)\widehat{\nu}(\theta_2) = \phi(\theta_1, \theta_2)$$

and P. Lévy's theorem, Theorem 3.20, completes the proof.

(b1) Immediate: $\mu * \nu$ and $\mu_n * \nu_n$ are the images of $\mu \otimes \nu$ and $\mu_n \otimes \nu_n$ respectively under the map $(x, y) \mapsto x + y$, which is continuous $\mathbb{R}^d \times \mathbb{R}^d \to \mathbb{R}^d$ (Remark 3.16).

(b2) We know (Example 3.9) that $\nu_n \to_{n\to\infty} \delta_0$ (the Dirac mass at 0). Hence thanks to (b1)

$$\mu * \nu_n \underset{n\to\infty}{\to} \mu * \delta_0 = \mu .$$

3.15 (a) As we assume that the partial derivatives of f are bounded, we can take the derivative under the integral sign (Proposition 1.21) and obtain, for $i = 1, \ldots, d$,

$$\frac{\partial}{\partial x_i} \mu * f(x) = \frac{\partial}{\partial x_i} \int_{\mathbb{R}^d} f(x-y) \, d\mu(y) = \int_{\mathbb{R}^d} \frac{\partial}{\partial x_i} f(x-y) \, d\mu(y) .$$

(b1) The $N(0, \frac{1}{n} I)$ density is

$$g_n(x) = \frac{n^{d/2}}{(2\pi)^{d/2}} e^{-\frac{1}{2} n|x|^2} .$$

The proof that the k-th partial derivatives of g_n are of the form

$$P_\alpha(x) e^{-\frac{1}{2} n|x|^2} , \qquad \alpha = (k_1, \ldots, k_d) \tag{7.43}$$

for some polynomial P_α is easily done by induction. Indeed the first derivatives are obviously of this form. Assuming that (7.43) holds for all derivatives up to the order $|\alpha| = k_1 + \cdots + k_d$, just note that for every i, $i = 1, \ldots, d$, we have

$$\frac{\partial}{\partial x_i} \left(P_\alpha(x) e^{-\frac{1}{2} n|x|^2} \right) = \left(\frac{\partial}{\partial x_i} P_\alpha(x) - nx_i P_\alpha(x) \right) e^{-\frac{1}{2} n|x|^2} ,$$

which is again of the form (7.43). In particular, all derivatives of g_n are bounded.
   (b2) If $\nu_n = N(0, \frac{1}{n} I)$, then (Exercise 2.5) the probability $\mu_n = \nu_n * \mu$ has density

$$f_n(x) = \int_{\mathbb{R}^d} g_n(x-y) \, d\mu(y)$$

with respect to the Lebesgue measure. The sequence $(\mu_n)_n$ converges in law to μ as

$$\widehat{\mu_n}(\theta) = \widehat{\nu_n}(\theta) \widehat{\mu}(\theta) = e^{-\frac{1}{2n} |\theta|^2} \widehat{\mu}(\theta) \underset{n\to\infty}{\to} \widehat{\mu}(\theta) .$$

Let us prove that the densities f_n are C^∞; let us assume $d = 1$ (the argument also holds in general, but it is a bit more complicated to write down). By induction: f_n is certainly differentiable, thanks to (a), as g_n has bounded derivatives. Let us assume next that Theorem 1.21 (derivation under the integral sign) can be applied m times and therefore that the relation

$$\frac{d^m}{dx^m} f_n(x) = \int_{\mathbb{R}} \frac{d^m}{dx^m} g_n(x-y) \, d\mu(y)$$

holds. As the integrand again has bounded derivatives, we can again take the derivative under the integral sign, which gives that f_n is $m + 1$ times differentiable. therefore, by recurrence, f_n is infinitely many times differentiable.

- This exercise, as well as Exercise 2.5, highlights the regularization properties of convolution.

3.16 (a1) We must prove that $f \geq 0$ ρ-a.e. and that $\int_E f(x) \, d\rho(x) = 1$. As $f_n \to_{n \to \infty} f$ in $L^1(\rho)$, for every bounded measurable function $\phi : E \to \mathbb{R}$ we have

$$\lim_{n \to \infty} \left| \int_E \phi(x) \, d\mu_n(x) - \int_E \phi(x) \, d\mu(x) \right|$$

$$= \lim_{n \to \infty} \left| \int_E \phi(x) \big(f_n(x) - f(x) \big) \, d\rho(x) \right|$$

$$\leq \lim_{n \to \infty} \int_E |\phi(x)| |f_n(x) - f(x)| \, d\rho(x)$$

$$\leq \|\phi\|_\infty \lim_{n \to \infty} \int_E |f_n(x) - f(x)| \, d\rho(x) = 0 \,,$$

i.e.

$$\lim_{n \to \infty} \int_E \phi(x) \, d\mu_n(x) = \int_E \phi(x) \, d\mu(x) \,. \tag{7.44}$$

By choosing first $\phi = 1$ we find $\int_E f(x) \, d\rho(x) = 1$. Next for $\phi = 1_{\{f < 0\}}$ (7.44) gives

$$0 \leq \lim_{n \to \infty} \int_{\{f < 0\}} f_n(x) \, d\rho(x) = \int_{\{f < 0\}} f(x) \, d\rho(x)$$

and therefore $\rho(\{f < 0\}) = 0$.

(a2) Choosing ϕ to be bounded continuous, (7.44) states that $\mu_n \to_{n \to \infty} \mu$ weakly. Also the relation $\mu_n(A) \to_{n \to \infty} \mu(A)$ for $A \in \mathcal{B}(E)$ follows from (7.44) with the choice $\phi = 1_A$.

(b1) Note that the functions f_n are ≥ 0, as cosine is always ≥ -1. Moreover

$$\int_0^1 f_n(x) \, dx = 1 + \frac{\sin(2n\pi x)}{2\pi n} \Big|_0^1 = 1 \,.$$

(b2) As now $E = \mathbb{R}$, in order to investigate weak convergence we can use the distribution function criterion, Proposition 3.23. The d.f. of μ_n is $F_n(x) = 0$ for $x \leq 0$, $F_n(x) = 1$ for $x \geq 1$ and

$$F_n(x) = \int_0^x \big(1 - \cos(2n\pi t)\big) \, dt = x + \frac{\sin(2n\pi x)}{2\pi n} \,, \qquad 0 \leq x \leq 1 \,,$$

hence

$$\lim_{n\to\infty} F_n(x) = x, \qquad 0 \le x \le 1,$$

(see the graphs of f_n and F_n in Figs. 7.7 and 7.8). We recognize the d.f. of a uniform law on $[0, 1]$. Therefore $(\mu_n)_n$ converges weakly to a uniform law on $[0, 1]$, i.e. having density $f = 1_{[0,1]}$ with respect to the Lebesgue measure.

(b3) By the periodicity of the cosine

$$\|f_n - f\|_1 = \int_0^1 |\cos(2n\pi x)|\, dx = \frac{1}{2n\pi} \int_0^{2n\pi} |\cos t|\, dt$$

$$= \frac{1}{2n\pi} \sum_{k=0}^{n-1} \int_{2k\pi}^{2(k+1)\pi} |\cos t|\, dt = \frac{1}{2n\pi} \sum_{k=0}^{n-1} \int_0^{2\pi} |\cos t|\, dt = \frac{C}{2\pi},$$

where $C = \int_0^{2\pi} |\cos t|\, dt > 0$ (actually $C = 4$). Therefore $\|f_n - f\|_1 \not\to 0$.

3.17 Let $f : E \to \mathbb{R}$ be a l.s.c. function bounded from below. By adding a constant we can assume $f \ge 0$. Then (Remark 2.1) we have

$$\int_E f(x)\, d\mu_n(x) = \int_0^{+\infty} \mu_n(f > t)\, dt.$$

As f is l.s.c., $\{f > t\}$ is an open set for every t, so that $\underline{\lim}_{n\to\infty} \mu_n(f > t) \ge \mu(f > t)$. By Fatou's Lemma

$$\underline{\lim}_{n\to\infty} \int_E f(x)\, d\mu_n(x) = \underline{\lim}_{n\to\infty} \int_0^{+\infty} \mu_n(f > t)\, dt$$

$$\ge \int_0^{+\infty} \mu(f > t)\, dt = \int_E f(x)\, d\mu(x).$$

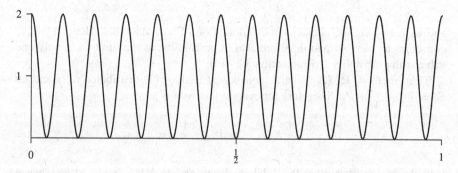

Fig. 7.7 The graph of f_n of Exercise 3.16 for $n = 13$. The rate of oscillation of f_n increases with n. It is difficult to imagine that it might converge in L^1

Fig. 7.8 The graph of the distribution function F_n of Exercise 3.16 for $n = 13$. The effect of the oscillations on the d.f. becomes smaller as n increases

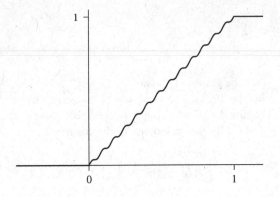

As this relation holds for every l.s.c. function f bounded from below, by Theorem 3.21(a) (portmanteau), $\mu_n \to_{n\to\infty} \mu$ weakly.

3.18 Recall that a $\chi^2(n)$-distributed r.v. has mean n and variance $2n$. Hence

$$E\left(\frac{1}{n} X_n\right) = 1, \qquad \operatorname{Var}\left(\frac{1}{n} X_n\right) = \frac{2}{n}.$$

By Chebyshev's inequality, therefore,

$$P\left(\left|\frac{X_n}{n} - 1\right| \geq \delta\right) \leq \frac{2}{\delta^2 n} \xrightarrow[n\to\infty]{} 0.$$

Hence the sequence $(\frac{1}{n} X_n)_n$ converges to 1 in probability and therefore also in law.

   Second method (possibly less elegant). Recalling the expression of the characteristic function of the Gamma laws (X_n is $\sim \operatorname{Gamma}(\frac{n}{2}, \frac{1}{2})$) the characteristic function of $\frac{1}{n} X_n$ is (Example 2.37(c))

$$\phi_n(\theta) = \left(\frac{\frac{1}{2}}{\frac{1}{2} - i\frac{\theta}{n}}\right)^{n/2} = \left(\frac{1}{1 - i\frac{2\theta}{n}}\right)^{n/2} = \left(1 - \frac{1}{n} 2\theta i\right)^{-n/2} \xrightarrow[n\to\infty]{} e^{i\theta}$$

and we recognize the characteristic function of a Dirac mass at 1. Hence $(\frac{1}{n} X_n)_n$ converges to 1 in law and therefore also in probability, as the limit takes only one value with probability 1 (Proposition 3.29(b)).

   Third method: let $(Z_n)_n$ be a sequence of i.i.d. $\chi^2(1)$-distributed r.v.'s and let $S_n = Z_1 + \cdots + Z_n$. Therefore for every n the two r.v.'s

$$\frac{1}{n} X_n \qquad \text{and} \qquad \frac{1}{n} S_n$$

have the same distribution. By Rajchman's strong law $\frac{1}{n} S_n \to_{n\to\infty} 1$ a.s., hence also in probability, so that $\frac{1}{n} X_n \xrightarrow[n\to\infty]{P} 1$.

- Actually, with a more cogent inequality than Chebyshev's, it is possible to prove that, for every $\delta > 0$,

$$\sum_{k=1}^{\infty} P\left(\left|\frac{X_n}{n} - 1\right| \geq \delta\right) < +\infty,$$

so that convergence also takes place a.s.

3.19 First method: distribution functions. Let F_n denote the d.f. of $Y_n = \frac{1}{n} X_n$: we have $F_n(t) = 0$ for $t < 0$, whereas for $t \geq 0$

$$F_n(t) = P(X_n \leq nt) = P(X_n \leq \lfloor nt \rfloor) = \sum_{k=0}^{\lfloor nt \rfloor} \frac{\lambda}{n}\left(1 - \frac{\lambda}{n}\right)^k$$

$$= \frac{\lambda}{n} \frac{1 - (1 - \frac{\lambda}{n})^{\lfloor nt \rfloor + 1}}{1 - (1 - \frac{\lambda}{n})} = 1 - \left(1 - \frac{\lambda}{n}\right)^{\lfloor nt \rfloor + 1}.$$

Hence for every $t \geq 0$

$$\lim_{n \to \infty} F_n(t) = 1 - e^{-\lambda t}.$$

We recognize on the right-hand side the d.f. of an exponential law of parameter λ. Hence $(\frac{1}{n} X_n)_n$ converges in law to this distribution.

Second method: characteristic functions. Recalling the expression of the characteristic function of a geometric law, Example 2.25(b), we have

$$\phi_{X_n}(\theta) = \frac{\frac{\lambda}{n}}{1 - (1 - \frac{\lambda}{n})e^{i\theta}} = \frac{\lambda}{n(1 - e^{i\theta}) + \lambda e^{i\theta}},$$

hence

$$\phi_{Y_n}(\theta) = \phi_{X_n}\left(\frac{\theta}{n}\right) = \frac{\lambda}{n(1 - e^{i\theta/n}) + \lambda e^{i\theta/n}}.$$

Noting that

$$\lim_{n \to \infty} n(1 - e^{i\theta/n}) = \theta \lim_{n \to \infty} \frac{1 - e^{i\theta/n}}{\frac{\theta}{n}} = -\theta \frac{d}{d\theta} e^{i\theta}\big|_{\theta=0} = -i\theta,$$

we have

$$\lim_{n \to \infty} \phi_{Y_n}(\theta) = \frac{\lambda}{\lambda - i\theta},$$

which is the characteristic function of an exponential law of parameter λ.

3.20 (a) The d.f. of X_n is, for $y \geq 0$,

$$F_n(y) = \int_0^y \frac{n}{(1+nx)^2} \, dx = 1 - \frac{1}{1+ny} \, .$$

As, of course, $F_n(y) = 0$ for $y \leq 0$,

$$\lim_{n \to \infty} F_n(y) = \begin{cases} 1 & y > 0 \\ 0 & y \leq 0 \, . \end{cases}$$

The limit is the d.f. of an r.v. X with $P(X = 0) = 1$. $(X_n)_n$ converges in law to X and, as the limit is an r.v. that takes only one value, the convergence takes place also in probability (Proposition 3.29(b)).

   (b) The a.s. limit, if it existed, would also be 0, but for every $\delta > 0$ we have

$$P(X_n > \delta) = 1 - P(X_n \leq \delta) = \frac{1}{1+n\delta} \, . \tag{7.45}$$

The series $\sum_{n=1}^{\infty} P(|X_n| > \delta)$ diverges and by the Borel-Cantelli Lemma (second half) $P(\overline{\lim}_{n \to \infty}\{X_n > \delta\}) = 1$ and the sequence does not converge to zero a.s. We have even that $X_n > \delta$ infinitely many times and, as δ is arbitrary, $\overline{\lim}_{n \to \infty} X_n = +\infty$.

   For the inferior limit note that for every $\varepsilon > 0$ we have

$$\sum_{n=1}^{\infty} P(X_n < \varepsilon) = \sum_{n=1}^{\infty} \left(1 - \frac{1}{1+n\varepsilon} \right) = +\infty \, ,$$

hence $P(\overline{\lim}_{n \to \infty}\{X_n < \varepsilon\}) = 1$. Therefore $X_n < \varepsilon$ infinitely many times with probability 1 and $\underline{\lim}_{n \to \infty} X_n = 0$.

3.21 Given the form of the r.v.'s Z_n of this exercise, it appears that their d.f.'s should be easier to deal with than their characteristic functions.

   (a) We have, for $0 \leq t \leq 1$,

$$P(Z_n > t) = P(X_1 > t, \ldots, X_n > t) = (1-t)^n \, ,$$

hence the d.f. F_n of Z_n is

$$F_n(t) = \begin{cases} 0 & \text{for } t < 0 \\ 1 - (1-t)^n & \text{for } 0 \leq t \leq 1 \\ 1 & \text{for } t > 1 \, . \end{cases}$$

Hence

$$\lim_{n\to\infty} F_n(t) = \begin{cases} 0 & \text{for } t \le 0 \\ 1 & \text{for } t > 0 \end{cases}$$

and we recognize the d.f. of a Dirac mass at 0, except for the value at 0, which however is not a continuity point of the d.f. of this distribution. We conclude that Z_n converges in law to an r.v. having this distribution and, as the limit is a constant, the convergence takes place also in probability. As the sequence $(Z_n)_n$ is decreasing it converges a.s.

(b) The d.f., G_n, of $n Z_n$ is, for $0 \le t \le n$,

$$G_n(t) = P(n Z_n \le t) = P\left(Z_n \le \tfrac{t}{n}\right) = F_n\left(\tfrac{t}{n}\right) = 1 - \left(1 - \tfrac{t}{n}\right)^n .$$

As

$$\lim_{n\to\infty} G_n(t) = G(t) := \begin{cases} 0 & \text{for } t \le 0 \\ 1 - e^{-t} & \text{for } t > 0 \end{cases}$$

the sequence $(n Z_n)_n$ converges in law to an exponential distribution with parameter $\lambda = 1$. Therefore, for n large,

$$P\left(\min(X_1, \ldots, X_n) \le \tfrac{2}{n}\right) \approx 1 - e^{-2} = 0.86 .$$

3.22 Let us compute the d.f. of M_n: for $k = 0, 1, \ldots$ we have

$$P(M_n \le k) = 1 - P(M_n \ge k+1) = 1 - P\left(U_1^{(n)} \ge k+1, \ldots, U_n^{(n)} \ge k+1\right)$$

$$= 1 - P\left(U_1^{(n)} \ge k+1\right)^n = 1 - \left(\frac{n-k}{n+1}\right)^n .$$

Now

$$\lim_{n\to\infty} \left(\frac{n-k}{n+1}\right)^n = \lim_{n\to\infty} \left(1 - \frac{k+1}{n+1}\right)^n = e^{-(k+1)} .$$

Hence

$$\lim_{n\to\infty} P(M_n \le k) = 1 - e^{-(k+1)} ,$$

which is the d.f. of a geometric law of parameter e^{-1}.

3.23 (a) The characteristic function of μ_n is

$$\widehat{\mu}_n(\theta) = (1 - a_n)\, e^{i\theta \cdot 0} + a_n\, e^{i\theta n} = 1 - a_n + a_n\, e^{i\theta n}$$

and if $a_n \to_{n \to \infty} 0$

$$\hat{\mu}_n(\theta) \underset{n \to \infty}{\to} 1 \qquad \text{for every } \theta ,$$

which is the characteristic function of a Dirac mass δ_0. It is possible to come to the same result also by computing the d.f.'s

(b) Let X_n, X be r.v.'s with $X_n \sim \mu_n$ and $X \sim \delta_0$. Then

$$E(X_n) = (1 - a_n) \cdot 0 + a_n \cdot n = n a_n ,$$

$$E(X_n^2) = (1 - a_n) \cdot 0^2 + a_n \cdot n^2 = n^2 a_n ,$$

$$\text{Var}(X_n) = E(X_n^2) - E(X_n)^2 = n^2 a_n (1 - a_n) .$$

If, for instance, $a_n = \frac{1}{\sqrt{n}}$ then $E(X_n) \to_{n \to \infty} +\infty$, whereas $E(X) = 0$. If $a_n = \frac{1}{n^{3/2}}$ then the expectations converge to the expectation of the limit but $\text{Var}(X_n) \to_{n \to \infty} +\infty$, whereas $\text{Var}(X) = 0$.

(c) By Theorem 3.21 (portmanteau), as $x \mapsto x^2$ is continuous and bounded below, we have, with $X_n \sim \mu_n$, $X \sim \mu$,

$$\lim_{n \to \infty} E(X_n^2) = \lim_{n \to \infty} \int x^2 \, d\mu_n \geq \int x^2 \, d\mu = E(X^2) .$$

The same argument applies for $\underline{\lim}_{n \to \infty} E(|X_n|)$.

3.24 (a) The d.f. of X_n is, for $t \geq 0$, $F_n(t) = 1 - e^{-\lambda_n t}$. As $F_n(t) \to_{n \to \infty} 0$ for every $t \in \mathbb{R}$, the d.f.'s of the X_n do not converge to any distribution function.

(b) Note in the first place that the r.v.'s Y_n take their values in the interval $[0, 1]$. We have, for every $t < 1$,

$$\{Y_n \leq t\} = \bigcup_{k=0}^{\infty} \{k \leq X_n \leq k + t\}$$

so that the d.f. of Y_n is, for $0 \leq t < 1$,

$$G_n(t) := P(Y_n \leq t) = \sum_{k=0}^{\infty} P(k \leq X_n < k + t) = \sum_{k=0}^{\infty} (e^{-\lambda_n k} - e^{-\lambda_n (k+t)})$$

$$= (1 - e^{-\lambda_n t}) \sum_{k=0}^{\infty} e^{-\lambda_n k} = \frac{1 - e^{-\lambda_n t}}{1 - e^{-\lambda_n}} = \frac{\lambda_n t + o(\lambda_n t)}{\lambda_n + o(\lambda_n)} .$$

Therefore

$$\lim_{n \to \infty} G_n(t) = t$$

and $(Y_n)_n$ converges in law to a uniform distribution on $[0, 1]$.

3.25 The only if part is immediate, as $x \mapsto \langle \theta, x \rangle$ is a continuous map $\mathbb{R}^d \to \mathbb{R}$. If $\langle \theta, X_n \rangle \to_{n \to \infty}^{\mathscr{L}} \langle \theta, X \rangle$ for every $\theta \in \mathbb{R}^d$, as both the real and the imaginary parts of $x \mapsto e^{ix}$ are bounded and continuous, we have

$$\lim_{n \to \infty} E(e^{i \langle \theta, X_n \rangle}) = E(e^{i \langle \theta, X \rangle})$$

and the result follows thanks to P. Lévy's Theorem 3.20.

3.26 By the Central Limit Theorem

$$\frac{X_1 + \cdots + X_n}{\sqrt{n}} \xrightarrow[n \to \infty]{\mathscr{L}} N(0, \sigma^2)$$

and the sequence $(Z_n)_n$ converges in law to the square of a $N(0, \sigma^2)$-distributed r.v. (Remark 3.16), i.e. to a Gamma$(\frac{1}{2}, \frac{1}{2\sigma^2})$-distributed r.v.

3.27 (a) By the Central Limit Theorem the sequence

$$S_n^* = \frac{X_1 + \cdots + X_n - nb}{\sqrt{n} \, \sigma}$$

converges in law to an $N(0, 1)$-distributed r.v., where b and σ^2 are respectively the mean and the variance of X_1. Here $b = E(X_i) = \frac{1}{2}$, whereas

$$E(X_1^2) = \int_0^1 x^2 \, dx = \frac{1}{3}$$

and therefore $\sigma^2 = \frac{1}{3} - \frac{1}{4} = \frac{1}{12}$. The r.v. W in (3.49) is nothing else than S_{12}^*.

It is still to be seen whether $n = 12$ is large enough for S_n^* to be approximatively $N(0, 1)$. Figure 7.9 and (b) below give some elements of appreciation.

(b) We have, integrating by parts,

$$E(X^4) = \frac{1}{\sqrt{2\pi}} \int_{-\infty}^{+\infty} x^4 e^{-x^2/2} \, dx$$

$$= \frac{1}{\sqrt{2\pi}} \left(-x^3 e^{-x^2/2} \Big|_{-\infty}^{+\infty} + 3 \int_{-\infty}^{+\infty} x^2 e^{-x^2/2} \, dx \right)$$

$$= 3 \underbrace{\frac{1}{\sqrt{2\pi}} \int_{-\infty}^{+\infty} x^2 e^{-x^2/2} \, dx}_{=\text{Var}(X)=1} = 3 \, .$$

The computation of the moment of order 4 of W is a bit more involved. If $Z_i = X_i - \frac{1}{2}$, then the Z_i's are independent and uniform on $[-\frac{1}{2}, \frac{1}{2}]$ and

$$E(W^4) = E[(Z_1 + \cdots + Z_{12})^4] . \tag{7.46}$$

Let us expand the fourth power $(Z_1 + \cdots + Z_{12})^4$ into a sum of monomials. As $E(Z_i) = E(Z_i^3) = 0$ (the Z_i's are symmetric), the expectation of many terms appearing in this expansion will vanish. For instance, as the Z_i are independent,

$$E(Z_1^3 Z_2) = E(Z_1^3)E(Z_2) = 0 .$$

A moment of reflection shows that a non-zero contribution is given only by the terms, in the development of (7.46), of the form $E(Z_i^2 Z_j^2) = E(Z_i^2)E(Z_j^2)$ with $i \neq j$ and those of the form $E(Z_i^4)$. The term Z_i^4 clearly has a coefficient $= 1$ in the expansion of the right-hand term in (7.46). In order to determine which is the coefficient of $Z_i^2 Z_j^2, i \neq j$, we remark that in the power series expansion around 0 of

$$\phi(x_1, \ldots, x_{12}) = (x_1 + \cdots + x_{12})^4$$

the monomial $x_i^2 x_j^2$, for $i \neq j$, has coefficient

$$\frac{1}{2!2!} \frac{\partial^4 \phi}{\partial x_i^2 \partial x_j^2}(0) = \frac{1}{4} \times 24 = 6 .$$

We have

$$E(Z_i^2) = \int_{-1/2}^{1/2} x^2 \, dx = \frac{1}{12} , \qquad E(Z_i^4) = \int_{-1/2}^{1/2} x^4 \, dx = \frac{1}{80} .$$

As all the terms of the form $E(Z_i^2 Z_j^2), i \neq j$, are equal and there are $11 + 10 + \cdots + 1 = \frac{1}{2} \times 12 \times 11 = 66$ of them, their contribution is

$$6 \times 66 \times \frac{1}{144} = \frac{11}{4} .$$

The contribution of the terms of the form $E(Z_i^4)$ (there are 12 of them), is $\frac{12}{80}$. In conclusion

$$E(W^4) = \frac{11}{4} + \frac{12}{80} = 2.9 .$$

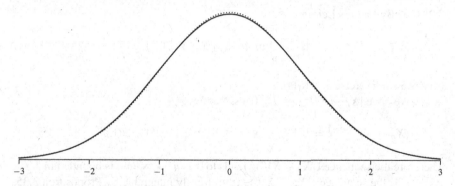

Fig. 7.9 Comparison between the densities of W (solid) and of a true $N(0, 1)$ density (dots). The two graphs are almost indistinguishable

The r.v. W turns out to have a density which is quite close to an $N(0, 1)$ (see Fig. 7.9). This was to be expected, the uniform distribution on $[0, 1]$ being symmetric around its mean, even if the value $n = 12$ seems a bit small.

However as an approximation of an $N(0, 1)$ r.v. W has some drawbacks: for instance it cannot take values outside the interval $[-6, 6]$ whereas for an $N(0, 1)$ r.v. this is possible, even if with a very small probability. In practice, in order to simulate an $N(0, 1)$ r.v., W can be used as a fast substitute of the Box-Müller algorithm (Example 2.19) for tasks that require a moderate number of random numbers, but one must be very careful in simulations requiring a large number of them because, then, the occurrence of a very large value is not so unlikely any more.

3.28 (a) Let $A := \overline{\lim}_{n \to \infty} A_n$. Recalling that $1_A = \overline{\lim}_{n \to \infty} 1_{A_n}$, by Fatou's Lemma

$$P\left(\overline{\lim_{n \to \infty}} A_n\right) = E\left(\overline{\lim_{n \to \infty}} 1_{A_n}\right) \geq \overline{\lim_{n \to \infty}} E(1_{A_n}) = \overline{\lim_{n \to \infty}} P(A_n) \geq \alpha .$$

(b) Let us assume *ad absurdum* that for some $\varepsilon > 0$ it is possible to find events A_n such that $P(A_n) \leq 2^{-n}$ and $Q(A_n) \geq \varepsilon$. If again $A = \overline{\lim}_{n \to \infty} A_n$ we have, by the Borel-Cantelli Lemma,

$$P(A) = 0$$

whereas, thanks to (a) with P replaced by Q,

$$Q(A) \geq \varepsilon ,$$

contradicting the assumption that $Q \ll P$.

- Note that (b) of this exercise is immediate if we admit the Radon-Nikodym Theorem: we would have $dQ = X\, dP$ for some density X and the result follows immediately thanks to Proposition 3.33, as $\{X\}$ is a uniformly integrable family.

3.29 (a) By Fatou's Lemma

$$M^r \geq \varliminf_{n \to \infty} E(|X_n|^r) \geq E(|X|^r)$$

(this is as in Exercise 1.15(a1)).

(b) We have $|X_n - X|^p \to_{n \to \infty} 0$ a.s. Moreover,

$$E\left[(|X_n - X|^p)^{r/p}\right] = E[|X_n - X|^r] \leq 2^{r-1}\big(E(|X_n|^r) + E(|X|^r)\big) \leq 2^r M^r.$$

Therefore the sequence $(|X_n - X|^p)_n$ tends to 0 as $n \to \infty$ and is bounded in $L^{r/p}$. As $\frac{r}{p} > 1$, the sequence $(|X_n - X|^p)_n$ is uniformly integrable by Proposition 3.35. The result follows thanks to Theorem 3.34.

If $X_n \to_{n \to \infty} X$ in probability only, just note that from every subsequence of $(X_n)_n$ we can extract a further subsequence $(X_{n_k})_{n_k}$ converging to X a.s. For this subsequence, by the result just proved we have

$$\lim_{k \to \infty} E(|X_{n_k} - X|^p) = 0$$

and the result follows thanks to the sub-sub-sequence Criterion 3.8.

3.30 (a) Just note that, for every R, ψ_R is a bounded continuous function.

(b) Recall that a uniformly integrable sequence is bounded in L^1; let $M > 0$ be such that $E(|X_n|) \leq M$ for every n. By the portmanteau Theorem 3.21(a) ($x \mapsto |x|$ is continuous and positive) we have

$$M \geq \varliminf_{n \to \infty} E(|X_n|) \geq E(|X|),$$

so that the limit X is integrable. Moreover, as $\psi_R(x) = x$ for $|x| \leq R$, we have $|X_n - \psi_R(X_n)| \leq |X_n| 1_{\{|X_n| > R\}}$ and $|X - \psi_R(X)| \leq |X| 1_{\{|X| > R\}}$. Let $\varepsilon > 0$ and R be such that

$$E(|X| 1_{\{|X| > R\}}) \leq \varepsilon \quad \text{and} \quad E(|X_n| 1_{\{|X_n| > R\}}) \leq \varepsilon$$

for every n. Then

$$\big|E(X_n) - E(X)\big|$$
$$\leq \big|E[X_n - \psi_R(X_n)]\big| + \big|E[\psi_R(X_n)] - E[\psi_R(X)]\big| + \big|E[\psi_R(X) - X]\big|$$
$$\leq E(|X_n| 1_{\{|X_n| > R\}}) + \big|E[\psi_R(X_n)] - E[\psi_R(X)]\big| + E(|X| 1_{\{|X| > R\}})$$
$$\leq 2\varepsilon + \big|E[\psi_R(X_n)] - E[\psi_R(X)]\big|.$$

Item (a) above then gives

$$\varlimsup_{n \to \infty} \left| E(X_n) - E(X) \right| \le 2\varepsilon$$

and the result follows thanks to the arbitrariness of ε.

• Note that in the argument of this exercise we took special care not to write quantities like $E(X_n - X)$, which might not make sense, as the r.v.'s X_n, X might not be defined on the same probability space.

3.31 (a) If $(Z_n)_n$ is a sequence of independent $\chi^2(1)$-distributed r.v.'s and $S_n = Z_1 + \cdots + Z_n$, then, for every n,

$$\frac{X_n - n}{\sqrt{2n}} \sim \frac{S_n - n}{\sqrt{2n}}$$

and, recalling that $E(Z_i) = 1$, $\mathrm{Var}(Z_i) = 2$, the term on the right-hand side converges in law to an $N(0, 1)$ law by the Central Limit Theorem. Therefore this is true also for the left-hand side.

(b1) We have

$$\lim_{n \to \infty} \frac{\sqrt{2n}}{\sqrt{2X_n} + \sqrt{2n - 1}} = \lim_{n \to \infty} \frac{1}{\sqrt{\frac{X_n}{n}} + \sqrt{\frac{2n-1}{2n}}}$$

and as, by the strong Law of Large Numbers,

$$\lim_{n \to \infty} \frac{X_n}{n} = 1 \quad \text{a.s.}$$

we obtain

$$\lim_{n \to \infty} \frac{\sqrt{2n}}{\sqrt{2X_n} + \sqrt{2n - 1}} = \frac{1}{2} \quad \text{a.s.} \tag{7.47}$$

(b2) We have

$$\sqrt{2X_n} - \sqrt{2n - 1} = \frac{2X_n - 2n + 1}{\sqrt{2X_n} + \sqrt{2n - 1}}$$

$$= \frac{2X_n - 2n}{\sqrt{2X_n} + \sqrt{2n - 1}} + \frac{1}{\sqrt{2X_n} + \sqrt{2n - 1}} \,.$$

The last term on the right-hand side is bounded above by $(2n-1)^{-1/2}$ and converges to 0 a.s., whereas

$$\frac{2X_n - 2n}{\sqrt{2X_n} + \sqrt{2n-1}} = 2\frac{X_n - n}{\sqrt{2n}} \times \frac{\sqrt{2n}}{\sqrt{2X_n} + \sqrt{2n-1}}.$$

We have seen in (a) that

$$\frac{X_n - n}{\sqrt{2n}} \xrightarrow[n\to\infty]{\mathscr{L}} N(0, 1)$$

and recalling (7.47), (3.50) follows by (repeated applications of) Slutsky's Lemma 3.45.

(c) From (a) we derive, denoting by Φ the d.f. of an $N(0, 1)$ law, the approximation

$$F_n(x) = P(X_n \le x) = P\Big(\frac{X_n - n}{\sqrt{2n}} \le \frac{x - n}{\sqrt{2n}}\Big) \sim \Phi\Big(\frac{x - n}{\sqrt{2n}}\Big) \tag{7.48}$$

whereas from (b)

$$\begin{aligned} F_n(x) = P(X_n \le x) &= P\big(\sqrt{2X_n} - \sqrt{2n-1} \\ &\le \sqrt{2x} - \sqrt{2n-1}\big) \sim \Phi\big(\sqrt{2x} - \sqrt{2n-1}\big). \end{aligned} \tag{7.49}$$

In order to deduce from (7.48) an approximation of the quantile $\chi^2_\alpha(n)$, we must solve the equation, with respect to the unknown x,

$$\alpha = \Phi\Big(\frac{x - n}{\sqrt{2n}}\Big).$$

Denoting by ϕ_α the quantile of order α of an $N(0, 1)$ law, x must satisfy the relation

$$\frac{x - n}{\sqrt{2n}} = \phi_\alpha,$$

i.e.

$$x = \sqrt{2n}\,\phi_\alpha + n.$$

Similarly, (7.48) gives the approximation

$$x = \frac{1}{2}\big(\phi_\alpha + \sqrt{2n-1}\big)^2.$$

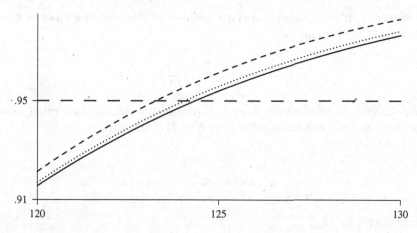

Fig. 7.10 The true d.f. of a $\chi^2(100)$ law in the interval $[120, 130]$, together with the CLT approximation (7.48) (dashes) and Fisher's approximation (7.49) (dots)

For $\alpha = 0.95$, i.e. $\phi_\alpha = 1.65$, and $n = 100$ we obtain respectively

$$x = 1.65 \cdot \sqrt{200} + 100 = 123.334$$

and

$$x = \frac{1}{2}(1.65 + \sqrt{199})^2 = 124.137 ,$$

which is a much better approximation of the true value 124.34. Fisher's approximation, proved in (b), remains very good also for larger values of n. Here are the values of the quantiles of order $\alpha = 0.95$ for some values of n and their approximations.

n	200	300	400	500
$\chi^2_\alpha(n)$	233.99	341.40	447.63	553.13
$\frac{1}{2}(\phi_\alpha + \sqrt{2n-1})^2$	233.71	341.11	447.35	552.84
$\sqrt{2n}\,\phi_\alpha + n$	232.90	340.29	446.52	552.01

see also Fig. 7.10.

3.32 (a) Recalling the value of the mean and variance of the Gamma distributions, $E(\frac{1}{n}X_n) = 1$ and $Var(\frac{1}{n}X_n) = \frac{1}{n}$. Hence by Chebyshev's inequality

$$P\left(\left|\frac{X_n}{n} - 1\right| \geq \delta\right) \leq \frac{1}{\delta^2 n} ,$$

so that $\frac{X_n}{n} \to_{n\to\infty} 1$ in probability and in law.

(b) Let $(Z_n)_n$ be a sequence of i.i.d. Gamma(1, 1)-distributed r.v.'s and let $S_n = Z_1 + \cdots + Z_n$. Then the r.v.'s

$$\frac{1}{\sqrt{n}} (X_n - n) \quad \text{and} \quad \frac{1}{\sqrt{n}} (S_n - n)$$

have the same distribution for every n. Now just note that, by the Central Limit Theorem, the latter converges in law to an $N(0, 1)$ distribution.

(c) We can write

$$\frac{1}{\sqrt{X_n}} (X_n - n) = \frac{\sqrt{n}}{\sqrt{X_n}} \frac{1}{\sqrt{n}} (X_n - n) .$$

Thanks to (a) and (b),

$$\frac{\sqrt{n}}{\sqrt{X_n}} \xrightarrow[n \to \infty]{\mathscr{L}} 1 ,$$

$$\frac{1}{\sqrt{n}} (X_n - n) \xrightarrow[n \to \infty]{\mathscr{L}} N(0, 1)$$

and by Slutsky's Lemma

$$\frac{1}{\sqrt{X_n}} (X_n - n) \xrightarrow[n \to \infty]{\mathscr{L}} N(0, 1) .$$

3.33 As the r.v.'s X_k are centered and have variance equal to 1, by the Central Limit Theorem

$$\sqrt{n}\, \overline{X}_n = \frac{X_1 + \cdots + X_n}{\sqrt{n}} \xrightarrow[n \to \infty]{\mathscr{L}} N(0, 1) .$$

(a) As the derivative of the sine function at 0 is equal to 1, the Delta method gives

$$\sqrt{n} \sin \overline{X}_n \xrightarrow[n \to \infty]{\mathscr{L}} N(0, 1) .$$

(b) As the derivative of the cosine function at 0 is equal to 0, again the Delta method gives

$$\sqrt{n} (1 - \cos \overline{X}_n) \xrightarrow[n \to \infty]{\mathscr{L}} N(0, 0) ,$$

i.e. the sequence converges in law to the Dirac mass at 0.

(c) We can write

$$n(1 - \cos \overline{X}_n) = \left(\sqrt{n} \sqrt{1 - \cos \overline{X}_n} \right)^2 .$$

Let us apply the Delta method to the function $f(x) = \sqrt{1 - \cos x}$. We have

$$f'(0) = \lim_{x \to 0} \frac{\sqrt{1 - \cos x}}{x} = \frac{1}{\sqrt{2}} \cdot$$

The Delta method gives

$$\sqrt{n}\sqrt{1 - \cos \overline{X}_n} \xrightarrow[n \to \infty]{\mathscr{L}} Z \sim N(0, \tfrac{1}{2}) ,$$

so that

$$n(1 - \cos \overline{X}_n) \xrightarrow[n \to \infty]{\mathscr{L}} Z^2 \sim \Gamma(\tfrac{1}{2}, 1) .$$

4.1 (a) The σ-algebra $\mathscr{G}$ is generated by the two-elements partition $A_0 = \{X + Y = 0\}$ and $A_1 = A_0^c = \{X + Y \geq 1\}$, i.e. $\mathscr{G} = \{\Omega, A_0, A_1, \emptyset\}$.

(b) We are as in Example 4.8: $E(X|\mathscr{G})$ takes on $A_i, i = 0, 1$, the value

$$\alpha_i = \frac{E(X 1_{A_i})}{P(A_i)} \cdot$$

As $X = 0$ on A_0, $E(X 1_{A_0}) = 0$ and $\alpha_0 = 0$.

On the other hand $X 1_{A_1} = 1_{\{X=1\}} 1_{\{X+Y \geq 1\}} = 1_{\{X=1\}}$ and therefore $E(X 1_{A_1}) = P(X = 1) = p$ and

$$\alpha_1 = \frac{p}{P(A_1)} = \frac{p}{1 - (1 - p)^2} \cdot$$

Hence

$$E(X|\mathscr{G}) = \frac{p}{1 - (1 - p)^2} 1_{\{X+Y \geq 1\}} \cdot \tag{7.50}$$

The r.v. $E(X|\mathscr{G})$ takes the values $p(1 - (1 - p)^2)^{-1}$ with probability $1 - (1 - p)^2$ and 0 with probability $P(A_0) = (1 - p)^2$. Note that $E[E(X|\mathscr{G})] = p = E(X)$.

By symmetry (the right-hand side of (7.50) being symmetric in X and Y) $E(X|\mathscr{G}) = E(Y|\mathscr{G})$.

As a non-constant r.v. cannot be independent of itself, $E(X|\mathscr{G})$ and $E(Y|\mathscr{G})$ are not independent.

4.2 (a) The r.v. $E(1_A|\mathscr{G})$ is $\mathscr{G}$-measurable so that $B = \{E(1_A|\mathscr{G}) = 0\} \in \mathscr{G}$ and, by the definition of conditional expectation,

$$E(1_A 1_B) = E\big[E(1_A|\mathscr{G}) 1_B\big] . \tag{7.51}$$

As $E(1_A|\mathcal{G}) = 0$ on B, (7.51) implies $E(1_A 1_B) = 0$. As $0 = E(1_A 1_B) = P(A \cap B)$ we have $B \subset A^c$ a.s.

(b) If $B = \{E(X|\mathcal{G}) = 0\}$ we have

$$E(X 1_B) = E\big[E(X|\mathcal{G}) 1_B\big] = 0 \,.$$

The r.v. $X 1_B$ is positive and its expectation is equal to 0, hence $X 1_B = 0$ a.s., which is equivalent to saying that X vanishes a.s. on B.

4.3 Statement (a) looks intuitive: adding the information $\mathcal{D}$, which is independent of X and of $\mathcal{G}$, should not provide any additional information useful to the prediction of X. But given how the exercise is formulated, the reader should have become suspicious that things are not quite as they seem. Let us therefore prove (b) as a start; we shall then look for a counterexample in order to give a negative answer to (a).

(b) The events of the form $G \cap D$, $G \in \mathcal{G}$, $D \in \mathcal{D}$, form a class that is stable with respect to finite intersections, generating $\mathcal{G} \vee \mathcal{D}$ and containing Ω. Thanks to Remark 4.3 we need only prove that

$$E\big[E(X|\mathcal{G}) 1_{G \cap D}\big] = E(X 1_{G \cap D})$$

for every $G \in \mathcal{G}$, $D \in \mathcal{D}$. As $\mathcal{D}$ is independent of $\sigma(X) \vee \mathcal{G}$ (and therefore also of $\mathcal{G}$),

$$E\big[E(X|\mathcal{G}) 1_{G \cap D}\big] = E\big[E(X 1_G|\mathcal{G}) 1_D\big]$$

$$= E(X 1_G)E(1_D) \overset{\downarrow}{=} E(X 1_G 1_D) = E(X 1_{G \cap D}) \,,$$

where $\downarrow$ denotes the equality where we use the independence of $\mathcal{D}$ and $\sigma(X) \vee \mathcal{G}$.

(a) The counterexample is based on the fact that it is possible to construct r.v.'s X, Y, Z that are pairwise independent but not independent globally and even such that X is $\sigma(Y) \vee \sigma(Z)$-measurable. This was seen in Remark 2.12. Hence if $\mathcal{G} = \sigma(Y)$, $\mathcal{D} = \sigma(Z)$, then

$$E(X|\mathcal{G}] = E(X)$$

whereas

$$E(X|\mathcal{G} \vee \mathcal{D}) = X \,.$$

4.4 (a) Every event $A \in \sigma(X)$ is of the form $A = \{X \in A'\}$ with $A' \in \mathcal{B}(E)$. Note that $\{X = x\} \in \sigma(X)$, as $\{x\}$ is a Borel set. In order for A to be strictly contained in $\{X = x\}$, A' must be strictly contained in $\{x\}$, which is not possible, unless $A' = \emptyset$.

(b) If A is an atom of $\mathcal{G}$ and X was not constant on A, then X would take on A at least two distinct values y, z. But then the two events $\{X = y\} \cap A$ and $\{X = z\} \cap A$

would be $\mathcal{G}$-measurable, nonempty and strictly contained in A, thus contradicting the assumption that A is an atom.

(c) $W = \mathrm{E}(Z|X)$ is $\sigma(X)$-measurable and therefore constant on $\{X = x\}$, as a consequence of (a) and (b) above. The value c of this constant is determined by the relation

$$cP(X = x) = \mathrm{E}(W1_{\{X=x\}}) = \mathrm{E}(Z1_{\{X=x\}}) = \int_{\{X=x\}} Z\, d\mathrm{P}\,,$$

i.e. (4.27).

4.5 (a) We have $\mathrm{E}[h(X)|Z] = g(Z)$, where g is such that, for every bounded measurable function ψ,

$$\mathrm{E}\big[h(X)\psi(Z)\big] = \mathrm{E}\big[g(Z)\psi(Z)\big]\,.$$

But $\mathrm{E}[h(X)\psi(Z)] = \mathrm{E}[h(Y)\psi(Z)]$, as $(X, Z) \sim (Y, Z)$, and therefore also $\mathrm{E}[h(Y)|Z] = g(Z)$ a.s.

(b1) The r.v.'s (T_1, T) and (T_2, T) have the same joint law. Actually (T_1, T) can be obtained from the r.v. $(T_1, T_2 + \cdots + T_n)$ through the map $(s, t) \mapsto (s, s + t)$. (T_2, T) is obtained through the same map from the r.v. $(T_2, T_1 + T_3 \cdots + T_n)$. As the two r.v.'s $(T_1, T_2 + \cdots + T_n)$ and $(T_2, T_1 + T_3 \cdots + T_n)$ have the same law (they have the same marginals and independent components), (T_1, T) and (T_2, T) have the same law. The same argument gives that $(T_1, T), \ldots, (T_n, T)$ have the same law.

(b2) Thanks to (a) and (b1) $\mathrm{E}(T_1|T) = \mathrm{E}(T_2|T) = \cdots = \mathrm{E}(T_n|T)$ a.s., hence a.s.

$$n\mathrm{E}(T_1|T) = \mathrm{E}(T_1|T) + \cdots + \mathrm{E}(T_n|T) = \mathrm{E}(T_1 + \cdots + T_n|T)$$
$$= \mathrm{E}(T|T) = T\,.$$

4.6 (a) (X, XY) is the image of (X, Y) under the map $\psi(x, y) := (x, xy)$. $(-X, XY)$ is the image of $(-X, -Y)$ under the same map ψ. As the Laplace distribution is symmetric, (X, Y) and $(-X, -Y)$ have the same distribution (independent components and same marginals), also their images under the same function have the same distribution.

(b1) We must determine a measurable function g such that, for every bounded Borel function ϕ

$$\mathrm{E}[X\,\phi(XY)] = \mathrm{E}[g(XY)\,\phi(XY)]\,.$$

Thanks to (a) $\mathrm{E}(X\,\phi(XY)) = -\mathrm{E}(X\,\phi(XY))$ hence $\mathrm{E}(X\,\phi(XY)) = 0$. Therefore $g \equiv 0$ is good and $\mathrm{E}(X|XY = z) = 0$.

(b2) Of course the argument leading to $\mathrm{E}(X|XY = z) = 0$ holds for every pair of independent integrable symmetric r.v.'s, hence also for $N(0, 1)$-distributed ones.

(b3) A Cauchy r.v. is symmetric but not integrable, nor l.s.i. as

$$E(X^-) = \int_{-\infty}^0 \frac{-x}{\pi(1+x^2)}\,dx = +\infty \,.$$

Conditional expectation for such an r.v. is not defined.

4.7 (a) Let $\phi : \mathbb{R}^+ \to \mathbb{R}$ be a bounded Borel function. We have, in polar coordinates,

$$E\big[\phi(|X|)\big] = \int_{\mathbb{R}^m} \phi(|x|)g(|x|)\,dx = \int_{\mathbb{S}_{m-1}} d\theta \int_0^{+\infty} \phi(r)g(r)r^{m-1}\,dr$$

$$= \omega_{m-1}\int_0^{+\infty} \phi(r)g(r)r^{m-1}\,dr \,,$$

where $\mathbb{S}_{m-1}$ is the unit sphere of $\mathbb{R}^m$ and ω_{m-1} denotes the $(m-1)$-dimensional measure of $\mathbb{S}_{m-1}$. We deduce that $|X|$ has density

$$g_1(t) = \omega_{m-1}g(t)t^{m-1} \,.$$

(b) Recall that every $\sigma(|X|)$-measurable r.v. W is of the form $W = h(|X|)$ (this is Doob's criterion, Proposition 1.7). Hence, for every bounded Borel function ψ we must determine a function $\widetilde{\psi} : \mathbb{R}^+ \to \mathbb{R}$ such that, for every bounded Borel function h, $E[\psi(X)h(|X|)] = E[\widetilde{\psi}(|X|)h(|X|)]$. We have, again in polar coordinates,

$$E\big[\psi(X)h(|X|)\big] = \int_{\mathbb{R}^m} \psi(x)h(|x|)g(|x|)\,dx$$

$$= \int_{\mathbb{S}_{m-1}} d\theta \int_0^{+\infty} \psi(r,\theta)h(r)g(r)r^{m-1}\,dr$$

$$= \frac{1}{\omega_{m-1}} \int_{\mathbb{S}_{m-1}} d\theta \int_0^{+\infty} \psi(t,\theta)h(t)g_1(t)\,dt$$

$$= \int_0^{+\infty} h(t)g_1(t)\Big(\frac{1}{\omega_{n-1}} \int_{\mathbb{S}_{m-1}} \psi(t,\theta)\,d\theta\Big)\,dt = E\big[\widetilde{\psi}(|X|)h(|X|)\big]$$

with

$$\widetilde{\psi}(t) := \frac{1}{\omega_{m-1}} \int_{\mathbb{S}_{m-1}} \psi(t,\theta)\,d\theta \,.$$

Hence

$$E\big[\psi(X)\,\big|\,|X|\big] = \widetilde{\psi}(|X|) \qquad \text{a.s.}$$

Note that $\widetilde{\psi}(t)$ is the average of ψ on the sphere of radius t.

4.8 (a) As $\{Z > 0\} \subset \{E(Z|\mathcal{G}) > 0\}$ we have

$$Z \geq Z1_{\{E(Z|\mathcal{G})>0\}} \geq Z1_{\{Z>0\}} = Z$$

and obviously

$$E(ZY|\mathcal{G}) = E(Z1_{\{E(Z|\mathcal{G})>0\}}Y|\mathcal{G}) = E(ZY|\mathcal{G})1_{\{E(Z|\mathcal{G})>0\}} \quad \text{a.s.}$$

(b1) As the events of probability 0 for P are also negligible for Q, $\{Z = 0\} \supset \{E(Z|\mathcal{G}) = 0\}$ also Q-a.s. Recalling that $Q(Z = 0) = E(Z1_{\{Z=0\}}) = 0$ we obtain $Q(E(Z|\mathcal{G}) = 0) \leq Q(Z = 0) = 0$.

(b2) First note that the r.v.

$$\frac{E(YZ|\mathcal{G})}{E(Z|\mathcal{G})}$$

of (4.29) is $\mathcal{G}$-measurable and well defined, as $E(Z|\mathcal{G}) > 0$ Q-a.s. Next, for every bounded $\mathcal{G}$-measurable r.v. W we have

$$E^Q\left[\frac{E(YZ|\mathcal{G})}{E(Z|\mathcal{G})}\,W\right] = E\left[Z\,\frac{E(YZ|\mathcal{G})}{E(Z|\mathcal{G})}\,W\right].$$

As in the mathematical expectation on the right-hand side Z is the only r.v. that is not $\mathcal{G}$-measurable,

$$\cdots = E\left[E\left(Z\,\frac{E(YZ|\mathcal{G})}{E(Z|\mathcal{G})}\,W\,\Big|\,\mathcal{G}\right)\right] = E\left[E(Z|\mathcal{G})\,\frac{E(YZ|\mathcal{G})}{E(Z|\mathcal{G})}\,W\right]$$

$$= E\left[E(YZ|\mathcal{G})W\right] = E(YZW) = E^Q(YW)$$

and the result follows.

• In solving Exercise 4.8 we have been a little on the sly on a delicate point that deserves more attention. Always recall that a conditional expectation (with respect to a probability P) *is not* an r.v., but a family of r.v.'s that differ among them only on P-negligible events. Therefore the quantity $E(Z|\mathcal{G})$ must be considered with caution when we argue with respect to a probability Q different from P, as a P-negligible event might not also be Q-negligible. In this case there are no such difficulties as $P \gg Q$.

4.9 (a) By the freezing lemma, Lemma 4.11, the Laplace transform of X is

$$L(z) = \mathrm{E}(e^{zZT}) = \mathrm{E}\big[\mathrm{E}(e^{zZT}\,|\,T)\big] = \mathrm{E}(e^{\frac{1}{2}z^2T^2}) = \int_0^1 2t\, e^{\frac{1}{2}z^2t^2}\, dt$$

$$= \frac{2}{z^2}\, e^{\frac{1}{2}z^2t^2}\Big|_{t=0}^{t=1} = \frac{2}{z^2}\,(e^{\frac{1}{2}z^2} - 1) = \sum_{n=0}^{\infty} \frac{1}{(n+1)!}\big(\tfrac{1}{2}z^2\big)^n \,. \tag{7.52}$$

L is defined on the whole of the complex plane so that the convergence abscissas are $x_1 = -\infty$, $x_2 = +\infty$. The characteristic function is of course

$$\phi(\theta) = L(i\theta) = \frac{2}{\theta^2}\,(1 - e^{-\frac{1}{2}\theta^2})\,.$$

See in Fig. 7.11 the appearance of the density having such a characteristic function.

(b) As its Laplace transform is finite in a neighborhood of the origin, X has finite moments of all orders. Of course $\mathrm{E}(X) = 0$ as ϕ is real-valued, hence X is symmetric. Moreover the power series expansion of (7.52) gives

$$\mathrm{E}(X^2) = L''(0) = \frac{1}{2}\,.$$

Alternatively, directly,

$$\mathrm{Var}(ZT) = \mathrm{E}(Z^2T^2) = \mathrm{E}(Z^2)\mathrm{E}(T^2) = \int_0^1 t^2 \cdot 2t\, dt = \frac{t^4}{2}\Big|_0^1 = \frac{1}{2}\,.$$

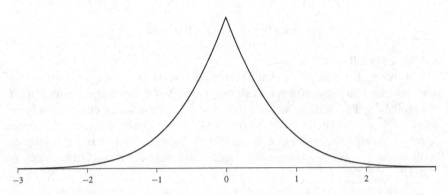

Fig. 7.11 The density of the r.v. X of Exercise 4.9, computed numerically with the formula (2.54) of the inversion Theorem 2.33. It looks like the graph of the Laplace density, but it tends to 0 faster at infinity

(c) Immediate with the same argument as in Exercise 2.44: by Markov's inequality, for every $R > 0$, $x > 0$,

$$P(X \geq x) = P(e^{RX} \geq e^{Rx}) \leq e^{-Rx} E(e^{RX}) = L(R) e^{-Rx}$$

and in the same way $P(X \leq -x) \leq L(-R) e^{-Rx})$. Therefore

$$P(|X| \geq x) \leq \big(L(R) + L(-R)\big)e^{-Rx} \, .$$

Of course property (c) holds for every r.v. X having both convergence abscissas infinite.

4.10 (a) Immediate, as X is assumed to be independent of $\mathscr{G}$ (Proposition 4.5(c)).
(b1) We have, for $\theta \in \mathbb{R}^m$, $t \in \mathbb{R}$,

$$\begin{aligned}
\phi_{(X,Y)}(\theta, t) &= E(e^{i\langle\theta,X\rangle}e^{itY}) = E\big[E(e^{i\langle\theta,X\rangle}e^{itY} \,|\, \mathscr{G})\big] \\
&= E\big[e^{itY}E(e^{i\langle\theta,X\rangle} \,|\, \mathscr{G})\big] = E\big[e^{itY}E(e^{i\langle\theta,X\rangle})\big] \quad\quad (7.53) \\
&= E(e^{i\langle\theta,X\rangle})E(e^{itY}) = \phi_X(\theta)\phi_Y(t) \, .
\end{aligned}$$

(b2) According to the definition, X and $\mathscr{G}$ are independent if and only if the events of $\sigma(X)$ are independent of those belonging to $\mathscr{G}$, i.e. if and only if, for every $A \in \mathscr{B}(\mathbb{R}^m)$ and $G \in \mathscr{G}$, the events $\{X \in A\}$ and G are independent. But this is immediate, thanks to (7.53): choosing $Y = 1_G$, the r.v.'s X and 1_G are independent thanks to the criterion of Proposition 2.35.

4.11 (a) We have (freezing lemma again),

$$E(e^{i\theta\sqrt{X}\,Y}) = E\big[E(e^{i\theta\sqrt{X}\,Y})|X\big] = E(e^{-\frac{1}{2}\theta^2 X})$$

and we land on the Laplace transform of the Gamma distributions. By Example 2.37 (c) or directly

$$E(e^{-\frac{1}{2}\theta^2 X}) = \lambda \int_0^{+\infty} e^{-\lambda t}e^{-\frac{1}{2}\theta^2 t}\, dt = \lambda \int_0^{+\infty} e^{-\frac{1}{2}(\theta^2 + 2\lambda)t}\, dt = \frac{2\lambda}{2\lambda + \theta^2} \, .$$

(b) The characteristic function of a Laplace distribution is computed in Exercise 2.43(a):

$$E(e^{i\theta W}) = \frac{\alpha^2}{\alpha^2 + \theta^2} \, .$$

(c) Comparing the results of (a) and (b) we see that Z has a Laplace law of parameter $\sqrt{2\lambda}$.

4.12 (a) We have

$$E(Z) = E[E(Z|Y)] = E[E(e^{-\frac{1}{2}\lambda^2 Y^2 + \lambda YX}|Y)].$$

By the freezing lemma $E(e^{-\frac{1}{2}\lambda^2 Y^2 + \lambda YX}|Y) = E[\Phi(Y)]$, where

$$\Phi(y) = E(e^{-\frac{1}{2}\lambda^2 y^2 + \lambda yX}) = e^{-\frac{1}{2}\lambda^2 y^2 + \frac{1}{2}\lambda^2 y^2} = 1.$$

Hence $E(Z) = 1$.

(b) Let us compute the Laplace transform of X under Q: for $t \in \mathbb{R}$

$$E^Q(e^{tX}) = E(e^{-\frac{1}{2}\lambda^2 Y^2 + \lambda YX}e^{tX}) = E(e^{-\frac{1}{2}\lambda^2 Y^2 + (\lambda Y + t)X})$$

$$= E[E(e^{-\frac{1}{2}\lambda^2 Y^2 + (\lambda Y + t)X}|Y)] = E[\Phi(Y)],$$

where now

$$\Phi(y) = E(e^{-\frac{1}{2}\lambda^2 y^2 + (\lambda y + t)X}) = e^{-\frac{1}{2}\lambda^2 y^2}e^{\frac{1}{2}(\lambda y + t)^2} = e^{\frac{1}{2}t^2 + \lambda ty},$$

so that

$$E^Q(e^{tX}) = e^{\frac{1}{2}t^2}E(e^{\lambda tY}) = e^{\frac{1}{2}t^2}e^{\frac{1}{2}\lambda^2 t^2} = e^{\frac{1}{2}(1+\lambda^2)t^2}.$$

Therefore $X \sim N(0, 1 + \lambda^2)$ under Q. Note that this law depends on $|\lambda|$ only and that the variance of X becomes larger under Q for every value of λ.

4.13 (a) The freezing lemma, Lemma 4.11, gives

$$E(e^{tXY}) = E[E(e^{tXY}|Y)] = E[e^{\frac{1}{2}t^2 Y^2}].$$

Hence, as $Y^2 \sim \Gamma(\frac{1}{2}, \frac{1}{2})$ (Remark 2.37 or Exercise 2.7), $E(e^{tXY}) = +\infty$ if $|t| \geq 1$ and

$$E(e^{tXY}) = \frac{1}{\sqrt{1 - t^2}} \qquad \text{if } |t| < 1.$$

(b) Thanks to (a) Q is a probability. Let $\phi : \mathbb{R}^2 \to \mathbb{R}$ be a bounded Borel function. We have

$$E^Q[\phi(X, Y)] = \sqrt{1 - t^2}\, E[\phi(X, Y)e^{tXY}]$$

$$= \frac{\sqrt{1 - t^2}}{2\pi} \int_{-\infty}^{+\infty} \int_{-\infty}^{+\infty} \phi(x, y)\, e^{txy}\, e^{-\frac{1}{2}(x^2 + y^2)}\, dx\, dy,$$

from which we derive that, under Q, the joint density with respect to the Lebesgue measure of (X, Y) is

$$\frac{\sqrt{1-t^2}}{2\pi} e^{-\frac{1}{2}(x^2+y^2-2txy)}.$$

We recognize a Gaussian law, centered and with covariance matrix C such that

$$C^{-1} = \begin{pmatrix} 1 & -t \\ -t & 1 \end{pmatrix},$$

i.e.

$$C = \frac{1}{1-t^2} \begin{pmatrix} 1 & t \\ t & 1 \end{pmatrix},$$

from which

$$\mathrm{Var}_Q(X) = \mathrm{Var}_Q(Y) = \frac{1}{1-t^2}, \qquad \mathrm{Cov}_Q(X, Y) = \frac{t}{1-t^2}.$$

4.14 Note that $S_{n+1} = X_{n+1} + S_n$ and that S_n is $\mathcal{F}_n$-measurable whereas X_{n+1} is independent of $\mathcal{F}_n$. We are therefore in the situation of the freezing lemma, Lemma 4.11, which gives that

$$E\big[f(X_{n+1} + S_n)|\mathcal{F}_n\big] = \Phi(S_n), \tag{7.54}$$

where, (recall that $X_n \sim \mu_n$)

$$\Phi(x) = E\big[f(X_{n+1} + x)\big] = \int f(y + x)\, d\mu_{n+1}(y). \tag{7.55}$$

The right-hand side in (7.54) is $\sigma(S_n)$-measurable (being a function of S_n) and this implies (4.31): indeed, as $\sigma(S_n) \subset \mathcal{F}_n$,

$$E\big[f(S_{n+1})|S_n\big] = E\big[E(f(S_{n+1})|\mathcal{F}_n)|S_n\big] = E(\Phi(S_n)|S_n) = \Phi(S_n)$$
$$= E\big[f(S_{n+1})|\mathcal{F}_n\big].$$

Moreover, by (7.55),

$$E\big[f(S_{n+1})|\mathcal{F}_n\big] = \Phi(S_n) = \int f(y + S_n)\, d\mu_{n+1}(y).$$

4.15 Recall that $t(1)$ is the Cauchy law, which does not have a finite mean. For $n \geq 2$ a look at the density that is computed in Example 4.17 shows that the mean exists, is finite, and is equal to 0 of course, as Student laws are symmetric.

As for the second order moment, let us use the freezing lemma, which is a better strategy than direct computation with the density that was computed in Example 4.17. Let $T = \frac{X}{\sqrt{Y}}\sqrt{n}$ be a $t(n)$-distributed r.v., i.e. with X, Y independent and $X \sim N(0, 1)$, $Y \sim \chi^2(n)$. We have

$$E(T^2) = E\left(\frac{X^2}{Y}n\right) = E\left[E\left(\frac{X^2}{Y}n \mid Y\right)\right] = E[\Phi(Y)],$$

where

$$\Phi(y) = E\left(\frac{X^2}{y}n\right) = \frac{n}{y},$$

so that

$$E(T^2) = E\left(\frac{n}{Y}\right) = \frac{n}{2^{n/2}\Gamma(\frac{n}{2})} \int_0^{+\infty} \frac{1}{y}\, y^{n/2-1} e^{-y/2}\, dy$$

$$= \frac{n}{2^{n/2}\Gamma(\frac{n}{2})} \int_0^{+\infty} y^{n/2-2} e^{-y/2}\, dy\,.$$

The integral diverges at 0 if $n \leq 2$. For $n \geq 3$ we can trace back the integral to a Gamma density and we have

$$\mathrm{Var}(T) = E(T^2) = \frac{n 2^{n/2-1}\Gamma(\frac{n}{2}-1)}{2^{n/2}\Gamma(\frac{n}{2})} = \frac{n}{2(\frac{n}{2}-1)} = \frac{n}{n-2}\,.$$

4.16 Thanks to the second freezing lemma, Lemma 4.15, the conditional law of W given $Z = z$ is the law of $zX + \sqrt{1-z^2}\, Y$, which is Gaussian $N(0, 1)$ and does not depend on z. This implies (Remark 4.14) that $W \sim N(0, 1)$ and that W is independent of Z.

4.17 By the second freezing lemma, Lemma 4.15, the conditional law of X given $Y = y$ is the law of $\frac{X}{\sqrt{y}}\sqrt{n}$, i.e. $\sim N(0, \frac{n}{y}C)$, hence with density with respect to the Lebesgue measure

$$\overline{h}(x; y) = \frac{y^{d/2}}{(2\pi n)^{d/2}\sqrt{\det C}}\, e^{-\frac{y}{2n}\langle C^{-1}x, x\rangle}\,.$$

Thanks to (4.19) the density of X is

$$h_X(x) = \int \overline{h}(x; y) h_Y(y)\, dy$$

$$= \frac{1}{2^{n/2}\Gamma(\frac{n}{2})(2\pi n)^{d/2}\sqrt{\det C}} \int_0^{+\infty} y^{d/2}\, y^{n/2-1} e^{-\frac{y}{2n}\langle C^{-1}x, x\rangle}\, e^{-y/2}\, dy$$

$$= \frac{1}{2^{n/2}\Gamma(\frac{n}{2})(2\pi n)^{d/2}\sqrt{\det C}} \int_0^{+\infty} y^{\frac{1}{2}(d+n)-1} e^{-\frac{y}{2}(1+\frac{1}{n}\langle C^{-1}x,x\rangle)}\, dy \; .$$

We recognize in the last integrand a Gamma($\frac{1}{2}(d+n)$, $\frac{1}{2}(1+\frac{1}{n}\langle C^{-1}x, x\rangle)$) density, except for the constant, so that

$$h_X(x) = \frac{1}{2^{n/2}\Gamma(\frac{n}{2})(2\pi n)^{d/2}\sqrt{\det C}} \; \frac{\Gamma(\frac{n}{2}+\frac{d}{2})2^{\frac{n+d}{2}}}{(1+\frac{1}{n}\langle C^{-1}x, x\rangle)^{\frac{n+d}{2}}}$$

$$= \frac{\Gamma(\frac{n}{2}+\frac{d}{2})}{\Gamma(\frac{n}{2})(\pi n)^{d/2}\sqrt{\det C}} \; \frac{1}{(1+\frac{1}{n}\langle C^{-1}x, x\rangle)^{\frac{n+d}{2}}} \; .$$

4.18 (a) Thanks to the second freezing lemma, Lemma 4.15, the conditional law of Z given $W = w$ is the law of the r.v.

$$\frac{X+Yw}{\sqrt{1+w^2}} \, ,$$

which is $N(0, 1)$ whatever the value of w, as $X + Yw \sim N(0, 1 + w^2)$.

(b) $Z \sim N(0, 1)$ thanks to Remark 4.14, which entails also that Z and W are independent.

4.19 (a) Let i be an index, $1 \le i \le n$. Let σ be a permutation such that $\sigma_1 = i$. The identity in law $X \sim X_\sigma$ of the vectors implies the identity in law of the marginals, hence $X_1 \sim X_{\sigma_1} = X_i$. Hence, $X_i \sim X_1$ for every $1 \le i \le n$.

If $1 \le i, j \le n, i \ne j$, then, just repeat the previous argument by choosing a permutation σ such that $\sigma_1 = i$, $\sigma_2 = j$ and obtain that $(X_i, X_j) \sim (X_1, X_2)$ for every $1 \le i, j \le n, i \ne j$.

(b) Immediate, as X and X_σ have independent components and the same marginals.

(c) The random vector $X_\sigma := (X_{\sigma_1}, \dots, X_{\sigma_n})$ is the image of $X = (X_1, \dots, X_n)$ under the linear map $A : (x_1, \dots, x_n) \to (x_{\sigma_1}, \dots, x_{\sigma_n})$. Hence (see (2.20)) X_σ has density

$$f_\sigma(x) = \frac{1}{|\det A|} f(A^{-1}x) \; .$$

Now just note that $f(A^{-1}x) = g(|A^{-1}x|) = g(|x|) = f(x)$ and also that $|\det A| = 1$, as the matrix A is all zeros except for exactly one 1 in every row and every column.

(d1) For every bounded measurable function $\phi : (E \times \dots \times E, \mathscr{E} \otimes \dots \otimes \mathscr{E}) \to \mathbb{R}$ we have

$$E[\phi(X_1, \dots, X_n)] = E\big[E[\phi(X_1, \dots, X_n)|Y]\big] \; . \tag{7.56}$$

As the conditional law of $(X_1, \ldots, X_n)$ given $Y = y$ is the product $\bar{\mu}_y \otimes \cdots \otimes \bar{\mu}_y$, hence exchangeable, we have $E[\phi(X_1, \ldots, X_n)|Y = y] = E[\phi(X_{\sigma_1}, \ldots, X_{\sigma_n})|Y = y]$ a.s. for every permutation σ. Hence

$$E[\phi(X_1, \ldots, X_n)] = E[E[\phi(X_1, \ldots, X_n)|Y]] = E[E[\phi(X_{\sigma_1}, \ldots, X_{\sigma_n})|Y]]$$
$$= E[\phi(X_{\sigma_1}, \ldots, X_{\sigma_n})] .$$

(d2) If $X \sim t(n, d, I)$, then $X \sim \frac{\sqrt{n}}{\sqrt{Y}}(Z_1, \ldots, Z_d)$, where $Z_1, \ldots, Z_d$ are independent $N(0, 1)$-distributed and $Y \sim \chi^2(n)$. Therefore, given $Y = y$, the components of X are independent and $N(0, \frac{n}{y})$ distributed, hence exchangeable thanks to (d1).

One can also argue that a $t(n, d, I)$ distribution is exchangeable because its density is of the form (4.32), as seen in Exercise 4.17.

4.20 (a) The law of $S = T + W$ is the law of the sum of two independent exponential r.v.'s of parameters λ and μ respectively. This can be done in many ways: by computing the convolution of their densities as in Proposition 2.18, or also by obtaining the density f_S of S as a marginal of the joint density of (T, S), which we are asked to compute anyway.

Let us follow the last path, taking advantage of the second freezing lemma, Lemma 4.15: we have $S = \Phi(T, W)$, where $\Phi(t, w) = t + w$, hence the conditional law of S given $T = t$ is the law of $t + W$, which has a density with respect to the Lebesgue measure given by $\bar{f}(s; t) = f_W(s - t)$.

Hence the joint density of T and S is

$$f(t, s) = f_T(t)\bar{f}(s; t) = \lambda\mu e^{-\lambda t} e^{-\mu(s-t)}, \qquad t > 0, s > t$$

and the density of S is, for $s > 0$,

$$f_S(s) = \int f(t, s)\, dt = \lambda\mu e^{-\mu s} \int_0^s e^{-(\lambda-\mu)t}\, dt$$

$$= \frac{\lambda\mu}{\lambda - \mu} e^{-\mu s}\left(1 - e^{-(\lambda-\mu)s}\right) = \frac{\lambda\mu}{\lambda - \mu}(e^{-\mu s} - e^{-\lambda s}) .$$

(b) The conditional density of T given $S = s$ is

$$\bar{f}(t; s) = \frac{f(t, s)}{f_S(s)}$$

and, replacing the expressions for f and f_S as computed in (a),

$$\bar{f}(t; s) = \begin{cases} \dfrac{(\lambda - \mu)\, e^{-\mu s}}{e^{-\mu s} - e^{-\lambda s}}\, e^{-(\lambda-\mu)t} & \text{if } 0 \le t \le s \\ 0 & \text{otherwise} . \end{cases}$$

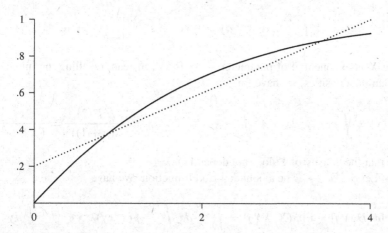

Fig. 7.12 The graph of the conditional expectation (solid) of Exercise 4.20 with the regression line (dots). Note that the regression line here is not satisfactory as, for values of s near 0, it lies above the diagonal, i.e. it gives an expected value of T that is larger than s, whereas we know that $T \leq S$

The conditional expectation of T given $S = s$ is the mean of this conditional density, i.e.

$$E(T \mid S = s) = \frac{(\lambda - \mu)\, e^{-\mu s}}{e^{-\mu s} - e^{-\lambda s}} \int_0^s t\, e^{-(\lambda - \mu)t}\, dt \ .$$

Integrating by parts and with some simplifications

$$E(T \mid S = s) = \frac{(\lambda - \mu)\, e^{-\mu s}}{e^{-\mu s} - e^{-\lambda s}} \left(-\frac{s}{\lambda - \mu}\, e^{-(\lambda - \mu)s} + \frac{1}{(\lambda - \mu)^2}(1 - e^{-(\lambda - \mu)s}) \right)$$

$$= \frac{s}{1 - e^{-(\mu - \lambda)s}} + \frac{1}{\lambda - \mu} \ .$$

In Exercise 2.30 we computed the regression line of T with respect to s, which was

$$s \mapsto \frac{\mu^2}{\lambda^2 + \mu^2}\, s + \frac{\lambda - \mu}{\lambda^2 + \mu^2} \ .$$

Figure 7.12 compares the graphs of these two estimates.

4.21 (a) For $x > 0$ we have

$$f_X(x) = \int_{-\infty}^{+\infty} f(x, y)\, dy = \int_0^{+\infty} \lambda^2 x\, e^{-\lambda x(y+1)}\, dy$$

$$= \lambda e^{-\lambda x} \int_0^{+\infty} \lambda x e^{-\lambda x y} \, dy = -\lambda e^{-\lambda x} \, e^{-\lambda x y} \Big|_{y=0}^{y=+\infty} = \lambda e^{-\lambda x} \, ,$$

hence X is exponential of parameter λ. As for Y, instead, recalling the integral of the Gamma densities, we have for $y > 0$

$$f_Y(y) = \int f(x, y) \, dx = \lambda^2 \int_0^{+\infty} x e^{-\lambda x (y+1)} \, dx = \frac{\lambda^2 \Gamma(2)}{(\lambda(y+1))^2} = \frac{1}{(y+1)^2} \, .$$

Note that the density of Y does not depend on λ.

(b) Let $\phi : \mathbb{R}^2 \to \mathbb{R}$ be a bounded Borel function. We have

$$E[\phi(U, V)] = E[\phi(X, XY)] = \int_0^{+\infty} dx \int_0^{+\infty} \phi(x, xy) \lambda^2 x \, e^{-\lambda x (y+1)} \, dy \, .$$

With the change of variable $z = xy$ in the inner integral, i.e. $x \, dy = dz$, we have

$$\cdots = \int_0^{+\infty} dx \int_0^{+\infty} \phi(x, z) \lambda^2 e^{-\lambda(z+x)} \, dz \, .$$

Hence the joint density of (U, V) is, for $u > 0$, $v > 0$,

$$g(u, v) = \lambda^2 e^{-\lambda(u+v)} = \lambda e^{-\lambda u} \cdot \lambda e^{-\lambda v} \, ,$$

so that U and V are independent and both exponential with parameter λ.

(c) The conditional density of X given $Y = y$ is, for $x > 0$,

$$\bar{f}(x; y) = \frac{f(x, y)}{f_Y(y)} = \lambda^2 x (y+1)^2 e^{-\lambda x (y+1)} \, ,$$

which is a Gamma$(2, \lambda(y + 1))$ density (as a function of x, of course). The conditional expectation $E(X | Y = y)$ is therefore the mean of this density, i.e.

$$E(X | Y = y) = \frac{2}{\lambda(y + 1)} \, .$$

Hence $E(X | Y) = \frac{2}{\lambda(Y+1)}$ and the requested squared L^2 distance is

$$E\big[(X - E(X | Y))^2 \big] \, .$$

By (4.6) this is equal to $E(X^2) - E[E(X | Y)^2]$. Now, recalling the expression of the moments of the exponential distributions, we have

$$E(X^2) = E(X)^2 + \text{Var}(X) = \frac{2}{\lambda^2}$$

and

$$E[E(X|Y)^2] = E\left(\frac{4}{\lambda^2(Y+1)^2}\right) = \frac{4}{\lambda^2}\int_0^{+\infty}\frac{1}{(y+1)^4}\,dy = \frac{4}{3\lambda^2},$$

from which the requested squared L^2 distance is equal to $\frac{2}{3\lambda^2}$.

• Note that (d) above states that, in the sense of L^2, the best approximation of X by a function of Y is $\frac{2}{\lambda(Y+1)}$. We might think of comparing this approximation with the regression line of X with respect to Y, which is the best approximation by an affine-linear function of Y. However we have

$$E(Y^2) = \int_0^{+\infty}\frac{y^2}{(1+y)^2}\,dy = +\infty.$$

Hence Y is not square integrable (not even integrable), so that the best approximation in L^2 of X by an affine-linear function of Y can only be a constant and this constant must be $E(X) = \frac{1}{\lambda}$, see the remark following Example 2.24 p.68.

4.22 (a) We know that $Z = X + Y \sim \text{Gamma}(\alpha + \beta, \lambda)$.

(b) As X and Y are independent, their joint density is

$$f(x,y) = \frac{\lambda^{\alpha+\beta}}{\Gamma(\alpha)\Gamma(\beta)}x^{\alpha-1}y^{\beta-1}e^{-\lambda(x+y)}$$

if $x, y > 0$ and $f(x,y) = 0$ otherwise. For every bounded Borel function $\phi : \mathbb{R}^2 \to \mathbb{R}$ we have

$$E[\phi(X, X+Y)]$$

$$= \frac{\lambda^{\alpha+\beta}}{\Gamma(\alpha)\Gamma(\beta)}\int_0^{+\infty}dx\int_0^{+\infty}\phi(x, x+y)\,x^{\alpha-1}y^{\beta-1}e^{-\lambda(x+y)}\,dy.$$

With the change of variable $z = x + y$, $dz = dy$ in the inner integral we obtain

$$\cdots = \frac{\lambda^{\alpha+\beta}}{\Gamma(\alpha)\Gamma(\beta)}\int_0^{+\infty}dx\int_x^{+\infty}\phi(x, z)\,x^{\alpha-1}(z-x)^{\beta-1}e^{-\lambda z}\,dz,$$

so that the density of $(X, X + Y)$ is

$$g(x, z) = \begin{cases} \dfrac{\lambda^{\alpha+\beta}}{\Gamma(\alpha)\Gamma(\beta)}x^{\alpha-1}(z-x)^{\beta-1}e^{-\lambda z} & \text{if } 0 < x < z \\ 0 & \text{otherwise}. \end{cases}$$

(c) Denoting by g_{X+Y} the density of $X + Y$, which we know to be Gamma($\alpha + \beta, \lambda$), the requested conditional density is

$$\overline{g}(x; z) = \frac{g(x, z)}{g_{X+Y}(z)} \, .$$

It vanishes unless $0 \le x \le z$. For x in this range we have

$$\overline{g}(x; z) = \frac{\frac{\lambda^{\alpha+\beta}}{\Gamma(\alpha)\Gamma(\beta)} x^{\alpha-1}(z-x)^{\beta-1}e^{-\lambda z}}{\frac{\lambda^{\alpha+\beta}}{\Gamma(\alpha+\beta)} z^{\alpha+\beta-1}e^{-\lambda z}} = \frac{\Gamma(\alpha+\beta)}{\Gamma(\alpha)\Gamma(\beta)} \frac{1}{z}(\tfrac{x}{z})^{\alpha-1}(1 - \tfrac{x}{z})^{\beta-1} \, .$$

(d) The conditional expectation $\mathrm{E}(X\,|\,X + Y = z)$ is the mean of this density. With the change of variable $t = \frac{x}{z}, dx = z\,dt$,

$$\int x\,\overline{g}(x; z)\,dx = \frac{\Gamma(\alpha+\beta)}{\Gamma(\alpha)\Gamma(\beta)} \int_0^z (\tfrac{x}{z})^{\alpha}(1 - \tfrac{x}{z})^{\beta-1}\,dx$$

$$= \frac{\Gamma(\alpha+\beta)}{\Gamma(\alpha)\Gamma(\beta)} z \int_0^1 t^{\alpha}(1-t)^{\beta-1}\,dt \, .$$

Recalling the expression of the Beta laws, the last integral is equal to $\frac{\Gamma(\alpha+1)\Gamma(\beta)}{\Gamma(\alpha+\beta+1)}$, hence, with the simplification formula of the Gamma function, the requested conditional expectation is

$$\frac{\Gamma(\alpha+\beta)}{\Gamma(\alpha)\Gamma(\beta)} \frac{\Gamma(\alpha+1)\Gamma(\beta)}{\Gamma(\alpha+\beta+1)} z = \frac{\alpha}{\alpha+\beta} z \, .$$

We know that the conditional expectation given $X+Y = z$ is the best approximation (in the sense of the L^2 distance) of X as a function of $X + Y$. The regression line is instead the best approximation of X as an affine-linear function of $X + Y = z$. As the conditional expectation in this case is itself an affine-linear function of z, the two functions necessarily coincide.

• Note that the results of (c) and (d) *do not* depend on the value of λ.

4.23 (a) We recognize that the argument of the exponential is, but for the factor $\frac{1}{2}$, the quadratic form associated to the matrix

$$M = \frac{1}{1 - r^2} \begin{pmatrix} 1 & -r \\ -r & 1 \end{pmatrix} \, .$$

M is strictly positive definite (both its trace and determinant are > 0, hence both eigenvalues are positive), hence f is a Gaussian density, centered and with covariance matrix

$$C = M^{-1} = \begin{pmatrix} 1 & r \\ r & 1 \end{pmatrix}.$$

Therefore X and Y are both $N(0, 1)$-distributed and $\text{Cov}(X, Y) = r$.

(b) As X and Y are centered and $\text{Cov}(X, Y) = r$, by (4.24),

$$E(X|Y = y) = \frac{\text{Cov}(X, Y)}{\text{Var}(Y)} y = ry.$$

Also the pair $X, X + Y$ is jointly Gaussian and again formula (4.24) gives

$$E(X|X + Y = z) = \frac{\text{Cov}(X, X + Y)}{\text{Var}(X + Y)} z = \frac{1 + r}{2(1 + r)} z = \frac{1}{2} z.$$

Note that $E(X|X + Y = z)$ does not depend on r.

4.24 (a) The pair (X, Y) has a density with respect to the Lebesgue measure given by

$$f(x, y) = f_X(x)\overline{f}(y; x) = \frac{1}{\sqrt{2\pi}} e^{-\frac{1}{2} x^2} \frac{1}{\sqrt{2\pi}} e^{-\frac{1}{2} (y - \frac{1}{2} x)^2}$$

$$= \frac{1}{2\pi} e^{-\frac{1}{2} (x^2 + y^2 - xy + \frac{1}{4} x^2)} = \frac{1}{2\pi} e^{-\frac{1}{2} (\frac{5}{4} x^2 + y^2 - xy)}.$$

At the exponential we note the quadratic form associated to the matrix

$$C^{-1} = \begin{pmatrix} \frac{5}{4} & -\frac{1}{2} \\ -\frac{1}{2} & 1 \end{pmatrix}.$$

We deduce that the pair (X, Y) has an $N(0, C)$ distribution with

$$C = \begin{pmatrix} 1 & \frac{1}{2} \\ \frac{1}{2} & \frac{5}{4} \end{pmatrix}.$$

(b) The answer is no and there is no need for computations: if the pair (X, Y) was Gaussian the mean of the conditional law would be as in (4.24) and necessarily an *affine-linear* function of the conditioning r.v.

(c) Again the answer is no: as noted in Remark 4.20(c), the variance of the conditional distributions of jointly Gaussian r.v.'s cannot depend on the value of the conditioning r.v.

5.1 As $(X_n)_n$ is a supermartingale, $U := X_m - E(X_n|\mathcal{F}_m) \geq 0$ a.s., for $n > m$. But $E(U) = E(X_m) - E[E(X_n|\mathcal{F}_m)] = E(X_m) - E(X_n) = 0$. The positive r.v. U having expectation equal to 0 is $= 0$ a.s., so that $X_m = E(X_n|\mathcal{F}_m)$ a.s. and $(X_n)_n$ is a martingale.

5.2 If $m < n$, as $\{M_m = 0\}$ is $\mathcal{F}_m$-measurable and $M_m 1_{\{M_m=0\}} = 0$, we have

$$E(M_n 1_{\{M_m=0\}}) = E(M_m 1_{\{M_m=0\}}) = 0 \,.$$

As $M_n \geq 0$, necessarily $M_n = 0$ a.s. on $\{M_m = 0\}$, i.e. $\{M_m = 0\} \subset \{M_n = 0\}$ a.s.

• Note that this is just Exercise 4.2 from another point of view.

5.3 We must prove that, if $m \leq n$,

$$E(M_n N_n 1_A) = E(M_m N_m 1_A) \tag{7.57}$$

for every $A \in \mathcal{H}_m$ or at least for every A in a subclass $\mathcal{C}_m \subset \mathcal{H}_m$ that generates $\mathcal{H}_m$, contains Ω and is stable with respect to finite intersections (Remark 4.3). Let $\mathcal{C}_m$ be the class of the events of the form $A_1 \cap A_2$ with $A_1 \in \mathcal{F}_m$, $A_2 \in \mathcal{G}_m$. $\mathcal{C}_m$ is stable with respect to finite intersections and contains both $\mathcal{F}_m$ (choosing $A_2 = \Omega$) and $\mathcal{G}_m$ (with $A_1 = \Omega$). As the r.v.'s $M_n 1_{A_1}$ and $N_n 1_{A_2}$ are independent (the first one is $\mathcal{F}_n$-measurable whereas the second one is $\mathcal{G}_n$-measurable) we have

$$E(M_n N_n 1_{A_1 \cap A_2}) = E(M_n 1_{A_1} N_n 1_{A_2}) = E(M_n 1_{A_1}) E(N_n 1_{A_2})$$

$$= E(M_m 1_{A_1}) E(N_m 1_{A_2}) = E(M_m 1_{A_1} N_m 1_{A_2}) = E(M_m N_m 1_{A_1 \cap A_2}) \,,$$

hence (7.57) is satisfied for every $A \in \mathcal{C}_m$ and therefore for every $A \in \mathcal{H}_m$.

5.4 (a) Z_n is $\mathcal{F}_{n-1}$-measurable whereas X_n is independent of $\mathcal{F}_{n-1}$, hence X_n and Z_n are independent and $Z_n^2 X_n^2$ is integrable, being the product of integrable independent r.v.'s. Hence Y_n is square integrable for every n. Moreover,

$$E(Y_{n+1}|\mathcal{F}_n) = E(Y_n + Z_{n+1} X_{n+1}|\mathcal{F}_n) = Y_n + Z_{n+1} E(X_{n+1}|\mathcal{F}_n)$$

$$= Y_n + Z_{n+1} E(X_{n+1}) = Y_n \,,$$

where we have taken advantage of the fact that Y_n and Z_{n+1} are $\mathcal{F}_n$-measurable, whereas X_{n+1} is independent of $\mathcal{F}_n$.

(b) As Z_k and X_k are independent, $E(Z_k X_k) = E(Z_k) E(X_k) = 0$ hence $E(Y_n) = 0$. Moreover,

$$E(Y_n^2) = E\Big[\Big(\sum_{k=1}^n Z_k X_k\Big)^2\Big] = E\Big(\sum_{k,h=1}^n Z_k X_k Z_h X_h\Big) = \sum_{k,h=1}^n E(Z_k X_k Z_h X_h) \,.$$

In the previous sum all terms with $h \neq k$ vanish: actually, let us assume $k > h$, then the r.v. $Z_k X_h Z_h$ is $\mathcal{F}_{k-1}$-measurable, whereas X_k is independent of $\mathcal{F}_{k-1}$. Hence

$$E(Z_k X_k Z_h X_h) = E(X_k)E(Z_k Z_h X_h) = 0 \,.$$

Therefore, as $E(Z_k^2 X_k^2) = E(Z_k^2)E(X_k^2) = \sigma^2 E(Z_k^2)$,

$$E(Y_n^2) = \sum_{k=1}^{n} E(Z_k^2 X_k^2) = \sigma^2 \sum_{k=1}^{n} E(Z_k^2) \,. \tag{7.58}$$

(c) The compensator $(A_n)_n$ of $(M_n^2)_n$ is given by the condition $A_0 = 0$ and the relations $A_{n+1} = A_n + E(M_{n+1}^2 - M_n^2 | \mathcal{F}_n)$. Now

$$E(M_{n+1}^2 - M_n^2 | \mathcal{F}_n) = E[(M_n + X_{n+1})^2 - M_n^2 | \mathcal{F}_n]$$

$$E(M_n^2 + 2M_n X_{n+1} + X_{n+1}^2 - M_n^2 | \mathcal{F}_n) = E(2M_n X_{n+1} + X_{n+1}^2 | \mathcal{F}_n)$$

$$= 2M_n E(X_{n+1} | \mathcal{F}_n) + E(X_{n+1}^2 | \mathcal{F}_n) \,.$$

As X_{n+1} is independent of $\mathcal{F}_n$,

$$E(X_{n+1} | \mathcal{F}_n) = E(X_{n+1}) = 0 \qquad \text{a.s.}$$

$$E(X_{n+1}^2 | \mathcal{F}_n) = E(X_{n+1}^2) = \sigma^2 \qquad \text{a.s.} \,,$$

hence $A_{n+1} = A_n + \sigma^2$ and, with the condition $A_0 = 0$, we have $A_n = n\sigma^2$. In order to compute the compensator of $(Y_n^2)_n$, $(\widetilde{A}_n)_n$ say, just repeat the same argument:

$$E(Y_{n+1}^2 - Y_n^2 | \mathcal{F}_n) = E[(Y_n + Z_{n+1} X_{n+1})^2 - Y_n^2 | \mathcal{F}_n]$$

$$= E(2Y_n Z_{n+1} X_{n+1} + Z_{n+1}^2 X_{n+1}^2 | \mathcal{F}_n)$$

$$= 2Y_n Z_{n+1} E(X_{n+1} | \mathcal{F}_n) + Z_{n+1}^2 E(X_{n+1}^2 | \mathcal{F}_n) = \sigma^2 Z_{n+1}^2 \,.$$

Therefore

$$\widetilde{A}_n = \sigma^2 \sum_{k=1}^{n} Z_k^2 \,.$$

5.5 (a) Let $m \leq n$: we have $E(M_n M_m) = E[E(M_n M_m | \mathcal{F}_m)] = E[M_m E(M_n | \mathcal{F}_m)] = E(M_m^2)$ so that

$$E[(M_n - M_m)^2] = E(M_n^2) + E(M_m^2) - 2E(M_n M_m) = E(M_n^2) - E(M_m^2) \,.$$

(b) Let us assume $M_0 = 0$ for simplicity: actually the martingales $(M_n)_n$ and $(M_n - M_0)_n$ have the same associated increasing process. Note that the suggested associated increasing process vanishes at 0 and is obviously predictable, so that it is sufficient to prove that $Z_n = M_n^2 - E(M_n^2)$ is a martingale (by the uniqueness of the associated increasing process). We have, for $m \leq n$,

$$
\begin{aligned}
E(M_n^2 | \mathscr{F}_m) &= E[(M_n - M_m + M_m)^2 | \mathscr{F}_m] \\
&= E[(M_n - M_m)^2 + 2(M_n - M_m)M_m + M_m^2 | \mathscr{F}_m] .
\end{aligned}
\tag{7.59}
$$

We have $E[(M_n - M_m)M_m | \mathscr{F}_m] = M_m E(M_n - M_m | \mathscr{F}_m) = 0$ and, as M has independent increments,

$$
E[(M_n - M_m)^2 | \mathscr{F}_m] = E[(M_n - M_m)^2] = E(M_n^2 - M_m^2) .
$$

Therefore, going back to (7.59), $E(M_n^2 | \mathscr{F}_m) = M_m^2 + E(M_n^2 - M_m^2)$ and

$$
E(Z_n | \mathscr{F}_m) = M_m^2 + E(M_n^2 - M_m^2) - E(M_n^2) = M_m^2 - E(M_m^2) = Z_m .
$$

(c) Let $m \leq n$. As M is a Gaussian family, $M_n - M_m$ is independent of $(M_0, \ldots, M_m)$ if and only if $M_n - M_m$ is uncorrelated with respect to M_k for every $k = 0, 1, \ldots, m$. But, by the martingale property,

$$
E[(M_n - M_m)M_k] = E\big[E[(M_n - M_m)M_k | \mathscr{G}_m]\big] = E\big[M_k \underbrace{E(M_n - M_m | \mathscr{G}_m)}_{=0}\big] ,
$$

so that $\mathrm{Cov}(M_n - M_m, M_k) = E[(M_n - M_m)M_k] = 0$.

5.6 (a) Let us denote by $(V_n)_{n \geq 0}$ the associated increasing process of $(S_n)_n$. As Y_{n+1} is independent of $\mathscr{F}_n$, $E(Y_{n+1} | \mathscr{F}_n) = E(Y_{n+1}) = 0$ and, recalling the definition of compensator in (5.3),

$$
\begin{aligned}
V_{n+1} - V_n &= E(S_{n+1}^2 - S_n^2 | \mathscr{F}_n) \\
&= E\big[(S_n + Y_{n+1})^2 - S_n^2 | \mathscr{F}_n\big] = E(Y_{n+1}^2 + 2Y_{n+1}S_n | \mathscr{F}_n) \\
&= E(Y_{n+1}^2 | \mathscr{F}_n) + 2S_n E(Y_{n+1} | \mathscr{F}_n) = E(Y_{n+1}^2) = 1 .
\end{aligned}
$$

Therefore $V_0 = 0$ and $V_n = n$. Note that this is a particular case of Exercise 5.5(b), as $(S_n)_n$ is a martingale with independent increments.

(b) We have

$$
E(M_{n+1} - M_n | \mathscr{F}_n) = E[\mathrm{sign}(S_n)Y_{n+1} | \mathscr{F}_n] = \mathrm{sign}(S_n)E(Y_{n+1} | \mathscr{F}_n) = 0 \qquad \text{a.s.}
$$

therefore $(M_n)_n$ is a martingale. It is obviously square integrable and its associated increasing process, $(A_n)_n$ say, is obtained as above:

$$A_{n+1} - A_n = E(M_{n+1}^2 - M_n^2 | \mathscr{F}_n)$$

$$= E\big(\text{sign}(S_n)^2 Y_{n+1}^2 + 2M_n \text{sign}(S_n) Y_{n+1} | \mathscr{F}_n\big)$$

$$= \text{sign}(S_n)^2 \underbrace{E(Y_{n+1}^2 | \mathscr{F}_n)}_{=1} + 2M_n \text{sign}(S_n) \underbrace{E(Y_{n+1} | \mathscr{F}_n)}_{=0} = \text{sign}(S_n)^2 = 1_{\{S_n \neq 0\}}$$

from which

$$A_n = \sum_{k=1}^{n-1} 1_{\{S_k \neq 0\}} .$$

Note that this is a particular case of Exercise 5.4(b).

(c1) On $\{S_n > 0\}$ we have $S_{n+1} \geq 0$, as $S_{n+1} \geq S_n - 1 \geq 0$; therefore $|S_{n+1}| - |S_n| = S_{n+1} - S_n = Y_{n+1}$ and

$$E[(|S_{n+1}| - |S_n|)1_{\{S_n > 0\}} | \mathscr{F}_n] = 1_{\{S_n > 0\}} E(Y_{n+1} | \mathscr{F}_n) = 0 .$$

The other relation is proved in the same way. Therefore

$$E\big(|S_{n+1}| - |S_n| \, \big| \, \mathscr{F}_n\big) = E[(|S_{n+1}| - |S_n|)1_{\{S_n=0\}} | \mathscr{F}_n]$$

$$= 1_{\{S_n=0\}} E\big(|Y_{n+1}| \, \big| \, \mathscr{F}_n\big) = 1_{\{S_n=0\}}$$

and

$$\widetilde{A}_{n+1} = \widetilde{A}_n + E\big(|S_{n+1}| - |S_n| \, \big| \, \mathscr{F}_n\big) = \widetilde{A}_n + 1_{\{S_n=0\}} .$$

Hence

$$\widetilde{A}_n = \sum_{k=0}^{n-1} 1_{\{S_k=0\}} .$$

(c2) We have

$$N_{n+1} - N_n = |S_{n+1}| - |S_n| - (\widetilde{A}_{n+1} - \widetilde{A}_n) = |S_{n+1}| - |S_n| - 1_{\{S_n=0\}} .$$

As

$$(|S_{n+1}| - |S_n|)1_{\{S_n > 0\}} = Y_{n+1} 1_{\{S_n > 0\}} ,$$

$$(|S_{n+1}| - |S_n|)1_{\{S_n < 0\}} = -Y_{n+1} 1_{\{S_n < 0\}} ,$$

$$(|S_{n+1}| - |S_n|)1_{\{S_n=0\}} = |Y_{n+1}| 1_{\{S_n=0\}} = 1_{\{S_n=0\}}$$

we have

$$\cdots = Y_{n+1}1_{\{S_n>0\}} - Y_{n+1}1_{\{S_n<0\}} = \text{sign}(S_n)Y_{n+1} = M_{n+1} - M_n \ .$$

Thus, as $M_0 = N_0 = 0$, $M_n = N_n = |S_n| - \sum_{k=0}^{n-1}1_{\{S_k=0\}}$ and M_n is $\sigma(|S_1|,\ldots,|S_k|)$-measurable.

Finally, recall that if $(M_n)_n$ is a martingale with respect to a given filtration, then it is also a martingale with respect to any smaller filtration (provided it is adapted to it) and it is immediate that $\mathcal{G}_n \subset \mathcal{F}_n$.

5.7 (a) As the sequence $(Z_n)_n$ is itself increasing, $E(Z_{n+1}|\mathcal{F}_n) \geq E(Z_n|\mathcal{F}_n) = Z_n$.
  (b) We have $A_0 = 0$ and

$$A_{n+1} = A_n + E(Z_{n+1}|\mathcal{F}_n) - Z_n \ . \tag{7.60}$$

Now

$$Z_{n+1}' = Z_n 1_{\{\xi_{n+1}\leq Z_n\}} + \xi_{n+1}1_{\{\xi_{n+1}>Z_n\}}$$

and by the freezing lemma, Lemma 4.11,

$$E(Z_{n+1}|\mathcal{F}_n) = E\big(Z_n 1_{\{\xi_{n+1}\leq Z_n\}} + \xi_{n+1}1_{\{\xi_{n+1}>Z_n\}}|\mathcal{F}_n\big) = \Phi(Z_n) \ ,$$

where

$$\Phi(z) = E\big(z 1_{\{\xi_{n+1}\leq z\}} + \xi_{n+1}1_{\{\xi_{n+1}>z\}}\big) \ ,$$

i.e.

$$\Phi(z) = z(1 - e^{-\lambda z}) + \lambda \int_z^{+\infty} y e^{-\lambda y}\,dy$$

$$= z(1 - e^{-\lambda z}) + \Big(-y e^{-\lambda y}\Big|_z^{+\infty} + \int_z^{+\infty} e^{-\lambda y}\,dy\Big)$$

$$= z(1 - e^{-\lambda z}) + z e^{-\lambda z} + \frac{1}{\lambda} e^{-\lambda z}$$

$$= z + \frac{1}{\lambda} e^{-\lambda z} \ ,$$

hence

$$E(Z_{n+1}|\mathcal{F}_n) - Z_n = \frac{1}{\lambda} e^{-\lambda Z_n}$$

and (7.60) becomes

$$A_{n+1} = A_n + \frac{1}{\lambda} e^{-\lambda Z_n} ,$$

so that

$$A_n = \frac{1}{\lambda} \sum_{k=0}^{n-1} e^{-\lambda Z_k} .$$

• Note that this gives the relation $\mathrm{E}(Z_n) = \mathrm{E}(A_n) = \frac{1}{\lambda} \sum_{k=0}^{n-1} \mathrm{E}(e^{-\lambda Z_k})$. The value of $\mathrm{E}(e^{-\lambda Z_k})$ was computed in (2.97), where we found

$$\mathrm{E}(e^{-\lambda Z_k}) = L_k(-\lambda) = k! \frac{\Gamma(2)}{\Gamma(k+2)} = \frac{1}{k+1} ,$$

so that we find again the value of the expectation $\mathrm{E}(Z_n)$ as in Exercise 2.48.

5.8 (a) The exponential function being convex we have, by Jensen's inequality,

$$\mathrm{E}(e^{M_n} | \mathscr{F}_{n-1}) \geq e^{\mathrm{E}(M_n | \mathscr{F}_{n-1})} = e^{M_{n-1}} ,$$

which implies (5.27).

(b) Recalling how Doob's decomposition was derived in Sect. 5.3, let us recursively define $A_0 = 0$ and

$$A_n = A_{n-1} + \log \mathrm{E}(e^{M_n} | \mathscr{F}_{n-1}) - M_{n-1} . \tag{7.61}$$

This defines an increasing predictable process and taking the exponentials we find

$$e^{A_n} = e^{A_{n-1}} \mathrm{E}(e^{M_n} | \mathscr{F}_{n-1}) e^{-M_{n-1}} ,$$

i.e., A_n being $\mathscr{F}_{n-1}$-measurable,

$$\mathrm{E}(e^{M_n - A_n} | \mathscr{F}_{n-1}) = e^{M_{n-1} - A_{n-1}} ,$$

thus proving (b).

(c1) We have

$$\begin{aligned}
\log \mathrm{E}(e^{M_n} | \mathscr{F}_{n-1}) &= \log \mathrm{E}(e^{W_1 + \cdots + W_n} | \mathscr{F}_{n-1}) \\
&= \log \left(e^{W_1 + \cdots + W_{n-1}} \mathrm{E}(e^{W_n} | \mathscr{F}_{n-1}) \right) \\
&= W_1 + \ldots + W_{n-1} + \log \mathrm{E}(e^{W_n} | \mathscr{F}_{n-1}) = M_{n-1} + \log L(1) ,
\end{aligned}$$

where we denote by L the Làplace transform of the W_k's (which is finite at 1 by hypothesis) and thanks to (7.61),

$$A_n = n \log L(1) .$$

(c2) Now we have

$$\log \mathrm{E}(e^{M_n} | \mathcal{F}_{n-1}) = \log \mathrm{E}\Big[\exp \Big(\sum_{k=1}^{n} Z_k W_k \Big) \Big| \mathcal{F}_{n-1} \Big]$$

$$= \sum_{k=1}^{n-1} Z_k W_k + \log \mathrm{E}(e^{Z_n W_n} | \mathcal{F}_{n-1}) = M_{n-1} + \log \mathrm{E}(e^{Z_n W_n} | \mathcal{F}_{n-1}) .$$

As W_n is independent of $\mathcal{F}_{n-1}$ and Z_n is $\mathcal{F}_{n-1}$-measurable, by the freezing lemma,

$$\mathrm{E}(e^{Z_n W_n} | \mathcal{F}_{n-1}) = \varPhi(Z_n) ,$$

where $\varPhi(z) = \mathrm{E}(e^{z W_n}) = L(z)$ and (7.61) gives $A_n = A_{n-1} + \log L(Z_n)$, i.e.

$$A_n = \sum_{k=1}^{n} \log L(Z_k) .$$

In particular $n \mapsto \exp \Big(\sum_{k=1}^{n} Z_k W_k - \sum_{k=1}^{n} \log L(Z_k) \Big)$ is an $(\mathcal{F}_n)_n$-martingale.

5.9 We already know that X_τ is $\mathcal{F}_\tau$-measurable (see the end of Sect. 5.4) hence we must just prove that, for every $A \in \mathcal{F}_\tau$, $\mathrm{E}(X 1_A) = \mathrm{E}(X_\tau 1_A)$. As $A \cap \{\tau = n\} \in \mathcal{F}_n$ we have $\mathrm{E}(X 1_{A \cap \{\tau=n\}}) = \mathrm{E}(X_n 1_{A \cap \{\tau=n\}})$ and, as τ is finite,

$$\mathrm{E}(X 1_A) = \sum_{n=0}^{\infty} \mathrm{E}(X 1_{A \cap \{\tau=n\}}) = \sum_{n=0}^{\infty} \mathrm{E}(X_n 1_{A \cap \{\tau=n\}})$$

$$= \sum_{n=0}^{\infty} \mathrm{E}(X_\tau 1_{A \cap \{\tau=n\}}) = \mathrm{E}(X_\tau 1_A) .$$

5.10 If X is a martingale the claimed property is a consequence of the stopping theorem (Corollary 5.11) applied to the stopping times $\tau_1 = 0$ and $\tau_2 = \tau$.

Conversely, in order to prove the martingale property we must prove that, if $n > m$,

$$\mathrm{E}(X_n 1_A) = \mathrm{E}(X_m 1_A) \qquad \text{for every } A \in \mathcal{F}_m . \tag{7.62}$$

The idea is to find two bounded stopping times τ_1, τ_2 such that the relation $E(X_{\tau_1}) = E(X_{\tau_2})$ implies (7.62). Let us choose, for $A \in \mathscr{F}_m$,

$$\tau_1(\omega) = \begin{cases} m & \text{if } \omega \in A \\ n & \text{if } \omega \in A^c \end{cases}$$

and $\tau_2 \equiv n$; τ_1 is a stopping time: indeed

$$\{\tau_1 \leq k\} = \begin{cases} \emptyset & \text{if } k < m \\ A & \text{if } m \leq k < n \\ \Omega & \text{if } k \geq n, \end{cases}$$

so that, in any case, $\{\tau_1 \leq k\} \in \mathscr{F}_k$. Now $X_{\tau_1} = X_m 1_A + X_n 1_{A^c}$ and the relation $E(X_{\tau_1}) = E(X_n)$ gives

$$E(X_m 1_A) + E(X_n 1_{A^c}) = E(X_{\tau_1}) = E(X_n) = E(X_n 1_A) + E(X_n 1_{A^c}),$$

and by subtraction we obtain (7.62).

5.11 (a) We must prove that, for $m \leq n$,

$$E(M_n 1_{\tilde{A}}) = E(M_m 1_{\tilde{A}}), \tag{7.63}$$

for every $\tilde{A} \in \widetilde{\mathscr{F}}_m$ or, at least for every $\tilde{A}$ in a class $\mathscr{C} \subset \widetilde{\mathscr{F}}_m$ of events that is stable with respect to finite intersections and generating $\widetilde{\mathscr{F}}_m$. Very much like Exercise 5.3, a suitable class $\mathscr{C}$ is that of the events of the form

$$\tilde{A} = A \cap B, \qquad A \in \mathscr{F}_m, \ B \in \mathscr{G}.$$

We have, $\mathscr{F}_n$ and $\mathscr{G}$ being independent,

$$E(M_n 1_{A \cap B}) = E(M_n 1_A 1_B) = E(M_n 1_A)E(1_B)$$

$$= E(M_m 1_A)E(1_B) = E(M_m 1_{A \cap B}).$$

which proves (7.63) for every $\tilde{A} \in \mathscr{C}$.

(b) Let $\widetilde{\mathscr{F}}_n = \sigma(\mathscr{F}_n, \sigma(\tau))$. Thanks to (a) $(M_n)_n$ is also a martingale with respect to $(\widetilde{\mathscr{F}}_n)_n$. Moreover we have $\{\tau \leq n\} \in \sigma(\tau) \subset \widetilde{\mathscr{F}}_n$ for every n, so that τ is a $(\widetilde{\mathscr{F}}_n)_n$-stopping time. Hence the stopped process $(M_{n \wedge \tau})_n$ is an $(\widetilde{\mathscr{F}}_n)_n$-martingale.

5.12 (a) By the Law of Large Numbers we have a.s.

$$\frac{1}{n} S_n = \frac{1}{n}(Y_1 + \cdots + Y_n) \xrightarrow[n \to \infty]{} E(Y_1) = p - q < 0.$$

Hence, for every δ such that $p - q < \delta < 0$, there exists, a.s., an n_0 such that $\frac{1}{n} S_n < \delta$ for $n \geq n_0$. It follows that $S_n \to_{n \to \infty} -\infty$ a.s.

(b) Note that $Z_n = (\frac{q}{p})^{Y_1} \dots (\frac{q}{p})^{Y_n}$, that the r.v.'s $(\frac{q}{p})^{Y_k}$ are independent and that

$$E[(\tfrac{q}{p})^{Y_k}] = \frac{q}{p} P(Y_k = 1) + \left(\frac{q}{p}\right)^{-1} P(Y_k = -1) = q + p = 1 ,$$

so that $(Z_n)_n$ are the cumulative products of independent r.v.'s having expectation $= 1$ and the martingale property follows from Example 5.2(b).

(c) As $n \wedge \tau$ is a bounded stopping time, $E(Z_{n \wedge \tau}) = E(Z_1) = 1$ by the stopping theorem, Theorem 5.10. Thanks to (a) $\tau < +\infty$ a.s., hence $\lim_{n \to \infty} Z_{n \wedge \tau} = Z_\tau$ a.s. As $-a \leq Z_{n \wedge \tau} \leq b$, we can apply Lebesgue's Theorem, which gives $E(Z_\tau) = \lim_{n \to \infty} E(Z_{n \wedge \tau}) = 1$.

(d1) As Z_τ can take only the values $-a$ or b, we have

$$1 = E(Z_\tau) = E[(\tfrac{q}{p})^{S_\tau}] = \left(\frac{q}{p}\right)^b P(S_\tau = b) + \left(\frac{q}{p}\right)^{-a} P(S_\tau = -a) .$$

As $P(S_\tau = -a) = 1 - P(S_\tau = b)$, the previous relation gives

$$1 - \left(\frac{q}{p}\right)^{-a} = P(S_\tau = b)\left(\left(\frac{q}{p}\right)^b - \left(\frac{q}{p}\right)^{-a}\right) ,$$

i.e.

$$P(S_\tau = b) = \frac{1 - (\tfrac{q}{p})^{-a}}{(\tfrac{q}{p})^b - (\tfrac{q}{p})^{-a}} ,$$

and, as $\frac{q}{p} > 1$

$$\lim_{a \to +\infty} P(S_{\tau - a, b} = b) = \lim_{a \to +\infty} \frac{1 - (\tfrac{q}{p})^{-a}}{(\tfrac{q}{p})^b - (\tfrac{q}{p})^{-a}} = \left(\frac{p}{q}\right)^b . \tag{7.64}$$

(d2) If $\tau_b(\omega) < n$, as the numerical sequence $(S_n(\omega))_n$ cannot reach $-n$ in less than n steps, necessarily $S_{\tau - n, b} = b$, hence $\{\tau_b < n\} \subset \{S_{\tau - n, b} = b\}$. Therefore by (7.64)

$$P(\tau_b < +\infty) = \lim_{n \to \infty} P(\tau_b < n) \leq \lim_{n \to \infty} P(S_{\tau - n, b} = b) = \left(\frac{p}{q}\right)^b .$$

On the other hand, thanks to the obvious inclusion $\{\tau_b < +\infty\} \supset \{S_{\tau - a, b} = b\}$ for every a, from (7.64) we have that the $=$ sign holds.

(d3) Obviously we have, for every n,

$$P(\tau_{-a} < +\infty) \geq P(S_{\tau_{-a,n}} = a)$$

and therefore

$$P(\tau_{-a} < +\infty) \geq \lim_{n \to \infty} P(S_{\tau_{-a,n}} = -a) = \lim_{n \to \infty} 1 - P(S_{\tau_{-a,n}} = n)$$

$$= \lim_{n \to \infty} 1 - \frac{1 - (\frac{q}{p})^{-a}}{(\frac{q}{p})^n - (\frac{q}{p})^{-a}} = \lim_{n \to \infty} \frac{(\frac{q}{p})^n - 1}{(\frac{q}{p})^n - (\frac{q}{p})^{-a}} = 1 .$$

• This exercise gives some information concerning the random walk $(S_n)_n$: it visits a.s. every negative integer but visits the strictly positive integers with a probability that is strictly smaller than 1. This is of course hardly surprising, given its asymmetry. In particular, for $b = 1$ (7.64) gives $P(\tau_b < +\infty) = \frac{p}{q}$, i.e. with probability $1 - \frac{p}{q}$ the random walk $(S_n)_n$ never visits the strictly positive integers.

5.13 (a) As X_{n+1} is independent of $\mathcal{F}_n$, $E(X_{n+1}|\mathcal{F}_n) = E(X_{n+1}) = x$ a.s. We have $Z_n = (X_1 - x) + \cdots + (X_n - x)$, so that $(Z_n)_n$ are the cumulative sums of independent centered r.v.'s, hence a martingale (Example 5.2(a)).

(b1) Also the stopped process $(Z_{n \wedge \tau})_n$ is a martingale, therefore $E(Z_{n \wedge \tau}) = E(Z_0) = 0$, i.e.

$$E(S_{n \wedge \tau}) = x E(n \wedge \tau) . \tag{7.65}$$

(b2) By Beppo Levi's Theorem $E(n \wedge \tau) \uparrow E(\tau)$ as $n \to \infty$. If we assume $X_k \geq 0$ a.s., the sequence $(S_{n \wedge \tau})_{n \geq 0}$ is also increasing, hence also $E(S_{n \wedge \tau}) \uparrow E(S_\tau)$ as $n \to \infty$ and from (7.65) we obtain

$$E(S_\tau) = x E(\tau) < +\infty . \tag{7.66}$$

As for the general case, if $x_1 = E(X_n^+)$, $x_2 = E(X_n^-)$ (so that $x = x_1 - x_2$), let $S_n^{(1)} = X_1^+ + \cdots + X_n^+$, $S_n^{(2)} = X_1^- + \cdots + X_n^-$ and

$$Z_n^{(1)} = S_n^{(1)} - nx_1 ,$$

$$Z_n^{(2)} = S_n^{(2)} - nx_2 .$$

As X_{n+1}^+ (resp. X_{n+1}^-) is independent of $\mathcal{F}_n$, $(Z_n^{(1)})_n$ (resp. $(Z_n^{(2)})_n$) is a martingale with respect to $(\mathcal{F}_n)_n$. By (7.66) we have

$$E(S_\tau^{(1)}) = x_1 E(\tau), \qquad E(S_\tau^{(2)}) = x_2 E(\tau) ,$$

and by subtraction, all quantities appearing in the expression being finite (recall that τ is assumed to be integrable),

$$E(S_\tau) = E(S_\tau^{(1)}) - E(S_\tau^{(2)}) = (x_1 - x_2)E(\tau) = xE(\tau) .$$

(c) The process $(S_n)_n$ can make, on $\mathbb{Z}$, only one step to the right or to the left. Therefore, recalling that we know that $\tau_b < +\infty$ a.s., $S_{\tau_b} = b$ a.s., hence $E(S_{\tau_b}) = b$. If τ_b were integrable, (c) would give instead

$$E(S_{\tau_b}) = E(X_1)E(\tau_b) = 0 .$$

5.14 (a) With the usual trick of splitting into the value at time n and the increment we have

$$E(W_{n+1}|\mathscr{F}_n) = E\big[(S_n + X_{n+1})^2 - (n+1)|\mathscr{F}_n\big]$$
$$= E(S_n^2 + 2S_n X_{n+1} + X_{n+1}^2|\mathscr{F}_n) - n - 1 .$$

Now S_n^2 is already $\mathscr{F}_n$-measurable, whereas

$$E(S_n X_{n+1}|\mathscr{F}_n) = S_n E(X_{n+1}) = 0 ,$$
$$E(X_{n+1}^2|\mathscr{F}_n) = E(X_{n+1}^2) = 1 ,$$

hence

$$E(W_{n+1}|\mathscr{F}_n) = S_n^2 + 1 - n - 1 = S_n^2 - n = W_n .$$

(b1) The stopping time $\tau_{a,b}$ is not bounded but, by the stopping theorem applied to $\tau_{a,b} \wedge n$,

$$0 = E(W_0) = E(W_{\tau_{a,b}\wedge n}) = E(S_{\tau_{a,b}\wedge n}^2) - E(\tau_{a,b} \wedge n) ,$$

hence

$$E(S_{\tau_{a,b}\wedge n}^2) = E(\tau_{a,b} \wedge n) .$$

Now $S_{\tau_{a,b}\wedge n}^2 \to_{n\to\infty} S_{\tau_{a,b}}$ a.s. and $E(S_{\tau_{a,b}\wedge n}^2) \to_{n\to\infty} E(S_{\tau_{a,b}}^2)$ by Lebesgue's Theorem as the r.v.'s $S_{\tau_{a,b}\wedge n}$ are bounded ($-a \le S_{\tau_{a,b}\wedge n} \le b$) whereas $E(\tau_{a,b} \wedge n) \uparrow_{n\to\infty} E(\tau_{a,b})$ by Beppo Levi's Theorem. Hence $\tau_{a,b}$ is integrable and

$$E(\tau_{a,b}) = E(S_{\tau_{a,b}}^2) = a^2 P(S_{\tau_{a,b}} = -a) + b^2 P(S_{\tau_{a,b}} = b)$$

$$= a^2 \frac{b}{a+b} + b^2 \frac{a}{a+b} = \frac{a^2 b + b^2 a}{a+b}$$

$$= ab .$$

(b2) We have, for every $a > 0$, $\tau_{a,b} < \tau_b$. Therefore $E(\tau_b) > E(\tau_{a,b}) = ab$ for every $a > 0$ so that $E(\tau_b)$ must be $= +\infty$.

5.15 (a) As $E(X_{n+1}) = E(X_{n+1}^3) = 0$, we have

$$
\begin{aligned}
E(Z_{n+1}|\mathscr{F}_n) &= E\big[(S_n + X_{n+1})^3 - 3(n+1)(S_n + X_{n+1})|\mathscr{F}_n\big] \\
&= E\big(S_n^3 + 3S_n^2 X_{n+1} + 3S_n X_{n+1}^2 + X_{n+1}^3 |\mathscr{F}_n\big) - 3(n+1)S_n \\
&= S_n^3 + 3S_n - 3(n+1)S_n = S_n^3 - 3nS_n \\
&= Z_n \, .
\end{aligned}
$$

(b1) By the stopping theorem, for every $n \geq 0$ we have $0 = E(Z_{n\wedge\tau})$, hence

$$
E(S_{n\wedge\tau}^3) = 3E[(n \wedge \tau)S_{n\wedge\tau}] \, . \tag{7.67}
$$

Note that $-a \leq S_{n\wedge\tau} \leq b$ so that $S_{n\wedge\tau}$ is bounded and that τ is integrable (Exercise 5.14). Then by Lebesgue's Theorem we can take the limit as $n \to \infty$ in (7.67) and obtain

$$
\begin{aligned}
E(\tau S_\tau) &= \frac{1}{3} E(S_\tau^3) = \frac{1}{3}\left(-a^3 \frac{b}{a+b} + b^3 \frac{a}{a+b}\right) \\
&= \frac{1}{3} \frac{-a^3 b + b^3 a}{a+b} = \frac{1}{3} ab(b-a) \, .
\end{aligned}
$$

As we know already that $E(S_\tau) = 0$, we obtain

$$
\mathrm{Cov}(S_\tau, \tau) = E(\tau S_\tau) = \frac{1}{3} ab(b-a) \, .
$$

If $b \neq a$ then S_τ and τ are correlated and cannot be independent, which is somehow intuitive: if b is smaller than a, i.e. the rightmost end of the interval is closer to the origin, then the fact that $S_\tau = b$ suggests that τ should be smallish.

(b2) Let us note first that, as $X_i \sim -X_i$, the joint distributions of $(S_n)_n$ and of $(-S_n)_n$ coincide. Moreover, we have

$$
P(S_\tau = a, \tau = n) = P(|S_0| < a, \ldots, |S_{n-1}| < a, S_n = a)
$$

and as the joint distributions of $(S_n)_n$ and of $(-S_n)_n$ coincide

$$
\begin{aligned}
P(S_\tau = a, \tau = n) &= P(|S_0| < a, \ldots, |S_{n-1}| < a, S_n = -a) \\
&= P(S_\tau = -a, \tau = n) \, .
\end{aligned} \tag{7.68}
$$

As $P(S_\tau = a, \tau = n) + P(S_\tau = -a, \tau = n) = P(\tau = n)$ and $P(S_\tau = a) = \frac{1}{2}$, from (7.68) we deduce

$$P(S_\tau = a, \tau = n) = \frac{1}{2} P(\tau = n) = P(S_\tau = a)P(\tau = n),$$

which proves that S_τ and τ are independent.

5.16 (a) Note that

$$E(e^{\theta X_k}) = \frac{1}{2} e^\theta + \frac{1}{2} e^{-\theta} = \cosh \theta$$

and that we can write

$$Z_n^\theta = \prod_{k=1}^n \frac{e^{\theta X_k}}{\cosh \theta}$$

so that the Z_n^θ are the cumulative products of independent positive r.v.'s having expectation equal to 1, hence a martingale as seen in Example 5.2(b).

Thanks to Remark 5.8 (a stopped martingale is again a martingale) $(Z_{n \wedge \tau}^\theta)_n$ is a martingale. If $\theta > 0$, it is also bounded: as S_n cannot cross level a without taking the value a, $S_{n \wedge \tau} \le a$ (this being true even on $\{\tau = +\infty\}$). Therefore, $\cosh \theta$ being always ≥ 1,

$$0 \le Z_{n \wedge \tau}^\theta \le e^{\theta a}.$$

(b1) Let $\theta > 0$. $(Z_{n \wedge \tau}^\theta)_n$ is a bounded martingale, hence bounded in L^2, and it converges in L^2 (and thus in L^1) and a.s. to an r.v. W^θ. On $\{\tau < \infty\}$ we have $W^\theta = \lim_{n \to \infty} Z_{n \wedge \tau}^\theta = Z_\tau^\theta = e^{\theta a}(\cosh \theta)^{-\tau}$; on the other hand $W^\theta = 0$ on $\{\tau = \infty\}$, since in this case $S_n \le a$ for every n whereas the denominator tends to $+\infty$. Therefore (5.28) is proved.

(b2) We have $W^\theta \to_{\theta \to 0+} 1_{\{\tau < +\infty\}}$ and, as for $\theta \le 1$,

$$W^\theta = \frac{e^{\theta a}}{(\cosh \theta)^\tau} 1_{\{\tau < +\infty\}} \le e^a, \tag{7.69}$$

by Lebesgue's Theorem

$$\lim_{\theta \to 0+} E(W^\theta) = P(\tau < +\infty). \tag{7.70}$$

Thanks to (b1) $E(W^\theta) = \lim_{n \to \infty} E(Z_{n \wedge \tau}^\theta) = E(Z_0^\theta) = 1$ for every $\theta \ge 0$, so that (7.70) gives $P(\tau < +\infty) = 1$.

Moreover, $1 = E(W^\theta) = E[e^{\theta a}(\cosh\theta)^{-\tau}]$ gives

$$E\left(\frac{1}{(\cosh\theta)^\tau}\right) = e^{-\theta a}. \tag{7.71}$$

(b3) For $\lambda > 0$ let $\theta \geq 0$ be such that $\cosh\theta = e^\lambda$, i.e. $\theta = \log\left(e^\lambda + \sqrt{e^{2\lambda} - 1}\right)$. Substituting this into (7.71) we find

$$E(e^{-\lambda\tau}) = \frac{1}{\left(e^\lambda + \sqrt{e^{2\lambda} - 1}\right)^a},$$

so that by analytical continuation, for $\Re z < 0$, we have

$$E(e^{z\tau}) = \frac{1}{\left(e^{-z} + \sqrt{e^{-2z} - 1}\right)^a}.$$

It is easy to check that

$$z \mapsto \frac{1}{\left(e^{-z} + \sqrt{e^{-2z} - 1}\right)^a}$$

does not have an analytic continuation on the half space $\Re z > 0$ (the square root is not analytic at 0), i.e. the right convergence abscissa of the Laplace transform of τ is $x_2 = 0$.

This is however immediate even without the computation above: τ being a positive r.v., its Laplace transform is finite on $\Re z \leq 0$. If the right convergence abscissa were > 0, τ would have finite moments of all orders, whereas we know (Exercises 5.13(d) and 5.14) that τ is not integrable.

5.17 (a) We have $E(e^{i\lambda X_k}) = \frac{1}{2}(e^{i\lambda} + e^{-i\lambda}) = \cos\lambda$ and, as X_{n+1} is independent of $\mathscr{F}_n$,

$$E\big[\cos(\lambda S_{n+1})|\mathscr{F}_n\big] = E\big(\Re e^{i\lambda(S_n + X_{n+1})}|\mathscr{F}_n\big) = \Re E\big(e^{i\lambda(S_n + X_{n+1})}|\mathscr{F}_n\big)$$
$$= \Re\big(e^{i\lambda S_n}E[e^{i\lambda X_{n+1}}]\big) = \Re\big(e^{i\lambda S_n}\cos\lambda\big) = \cos(\lambda S_n)\cos\lambda,$$

so that

$$E(Z_{n+1}|\mathscr{F}_n) = (\cos\lambda)^{-(n+1)}E\big[\cos(\lambda S_{n+1})|\mathscr{F}_n\big] = (\cos\lambda)^{-n}\cos(\lambda S_n) = Z_n.$$

The conditional expectation $E[\cos(\lambda(S_n + X_{n+1}))|\mathscr{F}_n]$ can also be computed using the addition formula for the cosine ($\cos(\alpha + \beta) = \cos\alpha\cos\beta - \sin\alpha\sin\beta$)), which leads to just a bit more complicated manipulations.

(b) As $n \wedge \tau$ is a bounded stopping time, $E(Z_{n \wedge \tau}) = E(Z_0) = 1$. Moreover, as

$$-\frac{\pi}{2} < -\lambda a \leq \lambda S_{n \wedge \tau} \leq \lambda a < \frac{\pi}{2},$$

we have $\cos(\lambda S_{n \wedge \tau}) \geq \cos(\lambda a)$ and

$$1 = E(Z_{n \wedge \tau}) = E\left[(\cos \lambda)^{-n \wedge \tau} \cos(\lambda S_{n \wedge \tau}) \right] \geq E[(\cos \lambda)^{-n \wedge \tau}] \cos(\lambda a) \,.$$

(c) The previous relation gives

$$E[(\cos \lambda)^{-n \wedge \tau}] \leq \frac{1}{\cos(\lambda a)} \, . \tag{7.72}$$

As $0 < \cos \lambda < 1$, we have $(\cos \lambda)^{-n \wedge \tau} \uparrow (\cos \lambda)^{-\tau}$ as $n \to \infty$, and taking the limit in (7.72), by Beppo Levi's Theorem we obtain

$$E[(\cos \lambda)^{-\tau}] \leq \frac{1}{\cos(\lambda a)} \, . \tag{7.73}$$

Again as $0 < \cos \lambda < 1$, $(\cos \lambda)^{-\tau} = +\infty$ on $\{\tau = +\infty\}$, and (7.73) entails $P(\tau = +\infty) = 0$. Therefore τ is a.s. finite.

(d1) We have $|S_{n \wedge \tau}| \to_{n \to \infty} |S_\tau| = a$ a.s. and therefore

$$Z_{n \wedge \tau} = (\cos \lambda)^{-n \wedge \tau} \cos(\lambda S_{n \wedge \tau}) \overset{\text{a.s.}}{\underset{n \to \infty}{\to}} (\cos \lambda)^{-\tau} \cos(\lambda a) = Z_\tau \, . \tag{7.74}$$

Moreover,

$$|Z_{n \wedge \tau}| = |(\cos \lambda)^{-n \wedge \tau} \cos(\lambda S_{n \wedge \tau})| \leq (\cos \lambda)^{-\tau}$$

and $(\cos \lambda)^{-\tau}$ is integrable thanks to (7.73). Therefore by Lebesgue's Theorem $E(Z_{n \wedge \tau}) \to_{n \to \infty} E(Z_\tau)$.

(d2) By Scheffé's Theorem $Z_{n \wedge \tau} \to_{n \to \infty} Z_\tau$ in L^1 and the martingale is regular.

(e) Thanks to (c) $1 = E(Z_\tau) = \cos(\lambda a) E[(\cos \lambda)^{-\tau}]$, so that

$$E[(\cos \lambda)^{-\tau}] = \frac{1}{\cos \lambda a} \, ,$$

which can be written

$$E[e^{\tau(-\log \cos \lambda)}] = \frac{1}{\cos \lambda a} \, . \tag{7.75}$$

Hence the Laplace transform $L(\theta) = E(e^{\theta\tau})$ is finite for $\theta < -\log\cos\frac{\pi}{2a}$ (which is a strictly positive number). (7.75) gives

$$\lim_{\theta \to -\log\cos\frac{\pi}{2a}-} L(\theta) = \lim_{\lambda \to \frac{\pi}{2a}} \frac{1}{\cos(\lambda a)} = +\infty$$

and we conclude that $x_2 := -\log\cos\frac{\pi}{2a}$ is the right convergence abscissa, the left one being $x_1 = -\infty$ of course. As the convergence strip of the Laplace transform contains the origin, τ has finite moments of every order (see (2.63) and the argument p.86 at the end of Sect. 2.7).

• In Exercises 5.16 and 5.17 it has been proved that, for the simple symmetric random walk, for $a > 0$ the two stopping times

$$\tau_1 = \inf\{n \geq 0, S_n = a\},$$

$$\tau_2 = \inf\{n \geq 0, |S_n| = a\}$$

are both a.s. finite. But the first one is not integrable (Exercise 5.13(d)) whereas the second one has a Laplace transform which is finite for some strictly positive values and has finite moments of all orders.

The intuition behind this fact is that before reaching the level a the random walk $(S_n)_n$ can make very long excursions on the negative side, therefore taking a lot of time before reaching a.

5.18 We know that the limit $\lim_{n\to\infty} U_n = U_\infty \geq 0$ exists a.s., $(U_n)_n$ being a positive supermartingale. By Fatou's Lemma

$$E(U_\infty) \leq \lim_{n\to\infty} E(U_n) = 0.$$

The positive r.v. U_∞ has mean 0 and is therefore $= 0$ a.s.

5.19 (a) By the strong Law of Large Numbers $\frac{1}{n}S_n \to_{n\to\infty} b < 0$, so that $S_n \to_{n\to\infty} -\infty$ a.s. Thus $(S_n)_n$ is bounded from above a.s.

(b) As $Y_1 \leq 1$ a.s. we have $e^{\lambda Y_1} \leq e^\lambda$ for $\lambda \geq 0$ and $L(\lambda) < +\infty$ on $\mathbb{R}^+$. Moreover $L(\lambda) \geq e^\lambda P(Y_1 = 1)$, which gives $\lim_{\lambda\to+\infty} \psi(\lambda) = +\infty$. As $L'(0+) = E(Y_i) = b$,

$$\psi'(0+) = \frac{L'(0+)}{L(0)} = b < 0.$$

ψ is continuous, vanishes at 0 with a right derivative that is strictly negative and converges to $+\infty$ as $\lambda \to +\infty$. Therefore, necessarily, it has another zero, λ_0, which is strictly positive (see Fig. 7.13). Thanks to the convexity of ψ this zero is unique.

(c) We have $Z_n = e^{\lambda_0 S_n} = e^{\lambda_0 Y_1} \cdots e^{\lambda_0 Y_n}$ and now just note that $E(e^{\lambda_0 Y_k}) = L(\lambda_0) = 1$ so that $(Z_n)_n$ are the cumulative products of independent positive

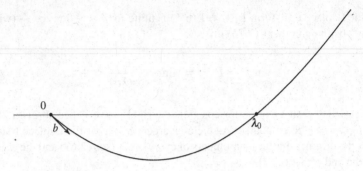

Fig. 7.13 A typical graph of ψ

r.v.'s having expectation equal to 1, hence a martingale (Example 5.2(b)). We noted already that $S_n \to_{n \to \infty} -\infty$ a.s., so that $\lim_{n \to \infty} Z_n = 0$ a.s.

(d) This is almost immediate as

$$\begin{cases} \lim_{n \to \infty} Z_{n \wedge \tau_K} = \lim_{n \to \infty} Z_n = 0 & \text{on } \{\tau_K = +\infty\} \\ \lim_{n \to \infty} Z_{n \wedge \tau_K} = Z_{\tau_K} = e^{\lambda_0 K} & \text{on } \{\tau_K < +\infty\}. \end{cases}$$

We use here the assumptions on the law of Y_n, which imply that S_n takes at most one step to the right and thus, necessarily, $S_{\tau_K} = K$.

(e) The stopped martingale $(Z_{n \wedge \tau_K})_n$ is bounded (it takes values between 0 and $e^{\lambda_0 K}$) and we can apply Lebesgue's Theorem in (5.30), which gives

$$1 = \mathrm{E}(Z_0) = \lim_{n \to \infty} \mathrm{E}(Z_{n \wedge \tau_K}) = e^{\lambda_0 K} \mathrm{P}(\tau_K < +\infty).$$

Therefore $\mathrm{P}(\tau_K < +\infty) = e^{-\lambda_0 K}$. Since obviously $\mathrm{P}(\tau_K < +\infty) = \mathrm{P}(W \geq K)$, W has a geometric law with parameter $p = 1 - e^{-\lambda_0}$.

With the given law for the Y_n, the Laplace transform is $L(\lambda) = q e^{-\lambda} + p e^{\lambda}$. The determination of the value $\lambda_0 > 0$ such that $L(\lambda_0) = 1$ reduces to the equation of the second degree

$$p e^{2\lambda} - e^{\lambda} + q = 0.$$

Its roots are $e^{\lambda} = 1$ (obviously, as $L(0) = 1$) and $e^{\lambda} = \frac{q}{p}$. Thus $\lambda_0 = \log \frac{q}{p}$ and in this case W has a geometric law with parameter $1 - e^{-\lambda_0} = 1 - \frac{p}{q}$.

5.20 (a) The S_n are the cumulative sums of independent centered r.v.'s, hence they form a martingale (Example 5.2(a)).

(b) The r.v.'s X_k are bounded, therefore $S_n \in L^2$. The associated increasing process, i.e. the compensator of the submartingale $(S_n^2)_n$, is defined by $A_0 = 0$ and

$$A_{n+1} = A_n + E(S_{n+1}^2 | \mathscr{F}_n) - S_n^2 = A_n + E(2S_n X_{n+1} + X_{n+1}^2 | \mathscr{F}_n)$$
$$= A_n + E(X_{n+1}^2) = A_n + 2^{-n}$$

hence, by induction,

$$A_n = \sum_{k=0}^{n-1} 2^{-k} = 2(1 - 2^{-n}) .$$

(Note that the increasing process $(A_n)_n$ is deterministic, as always with a martingale with independent increments, Exercise 5.5(b).)

(c) As the associated increasing process $(A_n)_n$ is bounded and

$$A_n = E(S_n^2) ,$$

we deduce that $(S_n)_n$ is bounded in L^2, so that it converges a.s. and in L^2 and is regular.

5.21 We have

$$E\left(\frac{p(X_k)}{q(X_k)}\right) = \sum_{x \in E} \frac{p(x)}{q(x)} q(x) = \sum_{x \in E} p(x) = 1 . \tag{7.76}$$

Y_n is therefore the product of positive independent r.v.'s having expectation equal to 1 and is therefore a martingale (Example 5.2(b)). Being a positive martingale it converges a.s. Recalling Remark 5.24(c), the limit is 0 a.s. and $(Y_n)_n$ cannot be regular.

5.22 (a) Let us argue by induction. Of course $X_0 = q \in [0, 1]$. Assume that $X_n^2 \in [0, 1]$. Then obviously $X_{n+1} \geq 0$ and also

$$X_{n+1} = \frac{1}{2} X_n^2 + \frac{1}{2} 1_{[0, X_n]}(U_{n+1}) \leq \frac{1}{2} + \frac{1}{2} = 1 .$$

(b) The fact that $(X_n)_n$ is adapted to $(\mathscr{F}_n)_n$ is also immediate by induction. Let us check the martingale property. We have

$$E(X_{n+1} | \mathscr{F}_n) = E\left[\frac{1}{2} X_n^2 + \frac{1}{2} 1_{[0, X_n]}(U_{n+1}) \,\middle|\, \mathscr{F}_n\right]$$

$$= \frac{1}{2} X_n^2 + \frac{1}{2} E\left[1_{[0, X_n]}(U_{n+1}) | \mathscr{F}_n\right] .$$

By the freezing lemma $E\left[1_{[0, X_n]}(U_{n+1}) | \mathscr{F}_n\right] = \Phi(X_n)$ where, for $0 \leq x \leq 1$,

$$\Phi(x) = E[1_{[0, x]}(U_{n+1})] = P(U_{n+1} \leq x) .$$

An elementary computation gives, for the d.f. of U_n, $P(U_n \le x) = 2x - x^2$ so that

$$E(X_{n+1}|\mathscr{F}_n) = \frac{1}{2}X_n^2 + X_n - \frac{1}{2}X_n^2 = X_n .$$

(c) $(X_n)_n$ is a bounded martingale, hence is regular and converges a.s. and in L^p for every $p \ge 1$ to some r.v. X_∞ and $E(X_\infty) = \lim_{n \to \infty} E(X_n) = E(X_0) = q$.

(d) (5.31) gives

$$X_{n+1}^2 - \frac{1}{2}X_n = \frac{1}{2}1_{[0,X_n]}(U_{n+1}) ,$$

hence $X_{n+1}^2 - \frac{1}{2}X_n$ can only take the values 0 or $\frac{1}{2}$ and, taking the limit, also $X_\infty - \frac{1}{2}X_\infty^2$ can only take the values 0 or $\frac{1}{2}$ a.s.

Now the equations $x - \frac{1}{2}x^2 = 0$ and $x - \frac{1}{2}x^2 = \frac{1}{2}$ together have the roots $0, 1, 2$. As $0 \le X_\infty \le 1$, X_∞ can only take the values 0 or 1, hence it has a Bernoulli distribution. As $E(X_\infty) = q$, $X_\infty \sim B(1,q)$.

5.23 Let us denote by E, E^Q the expectations with respect to P and Q, respectively.

(a) Recall that, by definition, for $A \in \mathscr{F}_m$, $Q(A) = E(Z_m 1_A)$. Let $m \le n$. We must prove that, for every $A \in \mathscr{F}_m$, $E(Z_n 1_A) = E(Z_m 1_A)$. But as $A \in \mathscr{F}_m \subset \mathscr{F}_n$, both these quantities are equal to $Q(A)$.

(b) We have $Q(Z_n = 0) = E(Z_n 1_{\{Z_n = 0\}}) = 0$ and therefore $Z_n > 0$ Q-a.s. Moreover, as $\{Z_n > 0\} \subset \{Z_m > 0\}$ a.s. if $m \le n$ (Exercise 5.2: the zeros of a positive martingale increase), for every $A \in \mathscr{F}_m$,

$$E^Q(1_A Z_n^{-1}) = E^Q(1_{A \cap \{Z_n > 0\}} Z_n^{-1}) = P(A \cap \{Z_n > 0\})$$
$$\le P(A \cap \{Z_m > 0\}) = E^Q(1_A Z_m^{-1}) \tag{7.77}$$

and therefore $(Z_n^{-1})_n$ is a Q-supermartingale.

(c) Let us assume $P \ll Q$: this means that $P(A) = 0$ whenever $Q(A) = 0$. Therefore also $P(Z_n = 0) = 0$ and

$$E^Q(Z_n^{-1}) = E(Z_n Z_n^{-1}) = 1 .$$

The Q-supermartingale $(Z_n^{-1})_n$ therefore has constant expectation and is a Q-martingale by the criterion of Exercise 5.1. Alternatively, just repeat the argument of (7.77) obtaining an equality.

5.24 (a) If $(M_n)_n$ is regular, then $M_n \to_{n\to\infty} M_\infty$ a.s. and in L^1 and $M_n = E(M_\infty | \mathcal{F}_n)$. Such an r.v. M_∞ is positive and $E(M_\infty) = 1$. Let Q be the probability on $\mathcal{F}$ having density M_∞ with respect to P. Then, if $A \in \mathcal{F}_n$, we have

$$Q(A) = E(1_A M_\infty) = E[1_A E(M_\infty | \mathcal{F}_n)] = E(1_A M_n) = Q_n(A) \,,$$

so that Q and Q_n coincide on $\mathcal{F}_n$.

(b) Conversely, let Z be the density of Q with respect to P. Then, for every n, we have for $A \in \mathcal{F}_n$

$$E(Z1_A) = Q(A) = Q_n(A) = E(M_n 1_A) \,,$$

which implies that

$$E(Z | \mathcal{F}_n) = M_n \,,$$

so that $(M_n)_n$ is regular.

5.25 (a) Immediate as M_n is the product of the r.v.'s $e^{\theta X_k - \frac{1}{2}\theta^2}$, which are independent and have expectation equal to 1 (Example 5.2(b)).

(b1) Let $n > m$. As X_n is independent of S_m, hence of M_m, for $A \in \mathcal{B}(\mathbb{R})$ we have

$$Q_m(X_n \in A) = E(1_{\{X_n \in A\}} M_m) = E(1_{\{X_n \in A\}}) E(M_m) = P(X_n \in A) \,.$$

X_n has the same law under Q_m as under P.

(b2) If $n \le m$ instead, X_n is $\mathcal{F}_m$-measurable so that

$$Q_m(X_n \in A) = E(1_{\{X_n \in A\}} M_m) = E\big[E(1_{\{X_n \in A\}} M_m | \mathcal{F}_n)\big]$$

$$= E\big[1_{\{X_n \in A\}} E(M_m | \mathcal{F}_n)\big] = E(1_{\{X_n \in A\}} M_n) = E\big(1_{\{X_n \in A\}} e^{\theta X_n - \frac{1}{2}\theta^2} M_{n-1}\big) \,.$$

As X_n is independent of $\mathcal{F}_{n-1}$ whereas M_{n-1} is $\mathcal{F}_{n-1}$-measurable,

$$\cdots = E\big(1_{\{X_n \in A\}} e^{\theta X_n - \frac{1}{2}\theta^2}\big) E(M_{n-1}) = \frac{1}{\sqrt{2\pi}} \int_A e^{\theta x - \frac{1}{2}\theta^2} e^{-x^2/2} \, dx$$

$$= \frac{1}{\sqrt{2\pi}} \int_A e^{-\frac{1}{2}(x-\theta)^2} \, dx \,.$$

If $n \le m$ then $X_n \sim N(\theta, 1)$ under Q_m.

5.26 (a) Follows from Remark 5.2(b), as the Z_n are the cumulative products of the r.v.'s $e^{X_k - \frac{1}{2}a_k}$, which are independent and have expectation equal to 1 (recall the Laplace transform of Gaussian r.v.'s).

(b) The limit $\lim_{n\to\infty} Z_n$ exists a.s., $(Z_n)_n$ being a positive martingale. In order to compute this limit, let us try Kakutani's trick (Remark 5.24(b)): we have

$$\lim_{n\to\infty} E(\sqrt{Z_n}) = \lim_{n\to\infty} E(e^{\frac{1}{2}S_n})e^{-\frac{1}{4}A_n} = \lim_{n\to\infty} e^{-\frac{1}{8}A_n} = 0. \qquad (7.78)$$

Therefore $\lim_{n\to\infty} Z_n = 0$ and $(Z_n)_n$ is not regular.

(c1) By (7.78) now

$$\lim_{n\to\infty} E(\sqrt{Z_n}) = e^{-\frac{1}{8}A_\infty} > 0.$$

Hence (Proposition 5.25) the martingale is regular.

Another argument leading directly to the regularity of $(Z_n)_n$ can also be obtained by noting that $(S_n)_n$ is itself a martingale (sum of independent centered r.v.'s) which is bounded in L^2, as $E(S_n^2) = A_n$. Hence $(S_n)_n$ converges a.s. and in L^2 to some limit S_∞, which is also Gaussian and centered (Proposition 3.36) as L^2 convergence entails convergence in law. Now if $Z_\infty := e^{S_\infty - \frac{1}{2}A_\infty}$ we have

$$E(Z_\infty | \mathscr{F}_n) = E\left[\exp\left(S_n - \frac{1}{2}A_n + \sum_{k=n+1}^{\infty} X_k - \frac{1}{2}\sum_{k=n+1}^{\infty} a_k \Big| \mathscr{F}_n \right)\right]$$

$$= e^{S_n - \frac{1}{2}A_n} E\left[\exp\left(\sum_{k=n+1}^{\infty} X_k - \frac{1}{2}\sum_{k=n+1}^{\infty} a_k \right)\right] = e^{S_n - \frac{1}{2}A_n} = Z_n,$$

again giving the regularity of $(Z_n)_n$. As a consequence of this argument the limit $Z_\infty = e^{S_\infty - \frac{1}{2}A_\infty}$ has a lognormal law with parameters $-\frac{1}{2}A_\infty$ and A_∞ (it is the exponential of an $N(-\frac{1}{2}A_\infty, A_\infty)$-distributed r.v.).

(c2) Let $f : \mathbb{R}^n \to \mathbb{R}$ be a bounded Borel function. Note that the joint density of $X_1, \ldots, X_n$ (with respect to P) is

$$\frac{1}{(2\pi)^{n/2}\sqrt{R_n}} e^{-\frac{1}{2a_1}x_1^2} \ldots e^{-\frac{1}{2a_n}x_n^2}$$

where $R_n = a_1 a_2 \ldots a_n$. Then we have

$$E^Q[f(X_1, \ldots, X_n)] = E[f(X_1, \ldots, X_n)Z_\infty]$$

$$= E\big[E[f(X_1, \ldots, X_n)Z_\infty | \mathscr{F}_n]\big] = E[f(X_1, \ldots, X_n)Z_n]$$

$$= E[f(X_1, \ldots, X_n) e^{S_n - \frac{1}{2}A_n}]$$

$$= \frac{1}{(2\pi)^{n/2}\sqrt{R_n}} \int_{\mathbb{R}^n} f(x_1, \ldots, x_n) e^{x_1 + \cdots + x_n - \frac{1}{2}A_n} e^{-\frac{1}{2a_1}x_1^2} \ldots$$

$$\ldots e^{-\frac{1}{2a_n}x_n^2} dx_1 \ldots dx_n$$

$$= \frac{1}{(2\pi)^{n/2}\sqrt{R_n}} \int_{\mathbb{R}^n} f(x_1, \ldots, x_n)\, e^{-\frac{1}{2}(a_1+\cdots+a_n)} e^{-\frac{1}{2a_1}(x_1^2-2a_1x_1)} \ldots$$

$$\ldots e^{-\frac{1}{2a_n}(x_n^2-2a_nx_n)}\, dx_1 \ldots dx_n$$

$$= \frac{1}{(2\pi)^{n/2}\sqrt{R_n}} \int_{\mathbb{R}^n} f(x_1, \ldots, x_n)\, e^{-\frac{1}{2a_1}(x_1^2-2a_1x_1+a_1^2)} \ldots$$

$$\ldots e^{-\frac{1}{2a_n}(x_n^2-2a_nx_n+a_n^2)}\, dx_1 \ldots dx_n$$

$$= \frac{1}{(2\pi)^{n/2}\sqrt{R_n}} \int_{\mathbb{R}^n} f(x_1, \ldots, x_n)\, e^{-\frac{1}{2a_1}(x_1-a_1)^2} \ldots e^{-\frac{1}{2a_n}(x_n-a_n)^2}\, dx_1 \ldots dx_n ,$$

so that under Q the joint density of $X_1, \ldots, X_n$ with respect to the Lebesgue measure is

$$g(x_1, \ldots, x_n) = \frac{1}{\sqrt{2\pi a_1}} e^{-\frac{1}{2a_1}(x_1-a_1)^2} \cdots \frac{1}{\sqrt{2\pi a_n}} e^{-\frac{1}{2a_n}(x_n-a_n)^2} ,$$

which proves simultaneously that $X_k \sim N(a_k, a_k)$ and that the r.v.'s X_n are independent. The same result can be obtained by computing the Laplace transform or the characteristic function of $(X_1, \ldots, X_n)$ under Q.

5.27 (a) By the freezing lemma, Lemma 4.11,

$$E[e^{\lambda X_n X_{n+1}}] = E[E(e^{\lambda X_n X_{n+1}}|\mathscr{F}_n)] = E(e^{\frac{1}{2}\lambda^2 X_n^2}) \tag{7.79}$$

and, recalling Exercise 2.7 (or the Laplace transform of the Gamma distributions),

$$E(e^{\lambda X_n X_{n+1}}) = \begin{cases} \dfrac{1}{\sqrt{1-\lambda^2}} & \text{if } |\lambda| < 1 \\[2mm] +\infty & \text{if } |\lambda| \geq 1 . \end{cases}$$

(b) We have

$$E(e^{Z_{n+1}}|\mathscr{F}_n) = e^{Z_n} E(e^{\lambda X_{n+1} X_n}|\mathscr{F}_n)$$

and by the freezing lemma again

$$\log E(e^{Z_{n+1}}|\mathscr{F}_n) = Z_n + \frac{1}{2}\lambda^2 X_n^2 . \tag{7.80}$$

Let $A_0 = 0$ and

$$A_{n+1} = A_n + \log E(e^{Z_{n+1}}|\mathscr{F}_n) - Z_n = A_n + \frac{1}{2}\lambda^2 X_n^2 , \tag{7.81}$$

i.e.

$$A_{n+1} = \frac{1}{2} \lambda^2 \sum_{k=1}^{n} X_k^2 .$$

$(A_n)_n$ is obviously predictable and increasing. Moreover, (7.81) gives

$$\log \mathrm{E}(e^{Z_{n+1}} | \mathcal{F}_n) - A_{n+1} = Z_n - A_n$$

and, taking the exponential and recalling that A_{n+1} is $\mathcal{F}_n$-measurable, we obtain

$$\mathrm{E}(e^{Z_{n+1}-A_{n+1}} | \mathcal{F}_n) = e^{Z_n - A_n} ,$$

so that $M_n = e^{Z_n - A_n}$ is the required martingale.

(c) Of course $(M_n)_n$ converges a.s., being a positive martingale. In order to investigate regularity, let us try Kakutani's trick: we have

$$\sqrt{M_n} = \exp \left(\frac{\lambda}{2} \sum_{k=1}^{n} X_{k-1} X_k - \frac{1}{4} \lambda^2 \sum_{k=1}^{n-1} X_k^2 \right).$$

One possibility in order to investigate the limit of this quantity is to write

$$\sqrt{M_n} = \exp \left(\frac{\lambda}{2} \sum_{k=1}^{n} X_{k-1} X_k - \frac{1}{8} \lambda^2 \sum_{k=1}^{n-1} X_k^2 \right) \exp \left(-\frac{1}{8} \lambda^2 \sum_{k=1}^{n-1} X_k^2 \right) := N_n \cdot W_n .$$

Now $(N_n)_n$ is a positive martingale (same as $(M_n)_n$ with $\frac{\lambda}{2}$ instead of λ) and converges a.s. to a finite limit, whereas $W_n \to_{n\to\infty} 0$ a.s., as $\mathrm{E}(X_k^2) = 1$ and, by the law of large numbers, $\sum_{k=1}^{n-1} X_k^2 \to_{n\to\infty} +\infty$ a.s. Hence $\sqrt{M_n} \to_{n\to\infty} 0$ a.s. and the martingale is not regular.

The courageous reader can also attempt to use Hölder's inequality in order to prove that $\mathrm{E}(\sqrt{M_n}) \to_{n\to\infty} 0$.

5.28 (a1) Just note that $\mathcal{B}_n = \sigma(S_n) \vee \sigma(X_j, j \geq n+1)$ and that $\sigma(X_j, j \geq n+1)$ is independent of $\sigma(S_n) \vee \sigma(X_k)$. The result follows thanks to Exercise 4.3(b).

(a2) Follows from the fact that the joint distributions of X_k, S_n and X_j, S_n are the same (see also Exercise 4.5).

(b1) Thanks to (a), as $\overline{X}_n = \frac{1}{n}(S_{n+1} - X_{n+1})$,

$$\mathrm{E}(\overline{X}_n | \mathcal{B}_{n+1}) = \mathrm{E}(\overline{X}_n | S_{n+1}) = \frac{1}{n} S_{n+1} - \frac{1}{n} \mathrm{E}(X_{n+1} | S_{n+1})$$

$$= \frac{1}{n} S_{n+1} - \frac{1}{n(n+1)} S_{n+1} = \frac{1}{n+1} S_{n+1} = \overline{X}_{n+1} .$$

(b2) By Remark 5.26, the backward martingale $(\overline{X}_n)_n$ converges a.s. to an r.v., Z say. As Z is measurable with respect to the tail σ algebra of the sequence $(X_n)_n$, as noted in the remarks following Kolmogorov's 0–1 law, p. 52, Z must be constant a.s. As the convergence also takes place in L^1, this constant must be $b = E(X_1)$.

6.1 (a) Thanks to Exercise 2.9 (b) a Weibull r.v. with parameters α, λ is of the form $X^{1/\alpha}$, where X is exponential with parameter λ. Therefore, recalling Example 6.3, if X is a uniform r.v. on $[0, 1]$, then $(-\frac{1}{\lambda}\log(1 - X))^{1/\alpha}$ is a Weibull r.v. with parameters α, λ.

(b) Recall that if $X \sim N(0, 1)$ then $X^2 \sim \text{Gamma}(\frac{1}{2}, \frac{1}{2})$. Therefore if $X_1, \ldots, X_k$ are i.i.d. $N(0, 1)$ distributed r.v.'s (obtained as in Example 6.4) then $X_1^2 + \cdots + X_k^2 \sim \text{Gamma}(\frac{k}{2}, \frac{1}{2})$ and $\frac{1}{2\lambda}(X_1^2 + \cdots + X_k^2) \sim \text{Gamma}(\frac{k}{2}, \lambda)$.

(c) Thanks to Exercise 2.20 (b) and (b) above if the r.v.'s $X_1, \ldots, X_k, Y_1, \ldots, Y_m$ are i.i.d. and $N(0, 1)$-distributed then

$$Z = \frac{X_1^2 + \cdots + X_k^2}{X_1^2 + \cdots + X_k^2 + Y_1^2 + \cdots + Y_m^2}$$

has a $\text{Beta}(\frac{k}{2}, \frac{m}{2})$ distribution.

(d) If the r.v.'s $X, Y_1, \ldots, Y_n$ are i.i.d. and $N(0, 1)$-distributed then

$$\frac{X}{Y_1^2 + \cdots + Y_n^2} \sqrt{n} \sim t(n) \,.$$

(e) In Exercise 2.43 it is proved that the difference of independent exponential r.v.'s of parameter λ has a Laplace law of parameter λ. Hence if X_1, X_2 are independent and uniform on $[0, 1]$, then $-\frac{1}{\lambda}\left(\log(1 - X_1) - \log(1 - X_2)\right)$ has the requested distribution.

(f) Thanks to Exercise 2.12(a), if X is exponential with parameter $-\log(1 - p)$, then $\lfloor X \rfloor$ is geometric with parameter p.

• Note that, for every choice of $\alpha, \beta \geq 1$, a $\text{Beta}(\alpha, \beta)$ r.v. can be obtained with the rejection method, Example 6.13.

6.2 For every orthogonal matrix $O \in O(d)$ we have

$$OZ = \frac{OX}{|X|} = \frac{OX}{|OX|} \,.$$

As $OX \sim X$ we have $OZ \sim Z$ so that the law of Z is the normalized Lebesgue measure of the sphere.

• Note that also in this case there are many possible ways of simulating the random choice of a point of the sphere with the normalized Lebesgue measure. Sarting from Exercise 2.14, for example, in the case of the sphere $\mathbb{S}_2$ of $\mathbb{R}^3$.

6.3 (a) We must compute the d.f., F say, associated to f and its inverse. We have, for $t \geq 0$,

$$F(t) = \int_0^t \frac{\alpha}{(1+s)^{\alpha+1}}\, ds = -\frac{1}{(1+s)^\alpha}\Big|_0^t = 1 - \frac{1}{(1+t)^\alpha}\,.$$

The equation

$$1 - \frac{1}{(1+t)^\alpha} = x$$

is easily solved, giving, for $0 < x < 1$,

$$\Phi(x) = \frac{1}{(1-x)^{1/\alpha}} - 1\,.$$

(b) The joint law of X and Y is, for $x, y > 0$,

$$h(x, y) = f_Y(y)\overline{f}(x; y) = \frac{1}{\Gamma(\alpha)} y^{\alpha-1} e^{-y} \times y\, e^{-yx} = \frac{1}{\Gamma(\alpha)} y^\alpha e^{-y(x+1)}$$

and the law of X has density with respect to the Lebesgue measure given by

$$f_X(x) = \int_{-\infty}^{+\infty} h(x, y)\, dy = \frac{1}{\Gamma(\alpha)} \int_0^{+\infty} y^\alpha e^{-y(x+1)}\, dy$$

$$= \frac{\Gamma(\alpha+1)}{\Gamma(\alpha)(1+x)^{\alpha+1}} = f(x)\,.$$

Therefore the following procedure produces a random number according with to law defined by f:

- first sample a number y with a Gamma$(\alpha, 1)$ distribution,
- then sample a number x with an exponential distribution with parameter y.

This provides another algorithm for generating a random number with density f, at least for the values of α for which we know how to simulate a Gamma$(\alpha, 1)$ r.v., see Exercise 6.1(b).

References

1. P. Baldi, L. Mazliak, P. Priouret, Solved exercises and elements of theory. *Martingales and Markov Chains* (Chapman & Hall/CRC, Boca Raton, 2002).
2. P. Billingsley, *Probability and Measure*. Wiley Series in Probability and Mathematical Statistics, 3rd edn. (John Wiley & Sons, New York, 1995)
3. M. Brancovan, T. Jeulin, *Probabilités Niveau M1* (Ellipses, Paris, 2006)
4. L. Breiman, *Probability* (Addison-Wesley, Reading, 1992)
5. P. Brémaud, *Probability Theory and Stochastic Processes* Universitext (Springer, Cham, 2020)
6. E. Çınlar, *Probability and Stochastics. Graduate Texts in Mathematics*, vol. 261 (Springer, New York, 2011)
7. L. Chaumont, M. Yor, A guided tour from measure theory to random processes, via conditioning. *Exercises in Probability*. Cambridge Series in Statistical and Probabilistic Mathematics, vol. 35, 2nd edn. (Cambridge University Press, Cambridge, 2012).
8. D. Dacunha-Castelle, M. Duflo, *Probability and Statistics*, vol. I (Springer-Verlag, New York, 1986)
9. C. Dellacherie, P.-A. Meyer, *Probabilités et Potentiel*, chap. I à IV (Hermann, Paris, 1975)
10. L. Devroye, *Nonuniform Random Variate Generation* (Springer-Verlag, New York, 1986)
11. R.M. Dudley, *Real Analysis and Probability*. Cambridge Studies in Advanced Mathematics, vol. 74 (Cambridge University Press, Cambridge, 2002). Revised reprint of the 1989 original
12. R. Durrett, *Probability–Theory and Examples*. Cambridge Series in Statistical and Probabilistic Mathematics, vol. 49, 5th edn. (Cambridge University Press, Cambridge, 2019)
13. W. Feller, *An Introduction to Probability Theory and Its Applications*, vol. II (John Wiley and Sons, New York, 1966)
14. G.S. Fishman, Concepts, algorithms, and applications. *Monte Carlo*. Springer Series in Operations Research (Springer-Verlag, New York, 1996)
15. J.E. Gentle, *Random Number Generation and Monte Carlo Methods*. Statistics and Computing, 2nd edn. (Springer, New York, 2003)
16. P.R. Halmos, *Measure Theory* (D. Van Nostrand Co., New York, 1950)
17. O. Kallenberg, *Foundations of Modern Probability*. Probability Theory and Stochastic Modelling, vol. 99, 3rd edn. (Springer, Cham, 2021)
18. D.E. Knuth, Seminumerical algorithms. *The Art of Computer Programming*, vol. 2, 3rd edn. (Addison-Wesley, Reading, 1998)
19. J.-F. Le Gall, *Measure Theory, Probability, and Stochastic Processes*. Graduate Texts in Mathematics, vol. 295 (Springer, Cham, 2022)
20. J. Neveu, *Mathematical Foundations of the Calculus of Probability* (Holden-Day, San Francisco/California/London/Amsterdam, 1965)

21. J. Neveu, *Discrete-Parameter Martingales*. North-Holland Mathematical Library, vol. 10, revised edn. (North-Holland Publishing/American Elsevier Publishing, Amsterdam/Oxford/New York, 1975)

22. W.H. Press, S.A. Teukolsky, W.T. Vetterling, B.P. Flannery, The art of scientific computing. *Numerical Recipes in C*, 2nd edn. (Cambridge University Press, Cambridge, 1992).

23. D.W. Stroock, S.R.S. Varadhan, *Multidimensional Diffusion Processes*. Grundlehren der Mathematisches Wissenschaften, vol. 233 (Springer, Berlin/Heidelberg/New York, 1979)

24. D. Williams, Foundations. *Diffusions, Markov Processes, and Martingales*, vol. 1. Probability and Mathematical Statistics (John Wiley & Sons, Chichester, 1979)

25. D. Williams, *Probability with Martingales*. Cambridge Mathematical Textbooks (Cambridge University Press, Cambridge, 1991)

Index

© The Author(s), under exclusive license to Springer Nature Switzerland AG 2023
P. Baldi, *Probability*, Universitext, https://doi.org/10.1007/978-3-031-38492-9

Printed in the United States
by Baker & Taylor Publisher Services